# U.S. CRIMINAL JUSTICE POLICY

## A CONTEMPORARY READER

# U.S. CRIMINAL JUSTICE POLICY

## A CONTEMPORARY READER

EDITED BY

## KARIM ISMAILI, PhD

Associate Dean, College of Humanities and Social Sciences

Professor, Department of Sociology and Criminal Justice

Kennesaw State University

Kennesaw, Georgia

JONES & BARTLETT
LEARNING

*World Headquarters*

Jones & Bartlett Learning
40 Tall Pine Drive
Sudbury, MA 01776
978-443-5000
info@jblearning.com
www.jblearning.com

Jones & Bartlett Learning
Canada
6339 Ormindale Way
Mississauga, Ontario L5V 1J2
Canada

Jones & Bartlett Learning
International
Barb House, Barb Mews
London W6 7PA
United Kingdom

Jones & Bartlett Learning books and products are available through most bookstores and online booksellers. To contact Jones & Bartlett Learning directly, call 800-832-0034, fax 978-443-8000, or visit our website, www.jblearning.com.

Substantial discounts on bulk quantities of Jones & Bartlett Learning publications are available to corporations, professional associations, and other qualified organizations. For details and specific discount information, contact the special sales department at Jones & Bartlett Learning via the above contact information or send an email to specialsales@jblearning.com.

**Production Credits**
Publisher, Higher Education: Cathleen Sether
Acquisitions Editor: Sean Connelly
Associate Editor: Megan R. Turner
Associate Production Editor: Lisa Cerrone
Associate Marketing Manager: Jessica Cormier
Manufacturing and Inventory Control Supervisor: Amy Bacus
Composition: Cape Cod Compositors, Inc.
Cover Design: Rena Schild/ShutterStock, Inc.
Photo Research and Permissions Manager: Kimberly Potvin
Photo and Permissions Associate: Emily Howard
Cover Image: © Rena Schild/ShutterStock, Inc.
Printing and Binding: Malloy, Inc.
Cover Printing: John P. Pow Company

**Library of Congress Cataloging-in-Publication Data**
U.S. criminal justice policy : a contemporary reader / [edited by] Karim
Ismaili.—1st ed.
      p.   cm.
   ISBN-13: 978-0-7637-4129-7
   ISBN-10: 0-7637-4129-9
   1. Criminal justice, Administration of—United States.   2. Crime—United States.   3. Crime prevention—United
States.   I. Ismaili, Karim.   II. Title: US criminal justice policy.
   HV9950.U15 2011
   364.973—dc22
                                                                            2010004960

6048

Printed in the United States of America
14  13  12  11  10    10  9  8  7  6  5  4  3  2  1

# Contents

Contributors    xi
Preface    xiii
Acknowledgments    xvii
About the Editor    xviii
About the Contributors    xix

**1    Thinking About Criminal Justice Policy:
Process, Players, and Politics** . . . . . . . . . . . . . . . . . . . . . . . . . . . **1**
by Karim Ismaili

Introduction    1
What Is Public Policy?    1
Criminology and the Policy-Making Process    3
Analyzing the Policy-Making Process    5
Step 1: Thinking About the "Peculiarities" of Criminal Justice Policy    7
Step 2: Thinking About the Criminal Justice Policy Community    9
Step 3: Thinking About Networks and Relationships Within the
    Policy Community    12
Conclusion    15
Discussion Questions    15
References    16

**Part I    Criminal Justice Policy and the Criminal
Justice System** . . . . . . . . . . . . . . . . . . . . . . . . . . . . . . . . . . . . **19**

**2    The Significance of Race in Contemporary Urban
Policing Policy** . . . . . . . . . . . . . . . . . . . . . . . . . . . . . . . . . . . . **21**
by Delores Jones-Brown and Erica King-Toler

Introduction    21
The History of Policing Race and Place    23

The Challenge of Urban Policing   27
Everything Old Is New Again: Assessing the Efficacy of
   Community Policing and the Broken Windows Approach   33
Recommendation: Making Multicultural Competency an Integral Part of
   Policing Policy and Practice   39
Conclusion   41
Endnotes   41
Discussion Questions   42
References   43
Court Cases Cited   48

**3   Combative and Cooperative Law Enforcement in
    Post-September 11th America** . . . . . . . . . . . . . . . . . . . . . . . . . . . **49**
by Ben Brown

Introduction   49
The Modern Evolution of American Law Enforcement:
   A Brief Overview   49
September 11th and the Emergence of Homeland Security   52
Combative Versus Cooperative Law Enforcement   55
The Rift of September 11th: Effects of New Policies on
   Immigration Control   60
Conclusion   63
Discussion Questions   68
References   68
Court Cases Cited   76

**4   Policies Promoting School–Police Partnerships** . . . . . . . . . . . . . **77**
by Nicole L. Bracy

Introduction   77
The School Resource Officer Program   79
A Federal Understanding of School Violence   80
Policies Supporting Placement of Police in Schools   81
Conclusion   90
Discussion Questions   90
References   91
Court Cases Cited   94

**5   Procedural Fairness, Criminal Justice Policy, and
    the Courts.** . . . . . . . . . . . . . . . . . . . . . . . . . . . . . . . . . . . . . . . . . **95**
by David B. Rottman

Introduction   95
What Is Procedural Fairness?   96
How Court Policy Is Made   101

A Balance Sheet: What Did Traditional Policy-Making Accomplish?    102
Procedural Fairness in Practice: Policies and Practices    103
Procedural Fairness and Statewide Court Policy-Making    108
The Public and Procedural Fairness    110
Conclusion    112
Endnotes    112
Discussion Questions    113
References    113
Court Cases Cited    116

## 6    Criminal Justice Policy and Problem-Solving Courts . . . . . . . . . .    117
by Susan T. Krumholz

Introduction    117
Background of Problem-Solving Courts    118
Drug Courts    120
Domestic Violence Courts    125
Discussion of the Policy Questions    128
Conclusion    131
Endnotes    131
Discussion Questions    132
References    132

## 7    U.S. Corrections Policy Since the 1970s . . . . . . . . . . . . . . . . . . . .    135
by Steven E. Barkan

Introduction    135
The Incarceration Boom    136
Why Has the Incarceration Boom Occurred?    139
The Incarceration Boom and the Crime Rate    143
Collateral Consequences of the Incarceration Boom    148
Conclusion    152
Discussion Questions    153
References    153

## 8    Reentry as a Process Rather Than a Moment . . . . . . . . . . . . . . . .    159
by Natasha A. Frost

Introduction    159
The Context for Reentry    161
The Challenge of Prisoner Reentry    163
Legislative Barriers to Successful Reintegration    167
Conclusion    176
Endnotes    177
Discussion Questions    177

References   178
Court Cases Cited   181

**9    Prison Privatization Turns 25** . . . . . . . . . . . . . . . . . . . . . . . . . . . **183**
by Richard Culp

Introduction   183
Why Privatization?   184
Research Synopsis   186
The Private Prison Market   188
Concentration in the Private Prison Industry   190
Competition and Innovation?   198
Conclusion   202
Endnotes   204
Discussion Questions   204
References   204
Court Cases Cited   209

**10    The U.S. Juvenile Justice Policy Landscape** . . . . . . . . . . . . . . . . . **211**
by Janeen Buck Willison, Daniel P. Mears, and Jeffrey A. Butts

Introduction   211
Juvenile Justice in the United States: Then and Now   213
Reform and Innovation in the 1900s   216
Analysis of Juvenile Justice Legislation   218
Is Juvenile Justice Punitive or Rehabilitative?   224
Practitioner Perspectives   232
Conclusion   235
Endnotes   236
Discussion Questions   237
References   237
Court Cases Cited   241

**Part II    Policy Intersections** . . . . . . . . . . . . . . . . . . . . . . . . . . . . . . . . . **243**

**11    Exploring the Relationship Between Contemporary
Immigration and Crime Control Policies** . . . . . . . . . . . . . . . . . . **245**
by Dana Greene

Introduction   245
U.S. Immigration Policy: Setting the Stage   246
Key Historic Benchmarks   248
Today's Immigration Landscape: Criminal Justice Redux   252
Does It Work? The Chimera of National Security   257
Consequences of a Crime Control Model   259

Conclusion   261
Discussion Questions   261
References   262

**12**   **Technology and Criminal Justice Policy** . . . . . . . . . . . . . . . . . . . **265**
by Phelan A. Wyrick

Introduction   265
From Technology Development to Saturation   268
Technology Profiles   276
Conclusion   283
Discussion Questions   283
References   284
Court Cases Cited   285

**13**   **White-Collar Crime and Public Policy:**
**The Sarbanes-Oxley Act and Beyond** . . . . . . . . . . . . . . . . . . . . . . **287**
by David O. Friedrichs (with Sarah Youshock)

Introduction   287
Conventional and White-Collar Offenders (and the Control
   of Conventional and White-Collar Crime) Compared   288
White-Collar Crime as a Focus of Public Policy   289
The Perceived Seriousness of White-Collar Crime   291
The Harms Caused by White-Collar Crime: Why It Merits More
   Public Policy Attention   292
Financier Crime and Public Policy   293
The Enron Case—And the Cases That Followed It   293
The Sarbanes-Oxley Act as a Public Policy Response to Enron   295
The Backlash to the Sarbanes-Oxley Act   297
Department of Justice Policy on Corporate Crime Cases   298
The Worst Economic Crisis Since the Great Depression and
   White-Collar Crime   299
Public Policy Responses to the Financier Crime at the Center
   of the Economic Crisis   302
Conclusion   302
Discussion Questions   303
References   303

**14**   **Criminal Justice Policy and Transnational Crime:**
**The Case of Anti-Human Trafficking Policy** . . . . . . . . . . . . . . . . . **307**
by Barbara Ann Stolz

Introduction   307
The Framework   308
Methods   311

U.S. Anti–Human Trafficking Legislation    312
Analysis: Stages in the Criminal Justice Policy-Making Process—
   The Case of Human Trafficking    319
Conclusion    324
Discussion Questions    326
Endnote    326
References    327
Laws Cited    331
Court Cases Cited    331

**15**   **When Is Crime a Public Health Problem?** . . . . . . . . . . . . . . . . . .   **333**
by Gregory J. DeLone and Miriam A. DeLone

Introduction    333
When Is Crime a Public Health Problem?    334
Integration of Two Intellectual Traditions    336
Conceptual Overlap of Two Disciplines    338
Conclusion    342
Discussion Questions    345
References    345

Subject Index    347

Author Index    369

# Contributors

Steven E. Barkan
University of Maine

Nicole L. Bracy
San Diego State University

Ben Brown
University of Texas at Brownsville

Jeffrey A. Butts
Public/Private Ventures in Philadelphia and New York City

Richard Culp
John Jay College
City University of New York

Gregory J. DeLone
University of Nebraska Omaha

Miriam A. DeLone
University of Nebraska Omaha

David O. Friedrichs
University of Scranton

Natasha A. Frost
Northeastern University

Dana Greene
New Mexico State University

Delores Jones-Brown
John Jay College

Erica King-Toler
John Jay College

Susan T. Krumholz
University of Massachusetts Dartmouth

Daniel P. Mears
Florida State University

David B. Rottman
National Center for State Courts

Barbara Ann Stolz
U.S. Government Accountability Office

Janeen Buck Willison
Justice Policy Center at the Urban Institute

Phelan A. Wyrick
Office of Justice Programs in the U.S. Department of Justice

Sarah Youshock
Senior Criminal Justice Major (Class of 2010) at the University of Scranton

# Preface

This collection of essays on contemporary U.S. criminal justice policy is a response to the significant recent growth of policy-oriented research in the fields of criminology and criminal justice. As this research has expanded, it has been accompanied by a proliferation in the number of college and university courses on crime policy offered in the United States. Influenced by both of these developments, the main goal of this book is to encourage students, instructors, and interested readers to engage in a dialogue about criminal justice policy and to think about the potential for criminal justice reform.

The authors invited to contribute to this book were asked to bring a policy focus to their areas of expertise. Specifically, they were asked to consider how criminal justice policy issues are framed; uncover who participates in the policy process; discuss how policy is made; consider the constraints and opportunities found in the policy process; link these findings to broader institutional, cultural, and global criminal justice trends; and grapple with what their research reveals about crime policy and democratic governance in the United States. Each essay approaches this challenge differently. Yet, as will become apparent in the pages that follow, all of the results are informative and thought provoking.

If there is one theme that connects each chapter, it is that criminal justice is an extremely broad and complex policy field. It features controversial trade-offs that attempt to balance security, equity, liberty, and efficiency. It consumes the time and energy of countless individuals, whether they are involved in developing, implementing, or evaluating crime policy. And it is emotionally charged, highly politicized, and often characterized by short-term fixes rather than by decisions based on sound policy analysis.

The publication of a "state-of-the-art" reader is especially timely considering the momentum currently building toward a full-scale review of U.S. criminal justice policy. Over the past 30 years, federal and state governments have been stiffening sanctions against criminal offenders. The United States now has the highest reported incarceration rate in the world, with 2.3 million (more than 1 in 100) American adults behind bars (Liptak, 2008). Minorities, especially black and Hispanic males, make up a disproportionately large share of prison populations. When the number of persons on probation or on parole is added to this number, 7.3 million (1 in 31) Americans are under some form of criminal justice supervision (S. 714, 2009). Each year, 700,000 prisoners are released from federal and state correctional facilities (Urban Institute, 2008, p. 4). The much publicized "crime drop" that has been documented by researchers in recent years has not reduced these numbers.

These trends have profound implications for the public policies pursued by governments in the United States. For example, according to a Pew Charitable Trusts report, in the last two decades state corrections spending increased 127%, while spending on higher education increased by a comparatively meager 21% (S. 714, 2009; *The New York Times*, 2009). On average, states spend almost 7% of their budgets on corrections, trailing only health care, education, and transportation (Liptak, 2008). Many of these facts and figures were recently cited in a bill introduced by Senator Jim Webb (D-VA) that calls for a comprehensive review of the criminal justice system (S. 714, March 26, 2009). The central contention of the bill, titled the National Criminal Justice Commission Act of 2009, is that U.S. criminal justice policy needs to be reformulated to "improve public safety, cost-effectiveness, overall prison administration, and fairness in the implementation of the Nation's criminal justice system." The essays included in this volume offer valuable insights that can inform this deliberative process as it moves forward.

## ■ ORGANIZATION OF THE BOOK

*U.S. Criminal Justice Policy: A Contemporary Reader* begins with an introduction that describes the context of criminal justice policy making. The chapter presents examples of what constitutes public policy and explores alternative images of the policy process. Special attention is devoted to the political process and to the various actors that attempt to influence the shape and direction of crime policy. Following the introduction, the book is organized into two major parts. Part I explores criminal justice policy as it relates to each "component" of the criminal justice system. Beginning with two chapters on contemporary policing policy, the section proceeds to analyze policy challenges and dilemmas facing the courts, corrections, and juvenile justice. Part II is designed to illustrate the breadth of criminal justice policy by exploring its intersection with other policy areas, including immigration policy, technology policy, regulatory policy, economic policy, international criminal justice policy, and public health policy. Each chapter provides insight into the past, present, and possible future of crime policy in America. It is those insights that become the raw material for the policy learning the book aims to convey to its readers.

## ■ References

Liptak, A. (2008, February 29). U.S. imprisons one in 100 adults, report finds. *The New York Times*. Retrieved March 11, 2010, from http://www.nytimes .com/2008/02/29/us/29prison.html

National Criminal Justice Commission Act of 2009, S. 714, 111th Cong., (2009).

Reviewing criminal justice. (2009, March 30). *The New York Times*. Retrieved March 11, 2010, from http://www.nytimes.com/2009/03/30/opinion/30mon1.html

Urban Institute. (2008). *The case for evidence-based policy*. Washington, DC: UI Press.

# Acknowledgments

This collection is the work of many people. I want to thank the authors of the various chapters included in this volume for sharing their time and knowledge. All have been gracious throughout the process, even when faced with responding to yet another communication from the editor. I am also grateful to my students for helping me think through the framework of the book while teaching courses on criminal justice policy.

The support and encouragement of numerous individuals at Jones & Bartlett Learning have been invaluable. I am especially grateful to Sean Connelly, Megan Turner, and Lisa Cerrone for their advice, assistance, and patience. I am also grateful to the reviewers who provided excellent comments and suggestions that led to a better final product: Kent R. Kerley, University of Alabama–Birmingham; Rebecca Paynich, Curry College; Anne Marie Rocheleau, Stonehill College; and Martha L. Shockey-Eckles, Saint Louis University.

This book has benefited immensely from conversations I have had with my colleagues over the years. I am especially grateful to Peter Cardalena, Keith Carrington, Giles Casaleggio, and Tom Ward, all of the Department of Criminal Justice and Legal Studies at St. John's University, for their wisdom and friendship.

This book was developed while I served as the Inaugural Chair of the Department of Criminal Justice and Criminology at Ryerson University in Toronto. I am deeply grateful to Dr. Carla Cassidy, former Dean of the Faculty of Arts and current Interim Vice President, Research, and Innovation, for providing me with the research support I needed to complete this project.

None of this could have happened without the love, patience, and understanding of my wife, Wendy. This is for you.

# About the Editor

**Karim Ismaili** is Associate Dean of the College of Humanities and Social Sciences, and Professor in the Department of Sociology and Criminal Justice at Kennesaw State University. He holds a PhD in political science from the University of Western Ontario. His areas of teaching, research, and publishing include criminal justice policy and analysis, penology, criminological theory, crime and inequality, and post-9/11 crime and security developments.

# About the Contributors

**Steven E. Barkan** is Professor of Sociology at the University of Maine. A former president of the Society for the Study of Social Problems, he is the author of *Criminology: A Sociological Understanding*, 4th edition (Prentice Hall), co-author of *Fundamentals of Criminal Justice: A Sociological View*, 2nd edition (Jones and Bartlett), and author of more than 30 journal articles and book chapters in the areas of criminology, law and society, and social movements.

**Nicole L. Bracy** is a Research Scientist at San Diego State University. She received a PhD in Criminology from the University of Delaware in 2009. Her research interests include police practices with youth, understandings and negotiations of juveniles' legal rights, and policing and punishment strategies in schools. Her research is published in *Youth and Society* and *Youth Violence and Juvenile Justice*.

**Ben Brown** is an Associate Professor of Criminal Justice at the University of Texas at Brownsville, where he teaches courses on law enforcement and research methods. He has conducted studies of crime, justice, and security issues in Mexico, South Korea, and the United States, and published his findings in numerous journals including *Crime and Delinquency*, *Police Practice and Research*, and *Journal of Interpersonal Violence*.

**Jeffrey A. Butts** is Executive Vice President for Research at Public/Private Ventures in Philadelphia and New York City. He has studied specialized courts for youth, drug treatment systems, and positive youth development as an intervention framework for youth justice. He is a graduate of the University of Oregon and holds a PhD from the University of Michigan.

**Richard Culp** is an Associate Professor of Public Administration at John Jay College of Criminal Justice and a member of the Doctoral Faculty in Criminal Justice at the City University of New York. Recent publications have appeared in *Journal of Contemporary Criminal Justice, Criminal Justice Policy Review, The Prison Journal,* and *Journal of Public Affairs Education.*

**Gregory J. DeLone** is an Assistant Professor in the School of Criminology and Criminal Justice at the University of Nebraska Omaha. He has been involved in numerous community service and research-oriented projects including an evaluation of the Nebraska State Patrol, strategic planning assistance to the Red Cross, and an evaluation of the Minor in Possession Diversion Program for the Health and Safety Council.

**Miriam A. DeLone** is an Associate Professor in the School of Criminology and Criminal Justice at the University of Nebraska Omaha. She co-authored *The Color of Justice* (Wadsworth Publishing) with colleagues Samuel Walker and Cassia Spohn. Her research interests include political economy and social control; race, ethnicity, gender, and sentencing; and corrections.

**David O. Friedrichs** is Professor of Sociology/Criminal Justice and Distinguished University Fellow at the University of Scranton (Pennsylvania). He is author of *Trusted Criminals: White Collar Crime in Contemporary Society*, 4th edition (Cengage, 2010) and *Law in Our Lives: An Introduction*, 2nd edition (Oxford University Press, 2006), and he is editor of *State Crime* (Ashgate, 1998). He served as President of the White Collar Crime Research Consortium (2002–2004).

**Natasha A. Frost** is an Assistant Professor in the College of Criminal Justice at Northeastern University. She holds a PhD in Criminal Justice from the City University of New York's Graduate School and University Center. Her primary research and teaching interests are in the areas of punishment and social control.

**Dana Greene** is an Assistant Professor in the Criminal Justice Department at New Mexico State University. Her interests include social change movements, restorative justice, penal history, penal abolition, and social control. She came to the study of punishment from a history of street activism and is deeply committed to applying her work beyond the academy.

**Delores Jones-Brown** is a Professor in the Department of Law and Police Science at John Jay College where she also directs the Center on Race, Crime, and Justice. Her research on race and policing has appeared in numerous journals and edited volumes. She is the author and co-editor of three books on race and justice. A former prosecutor, her current funded research examines the police use of confidential informants.

**Erica King-Toler** is an Assistant Professor at John Jay College in the SEEK Department. A licensed cross-cultural psychologist, her research focuses on helping professionals in education, nursing, and criminal justice improve their practice through the development of multicultural competence. Her EdM in higher education, MA in student personnel administration, and doctorate in counseling psychology were each earned at Teachers College, Columbia University.

**Susan T. Krumholz** received her JD from Seattle University and her PhD in Law, Policy and Society from Northeastern University. She is presently Director of the Crime and Justice Studies major at the University of Massachusetts Dartmouth. Her research interests include intimate violence, alternatives to the criminal/legal system, and women as students and practitioners of the law.

**Daniel P. Mears** is a Professor at Florida State University's College of Criminology and Criminal Justice. His research focuses on a range of theoretical and applied research topics in criminology, criminal justice, law, and sociology; his work has appeared in *Criminology, Journal of Research in Crime and Delinquency,* and *Law and Society Review.* Recent studies have focused on juvenile justice, homicide, prisoner reentry, and supermax prisons.

**David B. Rottman** is a Principal Court Research Consultant at the National Center for State Courts. His current research concerns public opinion on the courts, the evolution of court structures, and problem-solving courts. He has a PhD in sociology from the University of Illinois at Urbana and is co-author of books on community courts, social inequality, and modern Ireland.

**Barbara Ann Stolz**, Senior Analyst, U.S. Government Accountability Office, is a political scientist and criminologist who has worked in academia and government. She has published articles and two books on the making of criminal justice policy on capital punishment, juvenile delinquency, corrections, drug control, domestic violence, and intelligence surveillance. Her current area of research is human trafficking policy.

**Janeen Buck Willison** is a Research Associate in the Justice Policy Center at the Urban Institute where she evaluates innovative juvenile and criminal justice interventions. She has studied specialized courts, delinquency prevention initiatives, and faith-based reentry programs. Recent studies include the Assessing Policy Options (APO) project and implementation of an evidence-based jail transition strategy. She holds a master's degree in Justice Studies from American University.

**Phelan A. Wyrick** is a Senior Advisor in the Office of Justice Programs in the U.S. Department of Justice. He has over 15 years of experience conducting research and providing policy advice on federal, state, local, and tribal criminal and juvenile justice issues. From 2007 to 2009, he worked extensively on technology policy at the National Institute of Justice.

**Sarah Youshock** is a Senior Criminal Justice major (Class of 2010) at the University of Scranton, and an aspiring lawyer. She interned in the U.S. Attorney's Office in the Middle District of Pennsylvania, and is a member of Alpha Phi Sigma National Criminal Justice Honor Society.

# Thinking About Criminal Justice Policy: Process, Players, and Politics[1]

# CHAPTER

# 1

Karim Ismaili

## ■ Introduction

This chapter explores the relationship between criminology and public policy. It will argue that although criminology has been preoccupied with policy relevance since its inception (Gilsinan, 1991), contemporary criminal justice policy is all too often made without the benefit of criminological expertise. This problem is due largely to the fact that empirical evidence is easily overridden by the realities of the policy-making process. That process is complex, involves many actors, and functions under significant political and social pressure "to do something about crime." While criminologists have excelled in thinking about the causes and effects of crime, they have only very recently begun to think seriously about the policy-making process and their role within it. In order for research on the policy process to advance, the "real world" of criminal justice policy-making must be illuminated through the accumulation of knowledge both *of* and *in* the policy process (see Togerson, 1985).

## ■ What Is Public Policy?

Criminology is about controlling crime and achieving justice (Lafree, 2007; Welford, 1997). It is an applied social science in which criminologists conduct empirical research aimed at understanding, explaining, predicting, and preventing crime (Lanier & Henry, 1998, p. 2). What are the causes of terrorism? How can

[1]Portions of this chapter were published previously in K. Ismaili, *Crim. Justice Policy Rev.* 17 (2006): 255–269.

recidivism be reduced? Why is procedural fairness central to the legitimacy of the justice system? What constitutes an effective police practice? These are only some of the many varied questions explored by criminologists. Since one of the core responsibilities of governments in the United States is to provide for the safety and security of its citizens, it follows that criminological research findings have the potential to influence how this responsibility is defined and carried out. Yet sound research accompanied by results that have been replicated over time is no guarantee that Americans will benefit from this knowledge in their everyday lives. For improvements to occur, the results must be translated into public policy.

At its most general, public policy is a decision or action of government that addresses problems and issues (The Dirkson Congressional Center, 2006). In order for a policy to be considered public policy, it is widely accepted that "it must to some degree have been generated or at least processed within the framework of governmental procedures, influences and organizations" (Hogwood & Gunn, 1984, p. 24). Unlike the vast majority of private sector policies, public polices are funded through public resources and backed up by the legal system. This latter feature is particularly important to criminal justice policy since it is the threat of sanctions—including the loss of liberty (and possibly life)—that is a defining characteristic of this policy area.

The term *policy* has many uses. Some examples include (Hogwood & Gunn, 1984, pp. 12–19):

- Policy as a label for a field of activity (e.g., criminal justice policy, educational policy, welfare policy, military policy)
- Policy as a general purpose or desired state of affairs
- Policy as a specific proposal (put forward by an elected leader, government agency, or interest group)
- Policy as a decision of government (often arising from a crucial moment of choice)
- Policy as a formal authorization (e.g., a specific act, statute, or executive order)
- Policy as a program of activity (e.g., a legislative package that includes a mandate, organizational structure, and resources)
- Policy as an output (i.e., what governments actually deliver as opposed to what is promised or envisioned)
- Policy as a product of a particular activity
- Policy as a theory (if we do X, then Y will follow)
- Policy as a process that unfolds over a period of time

These descriptions not only provide some insight into the sheer range of activities that constitute policy, they also highlight the importance of considering the specific context in which the term policy is used.

## ■ Criminology and the Policy-Making Process

All public policy—including criminal justice policy—is the outcome of a process that is critical for criminologists to explore. Examining the policy process can facilitate an understanding of how the political process negotiates change, help to identify the constraints the process places upon the translation of ideas and analysis into action, describe the degree to which various actors influence the movement of criminal justice proposals through the policy process, and provide insight into how politics determines what is and can be implemented (Solomon, 1981, p. 5). This understanding, in turn, can shed light on how criminal justice policy differs from other policy realms, uncover whether crime policy tends to change more slowly than others, provide insight into whether there are greater obstacles to innovation in criminal justice, and describe how political structure and cultural traditions affect the policy process and account for variations between nations (Solomon, 1981, p. 6).

Despite the clear potential benefits of studying the criminal justice policy-making process, it is regrettable that very little is known about how criminal justice policy becomes the way it is (Jones & Newburn, 2002, p. 179). Trevor Jones and Tim Newburn have suggested two reasons for this state of affairs: First, criminologists have tended to focus their research on the *effects* of policy rather than on its *origins*; second, political scientists, while having a sophisticated notion of what policy is and how it comes about, have tended to neglect the field of crime control (Jones & Newburn, 2002, p. 179; emphasis added). To be fair, it is perfectly understandable why criminologists have been preoccupied with the effects of crime policy. Over the past several decades, these effects have been profound, particularly in the United States. Broad social transformations in governance (the rise of neo-liberalism) and politics (the emergence of a highly charged symbolic discourse about crime and the transformation of crime from a local and state issue into a federal one) have conspired to generate a "punitive turn" in criminal justice policy (Gest, 2001; Haggerty, 2004; Jacobson, 2006). The various policies associated with this turn—mandatory sentencing, private sector involvement in crime control, a victim-centered orientation to the criminal justice system, a dramatic increase in the use of prisons, etc.—have had an enormous impact on the nature of the policy research that has been undertaken by criminologists.

Unfortunately, the resulting research findings have not been influential in determining the course of criminal justice policy in the United States. Part of this failure has to do with the

> fact that in a society saturated with "crime talk," [criminologists] have utmost difficulty in communicating with politicians, policy makers, professionals and the public. Criminological reasoning is now mediated and contested by a range of vociferous interest groups, activists and a multitude of institutional actors and public opinions.

And criminologists are alienated from late modern political culture because crime, policing and punishment are defining electoral issues. (Chancer & McLaughlin, 2007, p. 157)

Moreover, little attention has been devoted to understanding the policy-making environment in which claims, counter-claims, and policy options are negotiated. Brownstein (2007) has described policy-making as a marketplace of claims wherein researchers compete and collaborate with lobbyists, politicians, government officials, and even other researchers. While each participant sets forth a favored position that will result in a desired outcome, criminologists—who are socialized to conduct objective research—find themselves "at a disadvantage in a marketplace filled with others more experienced and powerful than they are" (Brownstein, 2007, p. 126). It is important to note, however, that the estrangement of criminology from the "real world" of policy-making can be reversed if criminologists learn to be effective claims makers in the marketplace of claims, where policies are made and programs implemented (p. 130).

John Laub has recently stated that the belief that scholarly knowledge alone determines policy outcomes is naïve (Laub, 2004, p. 18). It is increasingly the case that empirical research about crime has an uneasy relationship with the values and needs that often dominate the world of politics and policy (Hawkesworth, 1988, chap. 3). If there is one lesson to be learned from the decline of criminological influence in the contemporary period, it is that pure reason competes with politics in shaping state responses to the crime "problem" (Zajac, 2002, p. 252). Put another way, both the process and environment of policy-making play a significant role in shaping policy outcomes. As Henry Ruth and Kevin Reitz (2003) have said:

> There is no doubt that data and empirical evidence supply only some of the inputs that influence the making of policy, and that they can be overridden by contrary moral sentiments, the tides of cultural change, the vagaries of politics, emotionalism, sensationalism, residual ignorance, and the inertial forces of laziness, habit, and vested interests. All of the messiness of real-world decision-making, even when fully acknowledged and experienced, does not diminish the importance of striving for an improved knowledge base. (pp. 39–40)

Understanding "the messiness of real-world decision-making" remains an important dimension of policy-making that could benefit from the cultivation of criminological expertise.

This book was conceived as a way to bring together a number of criminal justice experts based in universities, governments, and policy organizations devoted to meeting this challenge. It builds upon recent developments that have seen criminal justice policy emerge as a vital subfield within criminology. Over the past few years, new criminal justice policy journals have been founded to bridge "the gap between policy relevant findings and criminal justice policy" (Clear & Frost, 2007, p. 633); criminology and criminal justice conferences now regularly feature panels on a variety of policy issues; and colleges and universities have hired faculty members and designed policy-focused courses and programs based on the interest in and growth of this subfield. This intellectual and financial investment has generated a corresponding proliferation of research on a broad range of substantive policy issues. While the level of knowledge on the criminal justice policy-making process remains low, it too is evolving rapidly.

As the subfield of criminal justice policy attracts criminologists to engage in some form of policy analysis, understanding the policy-making environment in all of its complexity becomes more central to the enterprise of criminology. Building a theoretical infrastructure that supports the accumulation of knowledge about the criminal justice policy-making process would not only open an important field of inquiry, it might also help to reverse the decline in influence experienced by criminologists (Beckett, 1997; Currie, 2007; Garland, 2001; Gest, 2001; Ruth & Reitz, 2003). The pages that follow will present a "contextual approach" that can be used to examine the criminal justice policy-making environment and its accompanying policy-making process. As will be seen, one of the principal benefits of this approach is its emphasis on addressing the complexity inherent to policy contexts.

## ■ Analyzing the Policy-Making Process

Although some notable contributions to the policy-making literature have been made by criminologists (see Fairchild & Webb, 1985; Ismaili, 2006; Jones & Newburn, 2007; Rock, 1986, 1995; Stolz, 2002a), much of the influential thinking on this subject has originated in the disciplines of political science and public administration. The most influential view of the policy-making process to emerge is the classical perspective, which advanced rational principles as a way to rescue policy from the irrationality of politics. In this perspective, the policy-making process is depicted as a quasi-scientific enterprise that unfolds in a series of clearly defined steps (Sharkansky, 1992, p. 408):

1. Define the problem that is to be the object of policy-making.
2. Recognize the full collection of demands that are relevant to the decision that has to be made.
3. List all policy alternatives.
4. Define the resources necessary to achieve each alternative.
5. Calculate the benefits and costs associated with each alternative.
6. Make the decision on the basis of all relevant information in such a way as to achieve the most benefits at the least cost.

Deborah Stone (1988, p. 5) summarizes these steps in the following manner: "Identify objectives; identify alternative courses of action for achieving objectives; predict and evaluate the possible consequences of each alternative; select the alternative that maximizes the attainment of objectives."

While this rational, sequential, and mechanistic depiction of the policy process remains powerful today, it has been challenged by alternative models that highlight the "somewhat anarchic" nature of policy-making (Jones & Newburn, 2005, p. 61). For example, "the structured interaction perspective does not assume a single decision-maker, addressing a clear policy problem: it focuses on the range of participants in the game, the diversity of their understandings of the situation and the problem, the ways in which they interact with one another, and the outcomes of this interaction. It does not assume that this pattern of activity is a collective effort to achieve known and shared goals" (Colebatch, 1998, p. 102). Rather than being a rational enterprise, these models point to the volatility that is characteristic of the everyday policy-making process. Value differences, the role of interest groups, shifts in public mood, decisions based on political ambitions, and institutional constraints are only a few of the many challenges encountered in contemporary policy environments. Critics of the rational perspective contend that such forces make wholly rational policy action unlikely—especially in fields like criminal justice—and argue that policy research must move beyond a one-dimensional reliance on analytical/rational logic.

The late political sociologist Harold Laswell developed a contextual orientation to policy research that moves away from the narrow confines of the classical perspective by embracing the complexity found in various policy arenas. For Laswell, policy research was about making "sense of a vast and complex, often bewildering array of phenomena" (Marvick, 1977; Togerson, 1985, p. 245). He argued that the role of the policy analyst is to grapple with this "total configuration" and to recognize that the goal of understanding is never fully realized; it evolves and remains unfinished. Examples of the contextual orientation to policy research have ranged from formulations that are general to those that are very specific. Paul Sabatier (1991) presents a good example of the former:

> One of the conclusions emerging from the policy literature is that understanding the policy process requires looking at an intergovernmental policy community or sub-system—composed of bureaucrats, legislative personnel, interest group leaders, researchers, and specialist reporters within a substantive policy area—as the basic unit of study. The traditional focus of political scientists on single institutions, or single levels of government will help in understanding the effect of institutional rules on behaviour and, at times, in understanding specific decisions. But it is usually inadequate for understanding the policy process over any length of time. (p. 148)

A more specific formulation of the issues to be considered in a contextual analysis of public policy is offered by Fischer and Forester (1993). Their perspective emphasizes the context-specific character of analytical practises—the ways the symbolism of their language matters, the ways the consideration of their audience matters, the ways they construct problems before solving them. In political terms . . . [it teaches] us about the ways policy and planning arguments are intimately involved with relations of power and the exercise of power, including the concerns of some and excluding others, distributing responsibility as well as causality, imputing blame as well as efficacy, and employing particular strategies and problem framing and not others. (p. 7)

Ideally, a contextual analysis of public policy would include a combination of both of these levels of analysis. The pages that follow will outline and discuss three steps that should be taken in order to understand the criminal justice policy-making process. We will begin by identifying some of the peculiarities associated with this particular policy field.

## ■ Step 1: Thinking About the "Peculiarities" of Criminal Justice Policy

Observers of criminal justice have noted some peculiarities of the field that are important to consider before embarking on policy research and analysis. While these peculiarities are characteristic of criminal justice in most liberal democracies, the degree to which they influence public policy varies from jurisdiction to jurisdiction. Earlier in this chapter, the politicization of contemporary crime policy was presented as a challenge that has frustrated sound policy analysis. To this we can add two additional long-standing challenges: institutional fragmentation and the symbolic dimension of criminal justice policy and criminal law (Nagel, Fairchild, and Champagne, 1983a, p. 9).

An ongoing debate within the criminal justice literature involves the degree to which the state agencies and departments of the criminal justice process act as a coherent and unified system. Critics argue that the major components of the system—policing, courts, and corrections—carry out their respective mandates independently, generating systemic or institutional fragmentation. Diverse organizational objectives and differences in the use of discretionary powers exacerbate the fragmentation, leading many to view the criminal justice system as "a network of interrelated, yet independent, individuals and agencies, rather than as a system *per se*" (Griffiths & Verdun-Jones, 1994, p. 9). In such an environment, the development of coherent criminal justice policy that reflects the needs and expectations of each component of the system becomes a significant challenge. And in situations where a degree of consensus on policy is secured, difficulties can often arise at the implementation phase.

Nagel and his associates (1983a) contend that institutional fragmentation is the result of differences in training, status, and ideology among police, courts, and corrections personnel. They state:

These differences are reflected in the kinds of authority relationships that exist in the sub-system organizations. Positive sanction and normative power are more prevalent among legally-trained professionals in the system. Negative sanctions and coercive power tend to be stressed in law enforcement and, at least in relation to custodial work, in correctional organizations. All of these factors lead to diverse organizational climates that surround the different components of the system. (p. 9)

It is worth noting that institutional fragmentation can also exist within "sub-system organizations." For example, it is not uncommon to find executive and rank-and-file police officers, or prosecutors and defense lawyers, holding contrasting views on important policy issues.

Because criminal law and criminal justice policy are expected to embody fundamental principles of society, a distinctive characteristic of research in the area has been the attention devoted to analyzing the symbolic quality of crime and criminal justice (Gusfield, 1963; Hagan, 1983; Newburn & Jones, 2007). Crime and criminal justice are condensation symbols that have the potential to arouse, widen, and deepen public interest by appealing to ideological or moral concerns (Edelman, 1988; Scheingold, 1984, 1991). They condense a number of stresses that people experience in their day-to-day lives, and are powerful because they relate to the moral, ethical, and cultural concerns of the social order (Nagel et al., 1983a, p. 11). The subject matter of crime and criminal justice is a significant source of complexity. The potential for evoking strong responses suggests that crime and criminal justice symbols are especially vulnerable to transformation for strategic purposes. As Nagel et al. (1983a) have said:

> The symbolic implications of the criminal law and of law and order politics are particularly interesting because the emotional issues involved easily lend themselves to demagogic excesses. This is especially true in light of the fact that the workings of the criminal justice system are quite complex and not well understood by the public, which tends to oversimplify the issues that are involved. (pp. 11–12)

The variability of both the meaning of criminal justice and of the symbols underlying that meaning compels an analysis of the evolving material basis of the symbol, as well as the shifting socio-political environment in which it is lodged (Edelman, 1988; Hagan, 1983; Majchrzak, 1984). Newburn and Jones present an excellent example of this sort of policy research when they examine the recent popularity of the term *zero tolerance* in contemporary criminal justice. Zero tolerance, they argue, is now widely used by politicians, policy makers, and criminal justice officials "when there is a need to indicate strong measures and clear resolve" (2007, p. 222). In this sense, the term has been deployed to "convey a mood and to impress an audience rather than in any concrete way to describe a set of policies or to frame particular objectives" (p. 236).

The preceding discussion serves as a reminder of the complexity inherent to both the policy environment and the policy-making process in criminal justice (see **Figure 1–1**). It is not, however, a complete description of the various forces that influence the nature and direction of criminal justice policy. In order to enhance our understanding of this environment, it is important to organize for analysis the various institutions, groups, and individuals that participate in the policy-making process. Political scientists have developed just such an organiza-

■ **Political Culture**
  • Social and Economic Characteristics
  • Political Parties, Partisanship, and Ideology
  • Checks and Balances/Federalism

■ **Politicization of Crime**
  • Symbolic Dimension of Crime and Criminal Justice
  • The Definition and Construction of Policy "Problems"
  • Campaigns and Elections
  • Public Opinion
  • Policy Networks Within the Policy Community (*see Figure 1–2*)
  • Policy Trends in Other Policy Sectors and in Other Jurisdictions

■ **Institutional (Criminal Justice System) Cohesiveness/Fragmentation**

**Figure 1–1** Contextual Features of the Policy Environment
**Source:** Reproduced from K. Ismaili, *Crim. Justice Policy Rev.* 17 (2006): 255–269.

tional framework—the policy community—which captures the constellation of forces that influence the development of public policy in a variety of sectors. As will be discussed next, it is a framework that can be used to further contextualize the development of criminal justice policy.

## ■ Step 2: Thinking About the Criminal Justice Policy Community

Understanding the evolution of public policy requires researchers to pay special attention to the policy-making process, to the actors involved in that process (the public, professionals, and politicians), and to the sites where participants interact and policy decisions are made. A policy community

> is that part of a political system that—by virtue of its functional responsibilities, its vested interests, and its specialized knowledge—acquires a dominant voice in determining government decisions in a specific field of public activity, and is generally permitted by society at large and the public authorities in particular to determine public policy in that field. (Pross, 1986, p. 98)

It includes all actors or potential actors with a direct interest in the particular policy field, along with those who attempt to influence it—government agencies, pressure groups, media people, and individuals, including academics, consultants, and other "experts" (Pross, 1986).

## The Subgovernment and the Attentive Public

The policy community subdivides into two segments: the subgovernment and the attentive public. The subgovernment is composed of government agencies and institutionalized associations that actually make policy within the sector (Coleman & Skogstad, 1990, p. 25). It normally consists of a very small group of people who work at the core of the policy community. The attentive public, in contrast, is less tightly knit and more loosely defined. Its composition varies, but it usually contains important, though less central, government agencies, private institutions, pressure groups, specific interests, and individuals. As Pross states, "the attentive public lacks the power of the sub-government but still plays a vital role in policy development" (1986, p. 99).

Research has demonstrated that the structure and functions of policy communities vary from policy field to policy field (Pross, 1986, p. 106; see Coleman & Skogstad, 1990). Similarly, the relationships between actors in policy communities also vary. This reality has been captured in the concept "policy network." It refers to the relationships that emerge between both organizations and individuals who are in frequent contact with one another around issues of importance to the policy community (Atkinson & Coleman, 1992; Coleman & Skogstad, 1990). Atkinson and Coleman have commented that the concepts of policy community and policy network "appear to possess the required elasticity" to stretch across a variety of policy sectors (1992, p. 157). They are "encompassing and discriminating: encompassing because they refer to actors and relationships in the policy process that take us beyond political-bureaucratic relationships; discriminating because they suggest the presence of many communities and different types of networks" (p. 156).

Nancy Marion (2002) has developed a useful way to organize the various actors and significant forces that shape the criminal justice policy community. Her discussion begins with the observation that both the states and the federal government develop criminal justice policy in the United States. This concurrent power effectively means that "all levels of government have the right to act to control criminal behavior and create a safe society for its citizens" (2002, p. 30). With this in mind, it is more precise to state that 51 criminal justice policy communities coexist, each reflecting differences with respect to participants, cultural traditions, and political dynamics. As a system of government, federalism implies that intergovernmental relations represent a significant feature of the policy field, especially where crime control responsibilities overlap between the federal and state governments, or when the federal government attempts to create national criminal justice policy through its "power of the purse." While federalism represents a significant source of complexity for the criminal justice policy community, it is important to note that traits and experiences shared across jurisdictional and geographic boundaries can also serve to unite its diverse elements. As we will see, all criminal justice policy communities are encountering policy environments with increasingly diverse and rapidly maturing interest groups. All contain subgovernments dominated by a hierarchy of professional interests. And all are experiencing the pressures of an expanding attentive public.

Just as federalism provides an institutional context through which one can understand the development of criminal justice policy, so too does the system of checks and balances. Many elected officials of each branch of government are deeply involved in the criminal justice policy-making process; some are de facto members of the subgovernment. The executive branch actors of primary significance include the president (federal level), governors (state level), and mayors (local level) (Marion, 2002, p. 32). Elected members of the legislative branch, especially those serving on criminal justice legislative committees, can also be viewed as important members of the criminal justice policy community subgovernment. These individuals include elected members of congress, state legislatures, and town and city councils (Marion, 2002, p. 34). Finally, whether elected, as they are in 39 states (see Liptak, 2008), or appointed, judges shape criminal justice policy through the application, review, and interpretation of law. Often, underexamined judicial actors exert both a strong and steady influence on the work of the subgovernment. While not active participants in the policy-making work of the subgovernment in a conventional sense, they are, nonetheless, central to the criminal justice policy community.

Directly responsible for translating the political priorities of elected officials into public policy, appointed heads of criminal justice government departments and agencies are key participants in the policy community subgovernment. Also active in the initiation and screening of proposals for change in criminal justice policy are bureaucrats at all levels of government who have no operational responsibilities but who concern themselves with monitoring policies and advising elected officials (Solomon, 1991, p. 161). While not part of the formal criminal justice system, government departments and individual bureaucrats (or policy makers) are important to consider in terms of their efforts to reduce institutional fragmentation. The relationships that are struck between government policy-making agencies and the various components of the justice system are essential to its smooth functioning. Bureaucrats thus engage in a precarious balancing act, at once trying to accommodate the diverse needs and demands of the various system components, along with those of politicians, interest groups, and other interested members of the public.

The criminal justice policy community is also populated by a wide range of interest groups. Each group is committed to influencing the outcome of public policy, although the degree to which they are ultimately successful is subject to considerable variation. Stolz (2002b) has identified eight distinct types of interest groups that attempt to influence criminal justice policy: Professional-, business-, social welfare-, civic-, ad hoc-, victim-, ex-offender-, and offender-oriented interest groups mobilize resources and press their respective positions at various decision points in the policy process.

The interest groups that have traditionally had the most influence on criminal justice policy are those that represent professionals and other officials involved in the operation of the criminal justice system—police associations, bar associations, judicial organizations, and correctional associations (Fairchild, 1981). As active members of the policy community subgovernment, these interest groups represent professions with an institutionalized stake in the operation of the criminal law and the criminal justice system (Fairchild, 1981). Their influence is enhanced by the high degree of public deference accorded them, especially on policy matters concerning the day-to-day operation of the criminal justice system.

According to Nagel et al. (1983a; 1983b), the power of various professional groups reduces the influence of criminal justice clients (i.e., accused, convicted, and victims of crime) on the development of public policy. The largely closed, expert-driven criminal justice system, characterised by complicated laws and regulations, frustrates and disempowers clients, who often end up turning to underfunded and overworked third parties to represent their interests in the policy process. Considering the marginalization that the clients of criminal justice experience both in the system and in the wider society, it is not surprising that civil libertarian groups, prisoner advocacy groups, and other client-focused groups face an uphill battle in their attempts to influence the policy process. In such an environment, criminal justice policy often reflects little more than the bargaining and compromising of system players on matters of self-interest (Christie, 1993; Nagel et al., 1983a; 1983b).

### ■ Step 3: Thinking About Networks and Relationships Within the Policy Community

The relationships that develop between institutionalized interests and governments are considered crucial to the policy-making process. They are relationships based on a principle of reciprocal return. Governments, for example, can ill afford to develop policy that will be met with criticism from professionals. Close ties with professional groups are cultivated to preclude such occurrences. Views are solicited and perspectives shared. Their opinions are considered vital, and their support essential (Pross, 1986, p. 98). Similarly, professional organizations exist to ensure that their positions on issues are represented at various stages of the policy process. Their involvement goes a long way toward ensuring that this objective is met. It is this close and privileged access to the policy development process that often distinguishes the influence of professionally oriented groups from other interests in the criminal justice policy community.

In cases where a crime or criminal justice issue is in the public spotlight, elected officials are particularly responsive to public concerns and pay correspondingly less attention to views of policy professionals, including criminologists. It is therefore inaccurate to state that nonprofessional interest groups are completely shut out from the subgovernment in the criminal justice policy community. The views of a number of reform-oriented interest groups are considered vital to the policy process, particularly those with an established presence in the policy community. Indeed, if there is one trend that has characterized the policy community over the past three decades, it is the increasing number of maturing interests that have developed around issues of criminal justice. Institutionalized victims' groups, groups working toward the elimination of violence against women, and other groups focused on the needs of minorities are proliferating. All are seeking to influence the shape and direction of public policy.

The "attentive public" in the criminal justice policy community is also expanding. As Fairchild (1981) has stated, "Matters of criminal law go to the heart of questions about governmental legitimacy, state authority, and other popular conceptions of right and wrong, and are thus of closer concern to many individuals than are most other legislative issues" (p. 189). This explains, in large part, the increase in the number of ad hoc and single-issue reform interests that have recently been created around various criminal justice issues.

The attentive public obtains much of its information about crime and the criminal justice system from the media. Because the coverage of crime is so prominent in "all means of mass communications, including daily and weekly newspapers, television, radio, news magazines, and so on" (Marion, 2002, p. 39), both the quantity and nature of media imagery can have a significant influence on how crime is perceived and, ultimately, on which criminal justice policies are pursued. Katherine Beckett has argued that the coverage of crime in the media may influence the actions of both elected and unelected actors in the policy community, independent of any impact on public opinion. This theory is manifested when policy makers interpret heightened media coverage as an indication of public concern warranting public action, or as an opportunity for political exposure and/or direct political gain. The latter is especially true in the run-up to or during an election campaign. Beckett also argues that media coverage "is undoubtedly a crucial component of the context in which public opinions are formed" (Beckett, 1997, p. 78).

According to Gray Cavender (2004), depictions of crime and criminals began to change in the 1970s as a response to and reflection of rising crime rates in the United States. During this period, the media gave extensive attention to "a sense of social malaise that was defined by the belief that values were in decline and that modern life had become unpredictable and dangerous" (p. 345). Cavender asserts that this basic narrative structure has been in place since then and explains in large part why fear of crime continues to be high despite declines in the overall crime rate. Viewed in this light, the manner in which crime and criminal justice issues are framed by the mainstream media becomes a significant contextual feature of the policy community. As Marion (2002) has stated:

> The media is important because it educates the public about crime. Unfortunately, the media's coverage of crime events does not reflect reality. The media tend to cover crimes that occur less frequently such as mass shootings and extremely violent offenses.

Although they make for good media ratings, these types of crimes are rare. In addition to misrepresenting the types of crimes committed, television shows tend to depict stereotypes of criminals, prisoners, and victims that are usually not accurate. (p. 40)

Perceptions are, by their very nature, malleable. When those perceptions relate to crime, they are rarely grounded on sound, accurate information. What remains is an apparent gulf between two related but independent domains: attitudes about crime and knowledge about crime (Roberts, 1994, p. 1). It is difficult to narrow this gulf in a policy sector where political reaction to the public's anxiety over crime is commonplace. This is especially troubling since it is these very reactions that often reinforce and perpetuate the inaccurate perceptions held by a large segment of the population.

**Figure 1–2** summarizes the criminal justice policy community described thus far. It should be emphasized that the policy community is a dynamic entity,

## The Subgovernment

- **Elected (Executive) Actors**
  - President, Governors, Mayors
- **Key Elected Legistative Actors**
  - Senators, Representatives
- **Major Interest/Pressure Groups**
- **Appointed Heads of Government Departments and Agencies**
  - Cabinet Secretaries (U.S.)
- **Key Judicial Actors** (not active participants in the work of the subgovernment, but a major influence on the products of that work)

## The Attentive Public

- **Politicization of Crime**
  - The Media
  - Less Central Government Agencies
  - Experts, Academics, and Consultants
  - Interest/Pressure Groups
  - Elected Officials
  - Interested Members of the Public
  - Private Institutions and NGOs
  - Judicial Actors

**Figure 1–2** The Criminal Justice Policy Community
**Source:** Reproduced from K. Ismaili, *Crim. Justice Policy Rev.* 17 (2006): 255–269.

constantly shifting and evolving to meet the needs and expectations of diverse groups and individuals with an interest in both the process and substance of criminal justice policy.

## ■ Conclusion

This chapter began with the observation that the policy-making process has been largely neglected in studies of crime policy. It has argued that in order for a truly policy-oriented criminology to advance, it is vital that this process be explored in all of its complexity. Drawing on Harold Laswell's call for policy researchers to come to terms with the entire complex, shifting policy universe, this chapter has proposed a contextual approach to criminal justice policy analysis. It is a form of policy analysis that insists that the researcher grapple with the complexity inherent to the criminal justice policy-making process in order to accumulate knowledge both *of* and *in* the policy process. For those interested in the criminal justice policy-making process, this complexity is both an attraction and a significant challenge. Three steps have been identified in order to meet this challenge. The first step is identifying the various "peculiar" features that are more or less unique to the criminal justice policy environment. Step two involves organizing this environment into a policy community with a subgovernment and an attentive public. This organization not only provides shape to the policy sector, it also helps to expose patterns of power and influence. A final step requires the researcher to uncover the networks and relationships that evolve within the policy community and to consider the implications of these patterns for the development of policy. As the subfield of criminal justice policy matures, so too must its theoretical frameworks and conceptual tools. With the potential of significantly enhancing our understanding of democracy in action, the contextual approach presented here is a reflection of and response to this maturation.

## ■ Discussion Questions

1. Compare and contrast the classical and contextual orientations to policy research.
2. Discuss the relationship between politics and criminal justice policy.
3. Can criminal justice policies ever be value-free?
4. Why is it important to think about the "peculiarities" of criminal justice when discussing policy?
5. Select a controversial criminal justice issue and use the policy community framework to outline the participants, networks, and relationships that will likely influence the policy-making process.

## ■ References

Atkinson, M., & Coleman, W. D. (1992). Policy networks, policy communities and problems of governance. *Governance: An International Journal of Policy and Administration, 5*, 155–180.

Beckett, K. (1997). *Making crime pay: Law and order in contemporary American politics.* New York: Oxford University Press.

Brownstein, H. (2007). How criminologists as researchers can contribute to social policy and practice. *Criminal Justice Policy Review, 18*(2), 119–131.

Cavender, G. (2004). Media and crime policy. *Punishment and Society, 6*(3), 335–348.

Chancer, L., & McLaughlin, E. (2007). Public criminologies: Diverse perspectives on academia and policy. *Theoretical Criminology, 11*(2), 155–173.

Christie, N. (1993). *Crime control as industry: Towards gulags, Western style?* London: Routledge.

Clear, T., & Frost, N. (2007). Informing public policy. *Criminology and Public Policy, 6*(4), 633–640.

Colebatch, H. (1998). *Policy.* Buckingham: Open University Press.

Coleman, W. D., & Skogstad, G. (Eds.). (1990). *Policy communities and public policy in Canada.* Mississauga: Copp Clark Pitman Ltd.

Currie, E. (2007). Against marginality: Arguments for a public criminology. *Theoretical Criminology, 11*(2), 175–190.

The Dirksen Congressional Center. (2006). Congress link: Vocabulary. Retrieved February 3, 2010, from http://www.congresslink.org/print_lp_whyneed congress_vocab.htm

Edelman, M. (1988). *Constructing the political spectacle.* Chicago: University of Chicago Press.

Fairchild, E. S. (1981). Interest groups in the criminal justice process. *Journal of Criminal Justice, 9*, 181–194.

Fairchild, E. S., & Webb, V. J. (Eds.). (1985). *The politics of crime and criminal justice.* London: Sage Publications Ltd.

Fischer, F., & Forester, F. (Eds.). (1993). *The argumentative turn in policy analysis and planning.* Durham, NC: Duke University Press.

Garland, D. (2001). *The culture of control: Crime and social order in contemporary society.* Chicago: University of Chicago Press.

Gest, T. (2001). *Crime and politics: Big government's erratic campaign for law and order.* New York: Oxford University Press.

Gilsinan, J. F. (1991). Public policy and criminology: An historical and philosophical reassessment. *Justice Quarterly, 8*(2), 201–216.

Griffiths, C. T., & Verdun-Jones, S. N. (1994). *Canadian criminal justice.* Toronto: Harcourt, Brace and Company.

Gusfield, J. (1963). *Symbolic crusade.* Urbana, IL: University of Illinois Press.

Hagan, J. (1983). The symbolic politics of criminal sanctions. In S. Nagel, E. Fairchild, and A. Champagne (Eds.), *The political science of criminal justice.* (pp. 27–39). Springfield, Illinois: Thomas Books.

Haggerty, K. (2004). Displaced expertise: Three constraints on the policy relevance of criminological thought. *Theoretical Criminology, 8*(2), 211–231.

Hawkesworth, M. E. (1988). *Theoretical issues in policy analysis.* Albany: SUNY Press.

Hogwood, B. W., & Gunn, L. A. (1984). *Policy analysis for the real world.* New York: Oxford University Press.

Ismaili, K. (2006). Contextualizing the criminal justice policy-making process. *Criminal Justice Policy Review, 17*(3), 255–269.

Jacobson, M. (2006). Reversing the punitive turn: The limits and promise of current research. *Criminology and Public Policy, 5*(2), 277–284.

Jones, T., & Newburn, T. (2002). Policy convergence and crime control in the USA and UK: Streams of influence and levels of impact. *Criminal Justice, 2*(2), 173–203.

Jones, T., & Newburn, T. (2005). Comparative criminal justice policy-making in the United States and the United Kingdom: The case of private prisons. *British Journal of Criminology, 45*(1), 58–80.

Jones, T., & Newburn, T. (2007). *Policy transfer and criminal justice.* Maidenhead, UK: Open University Press.

Lafree, G. (2007). Expanding criminology's domain. *Criminology, 45*(1), 501–532.

Lanier, M., & Henry, S. (1998). *Essential criminology.* Boulder, CO: Westview Press.

Laub, J. H. (2004). The life course of criminology in the United States. *Criminology, 42*(1), 1–26.

Liptak, A. (May 25, 2008). Rendering justice, with one eye on re-election. *The New York Times.* Retrieved February 3, 2010, from http://www.nytimes.com/2008/05/25/us/25exception.html?_r=1

Majchrzak, A. (1984). *Methods for policy research.* London: Sage Publications.

Marion, N. E. (2002). *Criminal justice in America: The politics behind the system.* Durham, NC: Carolina Academic Press.

Marvick, D. (Ed.). (1977). *Harold D. Laswell on political sociology.* Chicago: The University of Chicago Press.

Nagel, S. et al. (1983a). Introduction. In S. Nagel, E. Fairchild, and A. Champagne (Eds.), *The Political Science of Criminal Justice.* (pp. 5–13). Springfield, Illinois: Thomas Books.

Nagel, S. et al. (1983b). Introduction: General relations between political science and criminal justice. In S. Nagel et al. (Eds.), *The Political Science of Criminal Justice.* (pp. ix–xii). Springfield, Illinois: Thomas Books.

Newburn, T., & Jones, T. (2007). Symbolizing crime control: Reflections on zero tolerance. *Theoretical Criminology, 11*(2), 221–243.

Pross, A. P. (1986). *Group politics and public policy.* Toronto: Oxford University Press.

Roberts, J. (1994). *Public knowledge of crime and justice: An inventory of Canadian findings.* Canada: Department of Justice.

Rock, P. (1986). *A view from the shadows: The Ministry of the Solicitor General of Canada and the making of the Justice for Victims of Crime initiative.* Oxford: Clarendon Press.

Rock, P. (1995). The opening stages of criminal justice policy-making. *British Journal of Criminology, 35,* 1–16.

Ruth, H., & Reitz, K. R. (2003). *The challenge of crime: Rethinking our response.* Cambridge, MA: Harvard University Press.

Sabatier, P. A. (1991). Toward better theories of the policy process. *PS: Political Science and Politics, June,* 147–156.

Scheingold, S. A. (1984). *The politics of law and order: Street crime and public policy.* New York: Longman.

Scheingold, S. A. (1991). *The politics of street crime: Criminal process and cultural obsession.* Philadelphia: Temple University Press.

Sharkansky, I. (1992). What a political scientist can tell a policy-maker about the likelihood of success or failure. *Policy Studies Review, 11*(3-4), 406–422.

Solomon, P. (1981). The policy process in Canadian criminal justice: A perspective and research agenda. *Canadian Journal of Criminology, 23,* 5–25.

Solomon, P. (1991). Politics and crime: A survey. In J. Gladstone, R. V. Ericson, and C. D. Shearing (Eds.), *Criminology: A readers guide* (pp. 157–176). Toronto: University of Toronto, Centre of Criminology.

Stolz, B. A. (2002a). *Criminal justice policymaking: Federal roles and processes.* Westport, CT: Praeger.

Stolz, B. A. (2002b). The roles of interest groups in US criminal justice policymaking: Who, when, and how. *Criminal Justice, 2*(1), 51–69.

Stone, D. (1988). *Policy paradox and political reason.* Boston: Scott, Foresman/ Little.

Togerson, D. (1985). Contextual orientation in policy analysis: The contribution of Harold D. Laswell. *Policy Sciences, 18,* 241–261.

Welford, C. (1997). Controlling crime and achieving justice. *Criminology, 35*(1), 35–79.

Zajac, G. (2002). Knowledge creation, utilization and public policy: How do we know what we know in criminology? *Criminology and Public Policy, 1*(2), 251–254.

# Criminal Justice Policy and the Criminal Justice System

# PART

# I

CHAPTER 2    The Significance of Race in Contemporary Urban Policing Policy

CHAPTER 3    Combative and Cooperative Law Enforcement in Post-September 11th America

CHAPTER 4    Policies Promoting School–Police Partnerships

CHAPTER 5    Procedural Fairness, Criminal Justice Policy, and the Courts

CHAPTER 6    Criminal Justice Policy and Problem-Solving Courts

CHAPTER 7    U.S. Corrections Policy Since the 1970s

CHAPTER 8    Reentry as a Process Rather Than a Moment

CHAPTER 9    Prison Privatization Turns 25

CHAPTER 10    The U.S. Juvenile Justice Policy Landscape

# The Significance of Race in Contemporary Urban Policing Policy

## CHAPTER 2

Delores Jones-Brown and Erica King-Toler

### ■ Introduction

The difficulty in writing a policing policy chapter lies in the fact that law enforcement in the United States consists of a complex network of sometimes overlapping or competing structures and functions. U.S. policing exists at the federal, state, and local levels. It includes specialized units such as park police, forest rangers, border patrols, drug enforcement agents, and immigration enforcement officers. It serves multiple, often not agreed-upon functions, including crime control, order maintenance, and service provision (Zhoa & Thurman, 1997). It exists in starkly different demographic and geographic contexts that are urban, suburban, rural, and international. It serves multiple constituents who may have significantly contradictory interests.

In addition, according to policing scholars, policing in the United States has changed over time and within and across various contexts. The political era, the reform era, and the community era have been described as the three major periods of U.S. policing reform (Kelling & Moore, 1988). This three-part typology has been criticized for its failure to acknowledge the significant role of law enforcement agents in maintaining racial dominance through slave patrols and the enforcement of discriminatory laws and practices even within its own ranks (Reichel, 1988; Williams & Murphy, 1990).

Any candid discussion of policing policy in the United States must acknowledge that all roads lead to considerations of race. From the 1967 presidential report on The Challenge of Crime in a Free Society and the 1968 Report of the National Advisory Commission on Civil Disorders to the 2007 RAND Corporation report on stop, question, and frisk practices in New York City, there is substantial evidence that, with few exceptions, virtually all implementation of police policy has significant racial effects. The causes of these race effects have been studied and debated for more than 40 years, producing volumes of theories, findings, and conclusions that are arrived at through varying personal orientations, theoretical paradigms, and social science methodologies. Few areas of consensus have emerged. There is a significant degree of divergence in thinking as to whether these race effects are warranted based on law-breaking behavior (see Russell, 1998; Joseph, 2000), and there is substantial disagreement as to whether anything can be done by criminal justice agencies to reduce or eliminate them.

From a practical standpoint, it cannot validly be contested that racial and ethnic minorities make up a substantial portion of the population in places that are heavily policed. Reported data indicate that these tend to be places where crime rates and calls for police service are high. Some have questioned whether the deployment of extra police resources to those places in some way increases the official crime rate, because these extra resources appear to result in a substantial number of arrests for behaviors that go undetected, unreported, or unprosecuted in other locations (see Beckett, Nyrop, & Pfingst, 2006). The counter to that suggestion has been that the high rate of violent crime, particularly homicide, would be a cause of public concern and police attention, regardless of where and among whom it occurs. But, as will be discussed later in this chapter, in an effort to interrupt patterns of violent crime commission, some urban police departments have adopted policing strategies that target low-level crime and that produce racially disparate rates of arrest (see Golub, Johnson, & Dunlap, 2007). The evidence regarding whether this practice actually reduces the occurrence of serious crime is equivocal at best.

Public policy has been defined as "the rules and regulations legislative bodies and agencies *choose* to establish" (Conser, Russell, Paynich, & Gingerich, 2005, p. 10; emphasis added) or, in short, "anything that the government chooses to do or to not do" (Conser et al., 2005, p. 342). This chapter examines the collection of governmental choices that have been made about how best to address issues of crime and disorder. It focuses on urban policing since a substantial portion of the nation's law enforcement resources is concentrated in cities. This is not to suggest that smaller police departments do not also utilize particular policing approaches when they are considered popular, effective, or potential sources of federal or state funding (Kraska & Cubellis, 1997).

Primarily, the chapter examines the historical origins of urban policing policy and its efficacy over the last 40 years. We give particular attention to the community policing and "broken windows" paradigms. While each paradigm involves somewhat divergent methods and ideologies, there are areas of significant overlap: both enjoy substantial academic and practical explication and debate; both make substantial claims to success; and both enjoy varying levels of support from a diverse array of constituents. We begin with a fuller discussion of the role that race has played in the development of policing policy from the colonial period to current day. We then connect race and ethnicity to the challenge of balancing effectiveness and fairness when developing and implementing contemporary police policy. We move to a critique of the efficacy of community policing and related approaches (e.g., problem solving), question their novelty, and use New York City's CompStat and broken windows-style policing to highlight problems that emerge from aggressive arrest-focused enforcement. By comparing the New York crime decline with that which occurred in San Diego, we raise the question of whether such policing is the most efficacious and humane means of producing public order. In the final section, we recommend that urban policing, in particular, can be substantially improved by the uniform adoption of a multicultural competence approach. We believe that multicultural competence training can help bridge the police–community gap, increase the fairness and effectiveness of policing, and generally improve the delivery of policing services in diverse and highly policed communities.

## ■ The History of Policing Race and Place

As mentioned, the strategic history of American policing has been described as falling into three general eras. During the political era, roughly the 1840s until the early 1900s, American police derived their authority and resources from local political leaders. In addition to their responsibility for maintaining order and preventing and controlling crime, police ran soup lines, provided lodging for immigrant workers, and assisted immigrants in finding work (Kelling & Moore, 1988).

These close ties to politicians and to citizens became a breeding ground for corruption, such that by the 1920s there was a call for police reform. The reform era, which lasted from about the 1930s to the 1970s (Cooper, 2005), called for the police to narrow their function to crime control and the apprehension of those who violated the criminal law. Policing was expected to become a politically disconnected profession (Kelling & Moore, 1988). Under this professional model, "officers were experts; they applied knowledge to their tasks and were the only ones qualified to do the job" (Dunham & Alpert, 2004, p. 31). In addition, "the department was [considered] autonomous from external influences"; it "made its own rules and regulated its [own] personnel." However, by the 1960s, the efficacy of the policing model touted during the reform era came under considerable attack. Outbreaks of black urban violence, most often sparked by police behavior, required yet another shift in police policy and function. As will be discussed in a later section, the community era of policing was ushered in during the 1970s and still claims considerable prominence today.

But well before these three eras, the American government had chosen to form "policing" agencies to interdict another type of crime. In 1619, the American colonies began importing African slaves who sometimes ran away. The southern slave patrols, identified as the precursors to modern American policing (Reichel, 1988; Williams & Murphy, 1990), were authorized by federal, state, and local laws to capture fugitive slaves and to detain others who facilitated the slaves' escape. The original federal constitution contained a fugitive slave clause in Article 4, Section 2. To enforce the slave-catching requirements, Congress enacted two versions of a fugitive slave law, one in 1793 and another in 1850 (Higginbotham, 1980; Jones-Brown, 2000; McIntyre, 1984). In the early colonial period, slave catchers were paid with private money, but as governments formed and solidified, lawmakers passed legislation authorizing payment from public treasuries.

In their official capacity, law enforcers did not provide assistance to these *in*voluntary immigrants as they would later provide for European immigrants. These early "police" agencies were purely mechanisms for maintaining racial dominance, containment, and control. In Boston, a city credited with having one of America's first organized urban police forces, as late as 1851, posters appeared warning blacks to stay away from watchman and police officers because they were employed in kidnapping, catching, and keeping slaves (Higginbotham, 1980). The American system of African slavery was maintained for more than two centuries. It was underpinned by myriad legal restrictions applied against people, both slave and free. American police were responsible for enforcing those restrictions along with the race-neutral mandates of the general criminal law.

The abolition of slavery did not much change the relationship between blacks and big-city policing authorities. Rather than choose to adopt inclusive policies that would counter the effects of more than two centuries of slavery, individual states and local governments in the North and the South enacted racially discriminatory laws that were enforced against "negroes" and other ethnic minorities. The public and the private enforcement of these laws, and the failure of the federal judiciary to condemn them, contributed to the creation of urban ghettos, defined by the 1968 Kerner Commission report as "area[s] within a city characterized by poverty and acute social disorganization, and inhabited by members of a racial or ethnic group under conditions of *in*voluntary segregation" (National Advisory Commission, 1968, p. 29; emphasis added).

Urban cities and their ghettos became the site of significant levels of crime and disorder. In order for the nation to maintain cohesion and legitimacy, there needed to be a governmental response. The professional policeman of the reform era might have been a viable solution if he had not been a myth. As noted by Dunham & Alpert (2004), "the reforms failed because the idea of policing could not be divorced from politics," and "the character of the big-city police was interconnected with policy-making agencies that helped to decide which laws were enforced, which public was served, and whose peace was kept" (p. 31). In the 1960s, the crime problems in major urban cities were no longer limited to individual and small group (i.e., gang) commission of interpersonal crime. Collective violence erupted in major cities including New York, Chicago, Detroit, Los Angeles, and Newark, New Jersey. All of these episodes, which have ironically and interchangeably been referred to by the contradictory terms "riot" and "civil disorder," were connected in some way to police behavior. The findings of the Kerner Commission candidly acknowledged that "in practically every city that . . . experienced racial disruption . . . abrasive relationships between police and Negroes and other minority groups have been a major source of grievance, tension and ultimately disorder" (National Advisory Commission, 1968, p. 157).

Part of the blame for urban violence and disorder was placed squarely on the shoulders of the police—not simply based on the conduct of individual officers, but also on the attributes of the attempted reform. During the reform era, police had been removed from foot patrol and placed in radio cars. They were told that as experts their goal was to make arrests and answer radio calls. Police had little direct contact with the people they policed. Rather than increasing their effectiveness, the loss of interaction with people resulted in decreases in case clearances (Cooper, 2005, pp. 2–3). When police did interact with the community, the interactions created rather than diminished problems. Black civilians became highly concerned with police harassment and police brutality. In a survey conducted by the Detroit Free Press in 1968, police brutality was named as the number one problem. The consistent series of riots evidenced that police reform had failed in significant ways. Instead of being politically disconnected professionals, examination of the incidents found police agents who were racially prejudiced and abusive (both physically abusive and abusive of their authority), or who simply lacked legitimacy[1] as a controlling force.

In New York City, demonstrators reacted violently to the fatal shooting of a 15-year-old black male by a white police officer. In Los Angeles, the reaction was to the arrest of a black motorist, his brother, and mother during what started as a traffic stop. In Newark, a black cab driver was subject to a severe police beating. In Detroit, the cause was a vice squad raid on an after-hours club in a predominantly black neighborhood. In Chicago, the 1966 "race riots" involved blacks, whites, and Latinos; the Division Street riot was in response to the police shooting of a young Puerto Rican man, and, in the suburb of Cicero, police who shared the racial views of white mob members failed to protect black marchers who entered the all-white community in search of access to better housing. The various riots lasted from a minimum of 2 days to a maximum of nearly a week and resulted in tremendous property damage, along with the death, injury, and arrest of many people.

These urban disturbances were indicators that the police had failed to contain and control those thought to be the most dangerous (National Commission on the Causes and Prevention of Violence, 1969). Violence had spilled over beyond their segregated and socioeconomically disadvantaged enclaves. A president and a nation wanted to know how this was possible and how to make it stop. The President's Commission on Law Enforcement and Administration of Justice (1967) and the Kerner Commission Report (National Advisory Commission, 1968) acknowledged the difficulties in attempting to enforce the letter of the law in areas of high poverty, low employment, poor housing, and overt racial and ethnic segregation and discrimination.

There was a call for retreat from the police policies and practices that had marked the era of reform. There was a push to renew partnerships with the community, to educate and train officers, and to measure success by the reduction of crime (Cooper, 2005). In keeping with the recommendations from the two prominent commissions, many efforts were made to organize and systematize the administration of American justice. It was believed that the system reforms, including those in policing, would improve its legitimacy and make it more effective. But outside the sphere of policing, few sustainable improvements occurred in the social structure of the ghetto. In fact, over time, poverty, social isolation, and racial/ethnic segregation increased (Sampson & Wilson, 1995; Wilson, 1987; Wilson, 1997); riots, primarily sparked by police behavior, continue to occur. Now, however, they are treated as isolated incidents, and there is no collective national or presidential response.[2]

We suggest that then and now many police officials attempt to respond to crime and disorder in ghetto neighborhoods with inadequate training regarding how best to effect and encourage community change. Instead, the prevailing response is one of containment and control. Many officers come from backgrounds that are starkly different from the communities they police. The social distance between the police and those they police disadvantages everyone in the relationship. We contend that for the most part, police continue to apply a middle-class measuring rod (Cohen, 1955) against people who live under circumstances that are far removed from such standards. This is no accident; most mainstream policing policies require them to do so. Police are expected to require ghetto residents to adhere to the general dictates of the law in the same way as nonghetto residents, even though the conditions of their lives are extremely different. Officers' strict adherence to what they believe to be the letter of the law may lead them to enforce standards of behavior that may be at odds with ghetto residents' economic capacity and self-interest—for example, arresting the homeless for vagrancy or trespassing; enforcing curfew laws against youth who may be attempting to avoid abusive or otherwise dysfunctional homes; enforcing gang ordinances in ways that fail to adequately distinguish between real gang members, wannabes, and non-delinquent youth (see Meares & Kahan, 1999; Meares & Skogan, 2004; Moskos, 2008; *City of Chicago v. Morales*, 1999). However, in so doing, the police reduce their own legitimacy as they depend on significant levels of coercion rather than cooperation in their efforts to reduce crime. And among those approaches that are designed to encourage police–community partnerships, rarely do they garner the participation of persons in the community from the most crime-prone groups. The history reported earlier in this chapter has taught that police coercion in such communities tends to contribute to crime and disorder problems rather than resolve them.

## ■ The Challenge of Urban Policing

One source has described the role of police in society as "to effectively control crime *and* ensure justice" (National Research Council, 2004, p. 2; emphasis added). Crime control may be objectively measured by crime statistics such as complaints, arrests, and clearance rates, but how do we measure justice? Legal philosophers, such as John Rawls (1971), have associated notions of justice with the principle of equal treatment under the law. But, contemporary police practices have raised important questions about in whose interest the police police. Ideally, the answer to this question is that the police police consistent with the best interest of all people. That is, that policing practice is distributed equitably in ways that benefit all of the law-abiding and that potentially leads to punishment of an equal distribution of those individuals who violate the law. However, conflict theory has always suggested that police act in the interest of the politically and socially powerful (see Chambliss & Mankoff, 1976) and that by its very nature policing is oppressive to those who are politically and socially powerless.

For historical reasons addressed earlier in this chapter, political and social powerlessness has traditionally been embedded in communities where the concentration of racial/ethnic nonwhite minorities is high. However, with the growth in community and grassroots organizing, those previously thought to be powerless (e.g., minority inhabitants of the ghetto) have theoretically become less so. In this, the post-civil rights era, minority members of society expect policing that will be different from the pre-civil rights era. Not unlike the 1960s, leaders within even the poorest high-crime minority communities ask for policing that is *fair and effective* (National Research Council, 2004).

### Is Urban Policing Policy Fair?

Among the common complaints leveled against policing policies that currently seek to address crime in urban neighborhoods, particularly those with high concentrations of racial/ethnic minorities, are that the policies are implemented in ways that target or criminalize the entire community—including the law abiding or minimally criminal (West, 2001)—and that the implementation of those policies has a significant negative impact on the life chances of community residents (Western, 2006). For example, during the 1980s, as part of its quality-of-life (QOL) approach, instead of issuing summonses, the New York City Police Department (NYPD) began to make custodial arrests for minor offenses. During the 1990s, smoking marijuana in public view (MPV) was particularly targeted. By 2000, MPV had become the leading arrest charge under QOL policing in the city (Golub, Johnson, & Dunlap, 2007). Between 1980 and 2003, blacks and Latinos combined made up between 74% and 91% of MPV arrests each year. These arrest figures were significantly disproportionate to their representation within the city's general population and significantly greater than the rate of similar arrests for their white counterparts.

Once arrested, blacks and Latinos were substantially more likely than whites to be detained pretrial, convicted, and sentenced to jail. While there is evidence that this arrest policy may have had an effect on civic norms—that is, it may have directly contributed to a reduction in the incident of smoking marijuana in public (but not in private) (Golub et al., 2006; Golub, Johnson, Taylor, & Eterno, 2003; Johnson, Golub, Dunlap, Sifaneck, & McCabe, 2006)—there is no definitive empirical evidence that these arrests led to the significant reductions in serious and violent crime that the NYPD's QOL approach has been credited with achieving.[3] However, there is plentiful evidence that minority males are at a greater disadvantage than their white counterparts when they have even minor criminal records (Pager, 2003; Pager, 2007; Pager, Western, & Sugie, 2009; Western, 2006) and that disadvantaged neighborhoods, in general, suffer more from policing that leads to higher rates of incarceration (Clear, 2009).

These questions arise: Should police agencies take these matters into consideration when setting policy? That is, should police agencies be required to consider the potential negative impact of their policies on vulnerable populations before implementing them?[4] Can an organization based on a paramilitary model effectively do so? Should the effectiveness of policing be based solely or primarily on evidence of reduced crime and disorder? Does law and morality require more?

Based on a comprehensive review of research on police policy and practice, a report from the National Research Council (2004) notes:

> Policing is primarily shaped by two public expectations. First . . . police are called on to deal with crime and disorder, preventing them when possible, and to bring to account those who disobey the law. Second, the public expects their police to be *impartial*, producing justice through the fair, effective, and *restrained* use of their authority. . . . The standards by which the public judges police success in meeting these expectations have become more exacting and challenging. . . . [P]olice agencies today must find ways to respond in an effective, affordable and legitimate way. (p. 1; emphasis added)

There is considerable question as to whether policing can effectively serve all of these masters. There are ways in which the development and implementation of police policy may create friction and contradictions when agencies attempt to achieve these potentially competing goals. For example, there is a question as to whether police exercise more or less restraint in carrying out intrusive policies (such as stop, question, and frisk) when they operate within predominantly minority versus predominantly white communities. This question is often intertwined with considerations of class and economic status; that is, do the police behave the same and/or develop similar or different policies and strategies for crime intervention in poorer communities than they do in more affluent communities? Citizen satisfaction with and trust in policing agencies will ultimately be determined by which of the priorities community members perceive as receiving the most attention and at whose behest (see Sharp & Johnson, 2009).

Because racial and ethnic minority group members have experienced discriminatory policing in the past, their need for equitable treatment under the law may be more acute than that envisioned by their white counterparts. Among those who live in high crime areas, with high concentrations of racial/ethnic minority residents, the policing challenge is to develop strategies that provide both public safety and clear consideration of individual constitutional rights. When police engage in targeted enforcement in the belief that they are maximizing costly police resources, minority communities seem to bear the brunt of enforcement shortcuts (see discussions of "rational discrimination" [Weitzer & Tuch, 2002] and "reasonable racism" [Kennedy, 1997] in reference to police use of racial profiling). The loss of legitimacy that can result from such enforcement may directly counteract the public safety gains expected from policy implementation.[5]

## Is Urban Policing Policy Effective?

Despite the well-publicized crime declines of the last 10 to 15 years, urban residents—minority residents in particular—still suffer levels of fear and victimization at rates substantially higher than their nonurban and nonminority counterparts (Jones-Brown, 2001). Also, more so than their counterparts, urban minority residents view and experience very public errors in policing practice and procedure (e.g., the 1993 beating of Rodney King by the Los Angeles Police Department, the 1999 41-shot killing of unarmed Amadou Diallo by members of the NYPD, and the January 2009 shooting of Oscar Grant by Oakland, California, transit police).

Over time, different overarching policies have been advocated and/or government-funded as *the* approach to best serve the policing needs of urban communities. In the implementation of these policies, the need for public safety has received considerable attention. Lesser attention is given to concerns for justice. In communities where violent crime rates are the highest, policing effectiveness is spoken of and carried out in ways that suggest that in order to gain the benefit of public safety, community members must naturally trade away their justice concerns. Reminiscent of the 1960s, some of the policies, when put in practice, have strained the relations between the police agents and the communities served, while simultaneously appearing to effectively reduce and control crime, at least for a time. Other policies have strained relations while only marginally reducing crime. Finally, there is anecdotal evidence suggesting that some policies have produced a perception of reduced crime and have reduced the fear of crime, but only for some segments of the population and not others. In New York City, for example, persons with higher incomes have been reported as feeling safer than persons of lower income (Jones-Brown, 2001).

In order to fully grasp the significance of policing policy and practice, it is necessary to recognize the magnitude of the policing apparatus in the United States and one's potential for having a police encounter. One source estimates that there are more than 17,000 law enforcement agencies, employing more than 700,000 police officers (Cooper, 2005). Other estimates place the figure closer to 800,000, including all full-time state, city, college, university, county, and federal law enforcement officers. In terms of structure and function, while Cooper (2005) describes law enforcement agencies as "modeled on a paramilitary hierarchy with defined roles and responsibilities and well established spans of control" (p. 15), others acknowledge that, even in modern times, the roles and responsibilities of law enforcement officers may be somewhat murky (to protect or to serve), con-tradictory (to control or to assist), and varied over time and place (to be the "law enforcer" in one era or area and the community "problem solver" in another) (Klockars & Mastrofski, 1991). In addition, private individuals' interests and con-cerns about policing policy may not be the same because people are not equally exposed to the possibility of a police encounter.

During 2006, the NYPD stopped over 500,000 pedestrians under its stop, question, and frisk (SQF) policy. Nearly 90% of the people stopped were black or Latino (Ridgeway, 2007), and 10% of the stops resulted in an arrest or summons. In addition to being stopped more, blacks and Latinos were more likely than whites to be frisked (i.e., patted down) during the stops (see Dunham, Alpert, & Smith, 2007, with similar findings for traffic stops in Miami). However, when whites were frisked, they were 70% more likely than blacks to be in possession of a weapon (Ridgeway, 2007).

Many explanations have been advanced to explain why the overwhelming majority of persons stopped, questioned, and (often) frisked by police were black and Latino in a city where the racial breakdown of the population is estimated at 35% white, 24% black, and 28% Latino. The explanations offered raise serious questions regarding both the effectiveness and fairness of this practice.

Some have attempted to explain the racial disparity in the stops by referenc-ing the fact that in the majority of calls for service, the "suspect" is described as a black or Latino male. Another explanation notes that the bulk of police atten-tion in the city is concentrated in neighborhoods where the residents are pre-dominantly black or Latino. The final, frequently recurring, explanation is that police stop people more in minority communities because violent crime rates are highest in those neighborhoods. Frequent stops are seen as a means of ensuring public safety.

The volume of the stops coupled with the failure to find evidence of a sum-monsable or arrestable charge during so many of them indicates there is something faulty about the information the police are using when deciding to stop, question, and frisk such a predominance of male blacks and Latinos. For example, the descriptions that the police are following up are not specific enough to rule out the people who are factually innocent; or, when following up a called-in description, the police are not confining their stops to persons narrowly described in the report. Instead, they may be simply rounding up all of the minority males in the area of the call. The latter behavior is not supported by the dictates of the U.S. Constitution (see Fagan & Davies, 2000, and Rice & Piquero, 2005, discussing the negative community impact of police operating pursuant to non-particularized suspicion).

Since the police are far more successful in discovering a chargeable offense while stopping fewer whites, the NYPD stop data also suggest that when "suspects" are white, the police may wait to gather more information to confirm their suspicions before approaching them to investigate the possibility of crime. Such a distinction would represent a classic violation of the Constitution's equal protection clause. White civilians are receiving different treatment by police than are blacks and Latinos. *Innocent* blacks and Latinos are being stopped and questioned at a rate higher than their white counterparts. Is it fair that substantial numbers of factually innocent blacks and Latinos are stopped and inconvenienced in order to find those who are guilty of some offense? On another note, what does it say about the policy's effectiveness overall that arrests or summonses occur in only 10% of the roughly half a million stops?

Because the policy was originally put in place to confiscate illegal guns, and many illegal guns are being found (reports have indicated roughly 2000 guns per year), there has been substantial support for the practice. But as the number of stops has increased substantially over time (Ridgeway, 2007), with some individuals being stopped multiple times without the recovery of a weapon or evidence of any other criminality, minority groups have come to see the practice as harassment, while many whites still see it as effective policing (Kennedy, 1997; Weitzer, 2000, 2002, 2005; Weitzer & Tuch, 2002, 2006). But the stop data, coupled with the arrest and summons data, indicate that when police are making decisions about who might be involved in criminality, they are only correct about 10% of the time. Would a 90% margin of error be accepted in any other profession? Should it be accepted in policing, especially when it is accompanied by such racial disparity?

The explanations offered earlier do not explain why the police are so often wrong when they stop black and Latino men on suspicion of weapons carrying or other criminality. They also do not seem to take into account the substantial role that personal discretion plays in policing. The RAND report acknowledged that the use of the racial distribution of individuals identified in crime suspect descriptions as an external benchmark for determining the efficacy of the NYPD stops was marked by "serious pitfalls" (Ridgeway, 2007, p. xi). This benchmark failed to take into account that officers can and do initiate SQF activity on their own and not in response to a specific description. They even initiate contact with a civilian without specifically knowing the crime he or she is suspected of committing. The report does note that "arrests may not reflect the types of suspicious activity that officers might observe . . . ," that "arrests can occur far from where the crime occurred," and, perhaps most importantly, that "since police make both the arrests and the stops, the arrest benchmark is *not independent of any biases that officers might have*" (Ridgeway, 2007, p. xii; emphasis added).

As long ago as 1966, sociologist Jerome Skolnick provided insight into how racial bias operates in urban policing. He describes a "perceptual shorthand" that police use to identify "symbolic assailants." He notes that the police see certain gestures, language, and attire as "preludes to violence" and that the police in most communities equate these cues and black men with danger. A substantial body of social psychological research has consistently confirmed Skolnick's original observations (see work by Correll, Park, Judd, & Wittenbrink, 2002; Eberhardt, Purdie, Goff, & Davies, 2004; Greenwald, Oakes, & Hoffman, 2003; Payne, 2001; Plant & Peruche, 2005; Plant, Peruche, & Butz, 2005). Such studies confirm the inability of police officers to make fine distinctions between minorities who are engaged in crime and those who are not. They also confirm police reliance on criminal "shorthand" based on the location, style of dress, language, gender, and racial/ethnic identity of "suspects." For example, to illustrate the influence of location on police decision making, the U.S. Supreme Court has ruled that police can use the fact that a person is running in a high crime area as an indicator that the runner is involved in (an unspecified) crime (*Illinois v. Wardlow*, 2000). As discussed in more detail at the end of this chapter, we suggest that by adopting police policies and practices that include attention to multicultural competence, police could come to know that many of the cues they associate with criminality—or more specifically, dangerousness—are instead normative cultural adaptations to very challenging social circumstances, which for historical and contemporary reasons are borne disproportionately by racial and ethnic minorities. This understanding, recognition, and acceptance of cultural differences can help police realize the need to gather more information about the behavior of individuals before acting on group cultural cues. Culturally competent policing may improve the success rate during stops and reduce the number of stops that occur based on faulty or unfounded suspicion. Again, we also hope that a positive collateral consequence of adapting this approach would be improved community relations.

To be clear, the racial disparity in those stopped and frisked in New York raises the question of fairness. The stop of such a substantial number of citizens of all races, without finding evidence of a chargeable crime, raises the question of effectiveness. And the concentration of unsuccessful stops (i.e., stops that fail to produce an arrest or summons) within the black and Latino population raises both fairness and effectiveness concerns.

# ■ Everything Old Is New Again: Assessing the Efficacy of Community Policing and the Broken Windows Approach

One irony of contemporary policing policy is its resemblance to the policies and practices of the past. As noted previously, according to Kelling and Moore (1988), the community era of policing began in about 1970 after the racial unrest over policing in the 1960s. In its ideal form, community policing conjures notions of police officers interacting with residents and being responsive to their needs, arguably addressing all three prongs of police function—crime control, order maintenance, and service provision. In contrast, during the reform era, which according to Kelling and Moore (1988) preceded the community era, policing was focused on arrests. Under contemporary notions of community policing, we have found considerable confusion in definition and practice, which arguably may have contributed to a return to arrest-focused policing in some locations. As will be discussed in greater detail, in New York City, for example, the return to arrest-focused policing was attributed to an adaptation of the "broken windows" thesis put forth by James Q. Wilson and George Kelling in 1982.[6]

## What Counts as Community Policing?

Overall, the terminology created over the last three decades to denote an expanding array of policing strategies and approaches uses confusing and ill-defined labels such as community policing, broken windows policing, problem-solving policing, hot spots policing, pulling levers policing, third-party policing, CompStat, and evidence-based policing (Weisburd & Braga, 2006). Some of these terms are known by multiple names. For example, community policing is also known as community-oriented policing. Problem-solving policing is also known as problem-oriented policing (Goldstein, 1979, 1990) and is sometimes used interchangeably with hot spots and third-party policing. Broken windows policing has been referred to as zero tolerance policing, quality-of-life policing, and order maintenance policing, all three of which have been written about in conjunction with CompStat and evidence-based police action.

The volumes of academic and practitioner literature produced about these various approaches does little to clarify their meaning or to establish the superiority of one over the other in accomplishing the tripartite goals of crime control, order maintenance, *and* service provision. At best, one approach or the other is heralded for its apparent crime control success, as evidenced by declining crime statistics. One or more have been credited with significant reductions in fear of crime, as evidenced by responses to community surveys or increased pedestrian activity in certain places (e.g., commercial centers) or at certain times (e.g., at night). Similarly, one or more claim responsibility for the improved physical conditions and heightened aesthetic appeal of specific locations, such as parks or retail and business areas. However, there are outstanding questions as to how much of these changes can be attributed directly to police intervention (as opposed to, or including, economic development and increased employment in marginal communities).

The community policing approach is perhaps the most ambitious of these policing policy and practice "innovations." Because it purports to include those who are policed within police decision making, it is a stark departure from the professional, arrest-focused model of the reform era where the police were considered autonomous and all-knowing about how best to detect and combat crime. Unlike some other innovations, the police–community collaboration that ideally characterizes the community-oriented policing approach suggests that, under this philosophy, citizens can expect that police will not only focus on catching criminals but can be counted on to provide unspecified services as well.

Providing other services to those in need is reminiscent of the positive aspects of policing under the political era (recall that police ran soup lines and assisted immigrants in finding lodging and work) and suggests a recognition that crime and disorder can be reduced or prevented by addressing potential root causes of crime, including some that might be unique to particular neighborhoods. For example, suppose abandoned buildings in a neighborhood are being used by substance abusers as crack houses or by unsupervised youth for drug dealing and other gang activity. Closing down one or more of these abandoned buildings would fall under problem-solving or problem-oriented policing. Closing down the building(s) by applying housing code regulations against an absent landlord or owner, or condemning and demolishing the building(s) under applicable civil law, would be examples of third-party policing (see Silverman's [1999] discussion of the NYPD's use of Civil Enforcement Units (CPUs); Buerger & Mazerolle, 1998). Any of these solutions might be viewed under the community policing umbrella and can take place without *anyone* being arrested.

It has been suggested that the attraction to the community policing movement lies in its potential to resolve the problems associated with conventional police–public relations (Sewell, 1999). However, community policing, identified by the National Research Council (2004) as "probably the most important development in policing in the past quarter century" (p. 85), did not become an official *federal* priority until 1994. In 1994, under the Violent Crime Control and Law Enforcement Act, police agencies at all levels were encouraged to adopt community policing as their primary policing mode. The legislation authorized the creation of the Office of Community Oriented Policing Services (COPS) within the Department of Justice (DOJ) under the auspices of the National Institute of Justice (NIJ). The agency that eventually became the NIJ was established by the Omnibus Crime Control and Safe Streets Act of 1968, a piece of legislation enacted in response to the turbulent times discussed earlier in this chapter.

COPS office funding resulted in the hiring of many new police officers, ushered in the age of technology to assist in record-keeping functions, and mandated the evaluation of policing programs. It has been noted that while federal funding did increase the number of officers with community-oriented assignments, the magnitude of the effect of community policing is uncertain (National Research Council 2004). One significant source of uncertainty is the lack of a uniform definition. Many different police agencies functioning in many different ways have claimed to be involved in community policing. During the 1990s, holding regular community meetings, conducting public surveys, forming community advisory committees, opening substations, and assigning patrol personnel to specific geographic areas were all indicators that community policing was taking place (National Research Council, 2004), even without proof of specific effects or outcomes.

By 1999, the federal government had awarded $4.3 billion for officer hiring (Roth, Ryan, & Koper, 2000, as cited in National Research Council, 2004). What were community-oriented police personnel expected to do differently from police personnel under the so-called professional model? Described as "both a philosophy and an organizational strategy that allows the police and residents to work closely together to address crime, physical and social disorders, and neighborhood decay" (Reiss, 2006, p. 8), community policing was expected to control crime while simultaneously increasing the relative value police placed on order maintenance and service provision (Eck & Rosenbaum, 1994; Zhao & Thurman, 1997).

From the beginning, some thought this new focus on community-oriented policing was doomed to fail. Klockars (1988) noted that the very nature of police work, which includes the authorization to use coercive force (Bittner, 1972), makes it impossible to adopt a community policing approach. The approach was criticized as nothing more than "'old wine in new bottles' developed more to manipulate public opinion than to reprioritize police functions" (Mastrofski, 1988; Zhao & Thurman, 1997, p. 355, citing Manning, 1988, 1989, 1995). Community policing was seen by some as a "soft" approach to crime control and by others as more rhetoric than reality (see Greene & Mastrofski, 1988). The approach called for a return to foot patrols—the beat officer concerned about his or her neighborhood (Sewell, 1999). In theory, the approach held promise—the promise of improving police–community relations *and* crime fighting through increased contact between police and the community.

In reality, although the community policing approach continues to garner support and requests for its "return" (Meares, 2002), there continue to be questions regarding what it is and what it does. Despite its popularity in a variety of quarters, the approach has been criticized for its ambiguity and accused of making inflated promises (Sewell, 1999). Included among its 10 principles are the following: a commitment to community empowerment; personalized policing; proactive problem solving; ethics, legality, responsibility, and trust; expansion of the police mandate; help for those with special needs; and grassroots creativity and support (Trojanowicz & Bacqueroux, 1990). But who is the "community" that the police are required to engage? In an increasingly diverse society, it has been said that the community policing approach rests on the "illusion of community" (Sewell, 1999, p. 219; see also McEvoy, 1976). For example, such approaches tend to attract the participation of the middle class, churchgoers, and the more economically and personally stable residents of a neighborhood (Meares, 2002), while the "crime problem" tends to be driven by substance abusers, unsupervised youth, the unemployed, and others existing on the margins of what we consider conventional life. Moreover, in large urban centers, where the police are not required to live within "the community," the "beat officer" is often a stranger sent to work within a particular geographic location for only a limited time (O'Donnell, 2004; Silverman, 1999).

For all of these reasons and more, it is difficult to quantify the success of community policing strategies. It is worth noting that, while not often publicized, the crime decline in New York City began under the community policing policies of former Mayor David Dinkins and former Police Commissioner Lee Brown. Dinkins and Brown introduced a vision of community policing to the NYPD in 1990 and set about making it the agency's dominant philosophy (Henry, 2002).[7] In 1991, at the urging of the mayor, the state legislature passed the Safe Streets, Safe City Act, which mandated that nearly 20,000 officers be assigned to patrol duties. To facilitate the community policing initiative, the Act also funded a host of social programs. However, before the community policing vision could reach its full potential in New York City, Mayor Dinkins was defeated in the 1993 mayoral race by Rudolph Giuliani. The new mayor and his new police commissioner, William Bratton, initiated a different approach to policing in the city.

### Broken Windows and Arrest-Focused Policing

With the election of the former prosecutor Rudolph Giuliani, the pendulum of policing in New York City swung back to the reform era. With the assistance of computerized crime statistics, which came to be known as CompStat, the police officer was once again expected to be the professional crime fighter, controlling and preventing serious crime. However, under Giuliani and Bratton's interpretation of the "broken windows" philosophy (see note 6), instead of engaging community members as potential partners in identifying and resolving crime-related concerns, the newly expanded patrol force aggressively approached members of the public who they *suspected* of being criminals or who they observed in the act of low-level offenses like fare beating, public urination, and even littering (Silverman, 1999). These low-level offenders were arrested, booked, and checked for warrants in hopes that the police would solve more serious crimes for which they might be guilty. Zero tolerance was shown for panhandling, loud radios, and the violation of transit rules. The middle class and tourists began to feel safer. The poor and racial/ethnic minorities began to feel like targets (O'Donnell, 2004, Silverman, 1999). Police practice was heralded as having improved the quality of life in the city, but for whom and at whose expense?

In addition to saddling low-level offenders with jail time and criminal records, the increase in misdemeanor marijuana arrests, mentioned earlier, created backlogs within the courts. A recent lawsuit in Camden, New Jersey, is evidence that courts and civilians are not the only ones negatively affected by police policies that call for increased arrests. A Camden city police union has filed suit against the Camden police department and the state's Attorney General's Office, alleging that the department has implemented an illegal quota system that rates officers on the number of citations they issue and arrests they make.

Two officers who are plaintiffs in the suit claim to have been transferred to lower paying positions when their arrest numbers were judged not to be high enough. While the department denies the use of arrest quotas, the officers note that the department's reliance on monthly accounts of stops, inquiries, summons, and arrests puts the rights of citizens at risk. They feel forced to issue tickets they would not normally issue, to make random stops of residents on the street, and to even knock on residential doors, all in the name of satisfying a numeric total. The president of the police union has stated, "we all want to stop violent crime, but not at the expense of the Constitution or to violate people['s] rights" (Mast, 2009, p. 18).

The controversy stems from a policy decision not unlike the NYPD's decision to implement CompStat. During July 2008, in response to violent crime, particularly the number of homicides, the Camden police instituted a weekly statistical reporting system. The statistical reporting was coupled with a program of "directed patrol" carried out by a newly expanded police force (Mast, 2009). The state's attorney general and the city's chief of police have responded by saying that the Camden police department is not directing its personnel to act illegally (there is a state law that prohibits the use of arrest quotas). The chief of police has stated that the lawsuit "will not deter us from holding people accountable to providing a service that makes our streets safer" (Mast, 2009, p. 38). The two sets of comments reflect profoundly different perspectives on what constitutes appropriate and effective police behavior and service provision. They also suggest that policing policy in this majority-minority city enforces a dichotomous relationship between residents' safety and justice concerns.

While New York City, Camden, and other locations (see Weisburd, Mastrofski, McNally, Greenspan, & Willis, 2003) have relied on more aggressive stop-and-arrest policies to bring about reductions in crime, San Diego has achieved double-digit crime reduction through tactics more closely aligned with Trojanowicz and Bucqueroux's (1990) community policing principles, although like CompStat, the San Diego police department uses computerized crime mapping in its neighborhood problem-solving approach. Since 1989, the department has collected snapshots of criminal activity. Patrol beats are organized by neighborhood rather than by precinct, and patrol officers are encouraged to look beyond an individual crime to try to determine its underlying causes. Using a data-driven, problem-solving model known as SARA (scanning, analysis, response, and assessment), the beat officers work closely with crime analysts and use sophisticated crime-mapping tools and laptop computers in the field to detect patterns of criminal activity and neighborhood hot spots, which are defined as specific blocks or locations with the highest concentration of crime or disorder (Grant & Terry, 2005). Residents and business owners are also included in this collaborative problem-solving process. The process has been noted for enhancing the communication between the law enforcement and civilian groups, and it helps the police to see neighborhood patterns more clearly (Mapping out Crime, 1999).

### Fear of Crime and Fear of the Police

Silverman (1999) warned that in the midst of the excitement about New York City's crime decline and its commitment to innovative policing, "Fear of police should not replace fear of crime" (p. 189). The aggressive tactics associated with New York–style policing, especially under former Mayor Rudolph Giuliani, have been contrasted with the outcomes of San Diego's less aggressive approach. Between 1992 and 1997, San Diego experienced a 61% decline in its serious crime—including murder, rape, assault, and burglary. San Diego's crime decline actually exceeded that of New York, which was 50.5% for the same period (Silverman, 1999). San Diego achieved this substantial crime reduction without using the New York–style tactics that have been criticized for their disparate negative impact on poor and minority residents (O'Donnell, 2004). Perhaps more importantly, with the exception of A. Demetrius Dubose's death, in achieving its crime reduction, San Diego seems to have avoided the series of questionable deaths that have been associated with New York's use of aggressive arrest-focused police tactics.

Between 1999 and 2004, four highly publicized shootings of factually innocent civilians by members of the NYPD all raised questions regarding the adequacy and appropriateness of the NYPD's training for policing in culturally and ethnically diverse neighborhoods. In each case, the victim was unarmed but was perceived to be dangerous by police officers engaged in special patrol or interdiction units. Each victim was black. Three had family origins from outside the United States. In all but one of the incidents, the police officers were white. A special Street Crimes Unit killed Amadou Diallo, mistaking him for a wanted rapist. Patrick Dorismond was shot by an officer from a sting unit investigating the sale of drugs in Manhattan nightclubs.[8] Ousmane Zongo was shot by an officer who was part of a team investigating possible product counterfeiting. Timothy Stansbury was shot by an officer during a "vertical sweep" of the public housing unit where he lived. Such sweeps were part of a routine patrol strategy designed to detect and reduce crime in the city's public housing complexes. The fact that each of these deaths occurred as the result of policing strategies intended to make communities safer raises serious efficacy questions. The fact that each also involves a black victim raises serious justice concerns.

Ironically, it was similar incidents and concerns that originally ushered in the community policing era.

Is it inevitable that racial and ethnic minorities must give up a greater portion of their constitutional rights in order to gain a greater level of personal safety in ghetto neighborhoods? Is it also unavoidable that innocent civilians from minority groups will die at the hands of the police under the mistaken belief that they are criminal and/or dangerous? Do utilitarian notions of safety justify police policies and practices that are associated with these outcomes? After nearly two centuries of "modern" policing and more than three decades of "innovations," is the current state of American policing as good as it gets? In the next and final section of this chapter we suggest that serious attention to the development of police agents and agencies that are multiculturally competent can address some of these concerns.

## ■ Recommendation: Making Multicultural Competency an Integral Part of Policing Policy and Practice

Ideally, there should be some level of consensus that fair and equitable policing is fundamental to a democratic society, if for no other reason than the recognition that attempts to secure public order in the absence of police legitimacy have proved treacherous in the past. Presently, it is clear that while major U.S. cities have enjoyed significant crime declines over the last decade or more, "America's assault on crime . . . has exacted a high price—more often than not, a price paid by communities of color" (West, 2001, p. v).

While it is true that some of the greatest declines have occurred in some cities previously thought to be among the most dangerous, among some of the "success stories," the declines have been associated with tactics that have been experienced or perceived as aggressive; racial and ethnic profiling of the innocent and the guilty; platforms or conduits for police misconduct; and, in its worst form, the cause of fatal police mistakes.

In the police section of the 1967 report The Challenge of Crime in a Free Society, the authors make the following observation:

At the very beginning of the [criminal justice] process or, more properly, before the process begins at all something happens that is scarcely discussed in lawbooks and is seldom recognized by the public: *law enforcement policy is made by the policeman.* (emphasis added)

The report continues:

Policemen cannot and do not arrest all the offenders they encounter. . . . A criminal code, in practice, is not a set of specific instructions to policemen but a more or less rough map of the territory in which policemen work. *How an individual policeman moves around that territory depends largely on his personal discretion.* (emphasis added)

It is also the case that department policing policies reflect the discretion of high-level police personnel (Bratton & Knobler, 1998).

Far more evidence than can be adequately presented in a chapter of this length suggests that while police agencies can be proud of the reforms and innovations that have contributed to crime declines, they must also take some responsibility (departmental and personal) for the collateral damage that policing strategies have caused in certain neighborhoods. We suggest that police departments, particularly those that serve in majority-minority locations, must move to perfect their policing skills.

One way of ensuring fair and effective policing is to make multicultural competence (MCC) an integral part of police policy and practice. Just as problem-oriented policing strategies recognize that not all problems of crime and disorder respond to the same set of predetermined and fixed techniques, multiculturally competent police leaders and line officers can learn to provide better policing services through a better understanding of those whom they police.

MCC refers to the ability to value diversity, demonstrate capacity for cultural assessment, show consciousness of the dynamics inherent when cultures interact, demonstrate institutionalized cultural knowledge, and develop adaptations to diversity (Jaffe-Ruiz, 2009). It has been used by numerous professional and academic disciplines, including education, psychology, and nursing, to address perceived and actual imbalances of privilege, power, and authority (Byars-Winston et al 2005).

MCC as applied in the field of policing has most often taken the form of diversity training (Barlow & Barlow, 1994). However, true MCC requires law enforcement professionals to develop ongoing self-awareness and reflective thinking regarding one's personal worldview as it relates to one's reference group memberships (e.g., race, socioeconomic class, culture, religion, gender, ability status, sexual orientation, and language ability) and how these reference group memberships impact reactions to others.

Purnell and Paulanka (1998) have developed a model of cultural competence that moves through four stages: (1) unconscious incompetence (arguably the worst or least developed); (2) conscious incompetence; (3) conscious competence; and (4) unconscious competence (most developed). The attainment of the most developed level of competence may prove difficult and labor-intensive, since despite diversity efforts, national statistics indicate that policing is still primarily a white male profession. In contrast to residents of highly policed areas, roughly 70% of full-time sworn police personnel are white men. Black men and women make up 11.7% of this police force, and Latinos represent 9.1% (Bureau of Justice Statistics, 2003). Although the percentages of minority officers in some urban cities are significantly greater than these national estimates, they do not significantly alter the national perception and experience of dominant police culture.

As noted in the chapter, for better or for worse, all of the police innovations of the last three decades promote or mandate high levels of police–civilian contact. That being the case, we consider MCC to be as essential a police skill as handling a firearm, making an arrest, taking a statement, writing a report, conducting an investigation, or testifying in court. In support of MCC as an essential component in policing policy and practice, Coderoni (2002) points out that it is extremely difficult to restore public trust once it has been interrupted, when citizens or police officers have been injured (or killed), or after property has been severely damaged as a result of police action (or inaction). Lawsuits typically follow such incidents—lawsuits that place a tremendous financial burden on the municipalities and cities that employ the officers involved in the cases. The loss of public revenues can negatively impact all members of the community, not just those directly involved with the incident.

## ■ Conclusion

While we used New York City as an example several times in this chapter, we reiterate that New York is not the only urban city experiencing police–community relations problems due to patterns, practices, and policies that reflect racially disparate policing. We suggest that although some of these cities continue to show reduced levels of crime,[9] they might gain more decreases (based on increased community cooperation, especially through crime reporting, information sharing, and self-policing) if they embrace a wider spectrum of their community through policing that focuses less on making low-level arrests and more on community problem solving that is innovative and culturally competent. Given the increasing diversity of urban populations and the challenges they face, developing MCC among police leaders and line officers is an essential step toward achieving effective *and* fair policing in contemporary urban settings.

## ■ Endnotes

1. The National Research Council (2004, p. 5) defines police legitimacy as "the degree to which citizens recognize the police as appropriate and justified representatives of government."

2. Since the Rodney King beating by Los Angeles police officers in 1993, which resulted in multimillion-dollar property damage following the acquittal of the police officers in state court, there have been riots in Cincinnati, Oakland, Pittsburgh, St. Petersburg, and other locations, leading to the assignment of federal monitors to access patterns and practice of policing in those specific locations.

3. There is a substantial portion of the mainstream public who believes these measures did work, and almost solely because the police said so (Silverman, 1999).

4. At least four states—Iowa, Minnesota, Wisconsin, and Connecticut—have enacted legislation that requires consideration of racial impact before implementing criminal justice policies (Mauer, 2009). Local governments within various states have also considered racial impact legislation. They include Seattle, Washington, and St. Paul, Minnesota.

5. See, for example, the Stop Snitching movement that encourages community members not to cooperate with the police and the reduced clearance rates for serious crime in some urban neighborhoods.

6. The term "broken windows" policing has been adopted and, in some cases, significantly modified from a concept developed in an article published by James Q. Wilson and George Kelling in 1982. The article suggested that by more aggressively enforcing minor criminal infractions and regulations aimed at preventing and controlling community disorder, police could prevent or reduce more serious crimes and effect community change. Wilson and Kelling suggested that the police had been complicit in the decline of urban cities by failing to engage in these practices over time (Wilson & Kelling, 1982).

7. However, it is noted that community policing in New York City dates back to 1984 when the NYPD introduced the Community Patrol Officer Program (CPOP) (Silverman, 1999).

8. The controversial death of Sean Bell in a hail of 50 NYPD bullets in 2006 was also related to a sting operation where officers were working undercover to detect possible prostitution, gambling, and drug sales in the club where Bell was having his bachelor party.

9. It is noted that despite its arrest-focused policies, Camden, New Jersey, continues to be rated as the most dangerous city in the state.

■ **Discussion Questions**

1. Why do critics of urban policing policy contend that it is often unfair and ineffective?

2. Discuss the evolution of urban policing policy in the United States.

3. Why is it important to think about race and place when examining urban policing policy?

4. Discuss some of the advantages and disadvantages of urban policing practices.

5. What is the relationship between multicultural competence and fair and equitable policing?

# References

Barlow, D., & Barlow, M. (1994). Cultural diversity training in criminal justice: A progressive or conservative reform? *Social Justice, 20,* 69–84.

Beckett, K., Nyrop, K., & Pfingst, L. (2006). Race, drugs, and policing: Understanding disparities in drug delivery arrests. *Criminology, 44*(1), 105–137.

Bittner, E. (1972). *The function of the police in modern society* (2nd ed.). Washington, DC: National Institute of Mental Health.

Bratton, W., & Knobler, P. (1998). *Turnaround: How America's top cop reversed the crime epidemic.* New York: Random House.

Buerger, M., & Mazerolle, L. (1998). Third-party policing: A theoretical analysis of an emerging trend. *Justice Quarterly, 15*(2), 301–327.

Bureau of Justice Statistics. (2003). *Local police departments, 2003.* Washington, DC: United States Department of Justice.

Byars-Winston, A. M., Akcali, O., Tao, K. W., Nepomuceno, C. A., Anctil, T., Acevedo, V., Benally, N., & Wilton, G. (2005). The challenges, impact, and implementation of critical multicultural pedagogies. In C. Z. Enns & A. L. Sinacore (Eds.), *Teaching and social justice: Integrating multicultural and feminist theories in the classroom* (pp. 125–41). Washington, DC: American Psychological Association.

Chambliss, W. J., & Mankoff, M. (1976). *Whose law? What order? A conflict approach to criminology.* New York: John Wiley & Sons.

Clear, T. R. (2009). *Imprisoning communities: How mass incarceration makes disadvantaged neighborhoods worse.* Oxford University Press.

Coderoni, G. R. (2002). The relationship between multicultural training for police and effective law enforcement. *FBI Enforcement Bulletin, November.* Washington, DC: Federal Bureau of Investigation.

Cohen, A. (1955). *Delinquent boys: The culture of the gang.* New York: Free Press.

Conser, J., Russell, G., Paynich, R., & Gingerich, T. (2005). *Law enforcement in the United States.* Sudbury, MA: Jones and Bartlett Publishers.

Cooper, W. E. (2005). *Leading beyond tradition.* Mukilteo, WA: Three-Star Publishing.

Correll, J., Park, B., Judd, C., & Wittenbrink, B. (2002). The police officer's dilemma: Using ethnicity to disambiguate potentially threatening individuals. *Journal of Personality and Social Psychology, 83,* 1314–1329.

Dunham, R., & Alpert, G. (2004). *Critical issues in policing: Contemporary readings* (5th ed). Long Grove, IL: Waveland Press.

Dunham, R., Alpert, G., & Smith, M. (2007). Investigating racial profiling by the Miami-Dade police department: A multimethod approach. *Criminology and Public Policy, 6*(1), 25–56.

Eberhardt, J., Purdie, V., Goff, P., & Davies, P. (2004). Seeing black: Race, crime and visual processing. *Journal of Personality and Social Psychology, 87,* 876–893.

Eck, J., & Rosenbaum, D. P. (1994). The new police order: Effectiveness, equity and efficiency in community policing. In D. Rosenbaum (Ed.), *The challenge of community policing: Testing the promises* (pp. 3–23). Thousand Oaks, CA: Sage.

Fagan, J., & Davies, G. (2000). Street stops and broken windows: Terry, race and disorder in New York City. *Fordham Urban Law Journal, 28,* 457–504.

Goldstein, H. (1979). Improving policing: A problem-oriented approach. *Crime and Delinquency, 25,* 236–258.

Goldstein, H. (1990). *Problem-oriented policing.* New York: McGraw-Hill.

Golub, A., Johnson, B., & Dunlap, E. (2006). Smoking marijuana in public: The spatial and policy shift in New York City arrests, 1992–2003. *Harm Reduction Journal, 3,* 22.

Golub, A., Johnson, B., & Dunlap, E. (2007). The race/ethnicity disparity in misdemeanor marijuana arrests in New York City. *Criminology and Public Policy, 6*(1), 131–164.

Golub, A., Johnson, B., Taylor, A., Eterno, J. (2003). Quality-of-life policing: Do offenders get the message? *Policing, 26,* 690–707.

Grant, H., & Terry, K. (2005). *Law enforcement in the 21st century.* New York: Pearson Education.

Greene, J., & Mastrofski, S. (1988). *Community policing: Rhetoric or reality?* New York: Praeger.

Greenwald, A., Oakes, M., & Hoffman, H. (2003). Targets of discrimination: Effects of race on responses to weapons holders. *Journal of Experimental Social Psychology, 39,* 399–405.

Henry, V. E. (2002). *The Compstat paradigm: Management accountability in policing, business and the public sector.* Flushing, NY: Looseleaf Law Publications, Inc.

Higginbotham, A. L. (1980). *In the matter of color: Race and the American legal process. 1: The Colonial Period.* New York: Oxford University Press.

Jaffe-Ruiz, M. (2009). *Cultural and linguistic competence for advanced nursing practice.* Paper presented for Nursing Grand Rounds at Mount Sinai Hospital, March 11, 2009.

Johnson, B., Golub, A., Dunlap, E., Sifaneck, S., & McCabe, J. (2006). Policing and social control of public marijuana use and selling in New York City. *Law Enforcement Executive Forum, 6*(5), 59-90.

Jones-Brown, D. (2000). *Race, crime and punishment.* Philadelphia: Chelsea House.

Jones-Brown, D. (2001). Fear of crime in a New York City neighborhood: The gap between perception and reality in Brownsville, Brooklyn. In A. Karmen (Ed.), *Crime and justice in New York City. Volume 1: New York City's Crime Problem.* Cincinnati, OH: Thomson Learning.

Joseph, J. (2000). Overrepresentation of minority youth in the juvenile justice system: Discrimination or disproportionality of delinquent acts? In M. Markowitz

& D. Jones-Brown (Eds.), *The system in black and white: Exploring the connections between race, crime and justice* (pp. 227–239). Westport, CT: Praeger.

Kelling, G. L., & Moore, M. H. (1988). The evolving strategy of policing. *Perspectives on Policing, No. 4.* Washington, DC: National Institute of Justice.

Kennedy, R. (1997). *Race, crime and the law.* New York: Pantheon.

Klockars, C. B. (1988). The rhetoric of community policing. In J. Greene & S. Mastrofski (Eds.), *Community policing: Rhetoric or reality?* (pp. 239–258). New York: Praeger.

Klockars, C. B., & Mastrofski, S. D. (1991). *Thinking about policing: Contemporary readings.* New York: McGraw-Hill.

Kraska, P. B., & Cubellis, L. J. (1997). Militarizing Mayberry and beyond: Making sense of American paramilitary policing. *Justice Quarterly, 14*(4), 607–629.

Mauer, M. (2009). Racial impact statements: Changing policies to address disparities. *Criminal Justice, 23*(4) (Winter).

Manning, P. (1988). Community policing as a drama of control. In J. Greene & S. Mastrofski (Eds.), *Community policing: Rhetoric or reality?* (pp. 27–45). New York: Praeger.

Manning, P. (1989). Community policing. In R. Dunham & G. Alpert (Eds.), *Critical issues in policing: Contemporary Readings* (pp. 395–405). Prospect Heights, IL: Waveland.

Manning, P. (1995). TQM and the future of policing. *Police Forum, April,* 1–5. Richmond, KY: Academy of Criminal Justice Sciences.

Mapping Out Crime: Highlights in Crime Mapping and Data-Driven Management (1999). Retrieved February 4, 2010, from http://govinfo.library.unt.edu/npr/library/papers/bkgrd/crimemap/section5.html

Mast, G. (2009). Quotas charged in police suit. *Camden Courier-Post.* April 7, 1B, 3B.

Mastrofski, S. (1988). Community policing as reform: A cautionary tale. In J. Greene & S. Mastrofski (Eds.), *Community policing: Rhetoric or reality?* (pp. 47–67). New York: Praeger.

McEvoy, D. W. (1976). *The police and their many publics.* Metuchen, NJ: Scarecrow Press, Inc.

McIntyre, C. (1984). *Criminalizing a race: Free blacks during slavery.* Queens, NY: Kayode.

Meares, T. (2002). Praying for community policing. *California Law Review, 90,* 1593.

Meares, T., & Kahan, D. (1999). *Urgent times: Policing and rights in inner city communities.* Boston: Beacon Press.

Meares, T., & Skogan, W. (2004). Lawful policing. *The Annals of the American Academy of Political and Social Science, 66*(18), 593.

Moskos, P. (2008). *Cop in the hood.* Princeton, NJ: Princeton University Press.

National Advisory Commission on Civil Disorder/Kerner Commission. (1968). Washington, DC: U.S. Government Printing Office.

National Commission on the Causes and Prevention of Violence. (1969). *To establish justice, to insure domestic tranquility.* New York: Universal Publishing and Distributing Corporation.

National Research Council. (2004). *Fairness and effectiveness in policing: The evidence.* Committee to Review Research on Police Policy and Practices. W. Skogan & K. Frydll (Eds.). Committee on Law and Justice, Division of Behavioral and Social Sciences and Education. Washington, DC: The National Academies Press.

O'Donnell, E. (2004). Fixing broken windows or fracturing fragile relationships? In D. Jones-Brown & K. Terry (Eds.), *Policing and minority communities: Bridging the gap* (pp. 72–84). Upper Saddle River, NJ: Pearson/Prentice Hall.

Pager, D. (2003). The mark of a criminal record. *American Journal of Sociology, 108*(5), 937–975.

Pager, D. (2007). *Marked: Race, crime and finding work in an era of mass incarceration.* Chicago: University of Chicago Press.

Pager, D., Western, B., & Sugie, N. (2009). Sequencing disadvantage: Barriers to employment facing young black and white men with criminal records. *Annals of the American Academy of Political and Social Sciences, 623*(May), 195–213.

Payne, K. (2001). Prejudice and perception: The role of automatic and controlled processes in misperceiving a weapon. *Journal of Personality and Social Psychology, 81,* 181–192.

Plant, E. A., & Peruche, B. M. (2005). The consequences of race for police officers' responses to criminal suspects. *Psychological Science, 16,* 180–183.

Plant, E. A., Peruche, B. M., & Butz, D. A. (2005). Eliminating automatic racial bias: Making race non-diagnostic for responses to criminal suspects. *Journal of Experimental Social Psychology, 41,* 141–156.

President's Commission on Law Enforcement and Administration of Justice. (1967). *Task force report: The police.* Washington, DC: U.S. Government Printing Office.

Purnell, L., & Paulanka, B. (1998). Transcultural health care: A culturally competent approach. Philadelphia: F.A. Davis.

Rawls, J. (1971). *A theory of justice.* Cambridge, MA: Harvard University Press.

Reichel, P. L. (1988). Southern slave patrols as a transitional police type. *American Journal of Police, 7,* 51–77.

Reiss, J. (2006). Community governance: An organized approach to fighting crime. *FBI Law Enforcement Bulletin, 75,* 8–11.

Rice, S., & Piquero, A. (2005). Perceptions of discrimination and justice in New York City. *Policing, 28,* 98–117.

Ridgeway, G. (2007). *Analysis of racial disparities in the New York Police Department's stop, question, and frisk practices.* Santa Monica, CA: RAND Corporation.

Roth, J., Ryan, J., & Koper, C. S. (2000). *National evaluation of the COPS program— Title I of the 1994 Crime Act.* Washington, DC: The Urban Institute.

Russell, K. (1998). *The color of crime.* New York: New York University Press.

Sampson, R., & Wilson, W. J. (1995). Toward a theory of race, crime and urban inequality. In J. Hagan & R. Peterson (Eds.), *Crime and inequality* (pp. 37–54). Stanford, CA: Stanford University Press.

Sewell, J. D. (1999). *Controversial issues in policing.* Boston: Allyn and Bacon.

Sharp, E. B., & Johnson, P. E. (2009). Accounting for variation in distrust of local police. *Justice Quarterly, 26*(1), 157–182.

Silverman, E. B. (1999). *NYPD battles crime: Innovative strategies in policing.* Boston: Northeastern University Press.

Skolnick, J. H. (1966). *Justice without trial.* New York: Wiley.

Trojanowicz, R., & Bacqueroux, B. (1990). *Community policing: A contemporary perspective.* Cincinnati, OH: Anderson Publishing.

Weisburd, D., & Braga, A. (2006). *Police innovation: Contrasting perspectives.* Cambridge: Cambridge University Press.

Weisburd, D., Mastrofski, S., McNally, A., Greenspan, R., & Willis, J. (2003). Reforming to preserve: Compstat and strategic problem solving in American policing. *Criminology and Public Policy, 2*(3), 421–456.

Weitzer, R. (2000). Racialized policing: Residents' perceptions in three neighborhoods. *Law and Society Review, 34,* 129–155.

Weitzer, R. (2002). Incidents of police misconduct and public opinion. *Journal of Criminal Justice, 30,* 397–408.

Weitzer, R. (2005). Can the police be reformed? *Contexts, 4*(Summer), 21–26.

Weitzer, R., & Tuch, S. (2002). Perceptions of racial profiling: Race, class and personal experience. *Criminology, 40,* 435–456.

Weitzer, R., & Tuch, S. (2006). *Race and policing in America: Conflict and reform.* Cambridge: Cambridge University Press.

West, M. H. (2001). *Community-centered policing: A force for change.* Oakland, CA: PolicyLink.

Western, B. (2006). *Punishment and inequality in America.* New York: Russell Sage Foundation.

Williams, H., & Murphy, P. (1990). *The evolving strategy of the police: A minority view.* Washington, DC: National Institute of Justice.

Wilson, J. Q., & Kelling, G. (1982). Broken windows: The police and neighborhood safety. *Atlantic Monthly, 249,* 29–38.

Wilson, W. J., Jr. (1987). *The truly disadvantaged: The inner city, the underclass, and public policy.* Chicago: University of Chicago Press.

Wilson, W. J., Jr. (1997). *When work disappears: The world of the new urban poor.* New York: Random House.

Zhoa, J., & Thurman, Q. C. (1997). Community policing: Where are we now? *Crime and Delinquency, 43*(3), 345–357.

## ■ Court Cases Cited

*City of Chicago v. Morales*, 527 U.S. 41 (1999).

*Illinois v. Wardlow*, 528 U.S. 119 (2000).

# Combative and Cooperative Law Enforcement in Post-September 11th America

CHAPTER

**3**

Ben Brown

## ■ Introduction

Law enforcement agencies, like all other public service organizations, must adapt to social change. As threats to public safety change, so too must law enforcement. More than two centuries have elapsed since law enforcement agencies were formed in the United States, and in that time the nation's law enforcement agencies have changed a great deal. This chapter provides a sketch of the modern evolution of law enforcement in the United States and an analysis of the impact the terrorist attacks of September 11th, 2001, had on the nation's law enforcement agencies. The central thesis advanced herein is that the events of September 11th sparked an ideological and practical rift in the law enforcement arena, with policy makers, scholars, law enforcement officials, and civil rights advocates fiercely debating whether combative or cooperative law enforcement strategies and tactics are the better means of enhancing public safety.

## ■ The Modern Evolution of American Law Enforcement: A Brief Overview

During the 1800s and early 1900s, American law enforcement officials were essentially political servants who obtained and retained their positions via political patronage. Law enforcement officials at all levels (local, state, and federal) had little if any training, limited knowledge of criminal law, and no real concern for the rights of the citizenry. Over the next half century, numerous reforms occurred, such as the creation of training academies and the utilization of databases and forensic science during criminal investigations. Little by little, the nation's law enforcement agencies evolved from an uncoordinated collection of poorly disciplined ruffians into a highly trained, technologically proficient, and well-organized network of professional public servants.

Despite the increase in professionalism among law enforcement officers, problems persisted. For instance, owing to the development of radio-dispatched patrols and the creation of the 911 emergency phone number, "incident-driven policing" became the norm, and local police agencies became reactive forces with police administrators being more concerned about the average response time than the causes of crime. In addition, violations of individual rights and police brutality remained problematic, and over time the public grew increasingly dissatisfied with law enforcement agencies. Public hostility toward the police was especially intense in impoverished minority neighborhoods and contributed to the rash of urban riots that plagued cities across the nation during the 1960s (Eck & Spelman, 1997; Fogelson, 1968; Kelling & Moore, 1988; Uchida, 1997; Wadman & Allison, 2004; Williams & Murphy, 1990).

In the latter decades of the 20th century, criminal justice scholars and practitioners sought to remedy the problems that plagued the nation's law enforcement system. For example, New York Police Commissioner Patrick Murphy emphasized the need for good police–community relations (Wadman & Allison, 2004), and Professor Herman Goldstein (1979) argued that law enforcement agencies needed to reduce reactive tactics and address social conditions at the local level that contributed to crime. These and other similar ideas took root and formed the basis of the community policing movement. Although there is no consensus in the field of law enforcement as to precisely what constitutes a "community policing" or "community-oriented policing" program, the essential elements are (1) quality relations between law enforcement officers and the public, and (2) concerted problem-solving efforts among local law enforcement officers and local denizens that focus on the identification and elimination of the causes of crime and disorder in the community (Community Policing Consortium, 1994).

During the 1980s support for community policing increased, and by the early 1990s police officials across the nation were trying a variety of amiable crime control methods such as bicycle patrol, the establishment of police ministations in decaying neighborhoods, and public clean-up projects. The police also began to seek input from the citizenry and increased their cooperation with other social service agencies (e.g., public schools, public housing authorities) in order to identify and address local problems that fostered crime (Roth, Roehl, & Johnson, 2004). The community policing movement was welcomed by the federal government, as evidenced by the creation of the Office of Community Oriented Policing Services (COPS) in 1994, a component of the U.S. Department of Justice charged with supporting community policing. Under the leadership of President Clinton, the COPS office was authorized to provide federal funding for the hiring of 100,000 new police officers with the stipulation that the officers be engaged in community policing. By the year 2000, more than two-thirds of all local police departments in the United States had some form of community policing program (Community Policing Consortium, 1994; Cordner, 1997; Hickman & Reaves, 2003; Moore, 1994; U.S. Department of Justice, 1994).

During the final decades of the 20th century, efforts were also made to rein in federal law enforcement and intelligence agencies, groups that had historically demonstrated little concern about the constitutionality of their tactics or objectives. For instance, during the Palmer Raids of 1919 and 1920, thousands of people were arrested and hundreds of them were deported based on little or no evidence that they posed any threat to public safety. Then there was the Federal Bureau of Investigation's (FBI) counterintelligence program carried out under the supervision of J. Edgar Hoover, which lasted from the 1950s until the 1970s and involved an assortment of nefarious activities including warrantless wiretaps, warrantless searches of offices and residences, and harassment designed to disrupt the activities of suspected socialists, civil rights advocates, and antiwar demonstrators. As the abuse of power by federal officials became public knowledge, the citizenry became increasingly disenchanted with the federal government. In response, during the mid-1970s the United States Senate Select Committee to Study Governmental Operations with Respect to Intelligence Activities (an assembly best known as The Church Committee) was formed to investigate and suggest reforms of the nation's federal law enforcement and intelligence operations. The combination of public distrust of the government and the findings of The Church Committee and other investigative bodies led to reforms such as the creation of the Foreign Intelligence Surveillance Court (Betts, 2004; Bulzomi, 2003; Chalmers, 1990; National Commission on Terrorist Attacks Upon the United States, 2004; Williams, 1981). Similar to the reforms of local police agencies, the reforms of federal law enforcement and intelligence agencies were designed to foster strategies and tactics that protected the public without unnecessarily intruding upon the private domain and violating the rights of the populace.

Throughout the mid-to-late 20th century, the courts also sought to bring law enforcement activities in line with the Constitution. For example, in the 1967 case of *Katz v. United States*, the Supreme Court held that an individual's Fourth Amendment right to privacy is not confined to an individual's home but travels with the individual. The ruling required that law enforcement officials obtain a warrant to conduct even noninvasive searches (i.e., electronic eavesdropping). And in the 1980 case of *Payton v. New York*, the high court ruled that law enforcement officers may not make a warrantless entry of a residence for the purposes of a felony arrest. Concisely stated, in a number of rulings the courts limited the authority of law enforcement and emphasized the rights of the citizenry.

Then came the horrific events of September 11th, 2001. It is impossible to adequately describe the carnage of the terrorist assault of September 11th, but suffice it to say that the attacks on the Pentagon and the World Trade Center killed more people than any other terrorist act in history, temporarily crippled the U.S.

economy, and shattered the tranquil sense of security that the American people had previously enjoyed. Immediately thereafter legislators and law enforcement professionals sought answers to the questions of how and why the attacks were carried out and began debating how to secure the nation against future attacks. The prevailing sentiment at the time was not that there was a need for more community involvement by the police or greater restrictions on the powers of law enforcement, but rather that there was a need to unleash law enforcement.

## ■  September 11th and the Emergence of Homeland Security

Organized terrorism (defined in this chapter as the use of violence against the general public and/or the symbols of a nation's strength by an organized group to achieve a political objective) has not historically been a major threat within the United States. Throughout the 1900s there were only a handful of organized terrorist attacks on U.S. soil, the most notable being the series of anarchist bombings in 1919 in cities across the nation, the 1975 bombing of Fraunces Tavern and the Anglers' Club in New York City by the Puerto Rican nationalist group *Fuerzas Armadas de Liberación Nacional* (in English, Armed Forces of National Liberation), and the 1993 bombing of the World Trade Center by a team with links to Al Qaeda. There were a number of other terrorist incidents in the United States during the late 1900s, but they were carried out by lone individuals or small groups rather than sizeable organizations with a political agenda. To provide a few examples, there was the sporadic mail bombing campaign of Theodore Kaczynski ("the Unabomber"), which lasted from 1978 to 1995, the bombing of the Alfred P. Murrah Federal Building in Oklahoma City by Timothy McVeigh in 1995, and Eric Robert Rudolph's bombing campaign of the early-to-mid 1990s in which he targeted gay and lesbian nightclubs, abortion clinics, and Centennial Olympic Park in Atlanta during the 1996 summer Olympics (Bodrero, 1999; Simonsen & Spindlove, 2004).

The aforementioned examples of terrorist activity notwithstanding, the majority of extraordinary organized terrorist strikes against the United States and its interests occurred overseas with terrorist groups focusing on American tourists, embassies, and military targets, as in the cases of the bombings of the U.S. embassy and a military barracks in Beirut, Lebanon, in 1983; the bombing of Pan Am Flight 103 bound for New York over Lockerbie, Scotland, in 1988; the 1998 attacks on U.S. embassies in Kenya and Tanzania; and the 2000 assault on the USS Cole while docked in Yemen. Because acts of organized terrorist violence in the United States were infrequent, American law enforcement agencies historically devoted few resources to counterterrorism (Bodrero, 1999; Lewis, 1999), a situation that began to change during the Clinton administration. In the 1990s there was substantial discussion among the nation's leaders about terrorism, and measures were taken to enhance homeland security. For example, in response to the 1993 bombing of the World Trade Center and the 1995 bombing of the Alfred P. Murrah Federal Building, President Clinton signed the Antiterrorism and Effective Death Penalty Act of 1996, which, among other things, restricted access to explosive materials and sought to decrease the funding of foreign organizations with ties to terrorist groups (Doyle, 1996).

President Clinton, Secretary of Defense William S. Cohen, and FBI Director Louis Freeh placed a greater emphasis on counterterrorism than did their predecessors. During the 1990s, federal spending on counterterrorism increased considerably, with a sizeable portion of the money being allocated to prepare first responders to handle a biological, chemical, or nuclear attack. In addition, it was during President Clinton's tenure that cooperative counterterrorism planning between defense, intelligence, and federal law enforcement agencies began and counterterrorism became a priority for the FBI (Dreyfuss, 2000; National Commission on Terrorist Attacks Upon the United States, 2004). However, before the September 11th attacks, there were no major reforms made to the nation's law enforcement and intelligence apparatuses. The bulk of discussion about terrorism among the nation's leaders resulted in little more than plans to handle hypothetical scenarios and recommendations for more preparedness. For instance, a few months prior to the attacks of September 11th, the U.S. Commission on National Security recommended that a National Homeland Security Agency be formed. It was also recommended that the director be assigned a Cabinet position and given the authority to unify and oversee the operations of numerous agencies including the Coast Guard, the Customs Service, and the Border Patrol (Mufson, 2001). The recommendations were not acted upon.

Then came the nightmarish events of September 11th. The kamikaze attacks on the Pentagon and World Trade Center that utilized hijacked commercial jets were a shocking demonstration of the extent to which terrorist organizations had evolved and showed the determination of such groups to bring the fight to the United States. Although the attacks of September 11th involved no biological, chemical, or nuclear weapons—the threats the Clinton administration had most feared—the attacks were nonetheless tactically brilliant and executed with remarkable efficiency. The ingenuity and magnitude of the assault caught U.S. officials off guard. More than 30 years had elapsed since there had been an incident of simultaneous multiple airplane hijackings anywhere in the world and such an incident had never previously occurred in the United States (National Commission on Terrorist Attacks Upon the United States, 2004).

Following the attack, the nation's highest ranking law enforcement officials vowed to get tough with terrorists, and the Pentagon/Twin Towers Bombing (PENTTBOM) Investigation was launched. Attorney General John Ashcroft (2001) clearly stated that the Justice Department would use "aggressive arrest and detention tactics in the war on terror" (p. 2), and FBI Director Robert S. Mueller (2001) commented that "The FBI is pouring its heart and soul into the investigation of the September 11 attacks. Every resource that can be deployed is being deployed" (p. 12). Federal agents sought anyone and everyone who might have had some involvement in the attacks and began making arrests based on little or no credible evidence, the primary targets being young Muslim males who had recently immigrated to the United States. In some cases the individuals were picked up and held as material witnesses, while in other cases immigration charges (e.g., overstaying a visa) were used to justify the arrests and detention. Within 2 months, more than 1000 people had been arrested (Fine, 2003).

There was also an intense effort among lawmakers to draft counterterrorism legislation. The drafting of legislation is normally a protracted process wherein legislators take months or even years to discuss, scrutinize, and debate proposed policies in an effort to forge sound legislation. In contrast, within days of the September 11th attacks, counterterrorism legislation was introduced in the Senate. A little over a month later, the United and Strengthening America by Providing Appropriate Tools Required to Intercept and Obstruct Terrorism Act of 2001 (USA PATRIOT Act) was signed into law by President George W. Bush, despite the legislation having been subjected to little scrutiny (Wong, 2006a, 2006b). The provisions of the USA PATRIOT Act went well beyond those of previous anti-terrorism legislation. For example, whereas a proposal to allow law enforcement agents to use roving wiretaps had been rejected during the debates over the Anti-Terrorism and Effective Death Penalty Act, the USA PATRIOT Act grants law enforcement officials such authority. The USA PATRIOT Act also included provisions that enhance electronic surveillance authority, allow nationwide execution of warrants for suspected terrorist activity, permit the use of "sneak and peek" search warrants, expand the surveillance and search authorities of the Foreign Intelligence Surveillance Act, and encourage cooperation between law enforcement and intelligence agencies (Bulzomi, 2003; Doyle, 2002; White, 2004; Whitehead & Aden, 2002).

There was also a reorganization of federal law enforcement agencies, including (1) the creation of the Department of Homeland Security and the Transportation Security Administration; (2) the amalgamation of the Immigration and Naturalization Service and Customs Service, and the subsequent formation of three new agencies—Citizenship and Immigration Services, Immigration and Customs Enforcement, and Customs and Border Protection; and (3) the restructuring of priorities among several federal law enforcement agencies such as the FBI and the Bureau of Alcohol, Tobacco, Firearms, and Explosives (ATF), with each agency receiving additional funding to enhance counterterrorism measures (Donohue, 2002; Oliver, 2002; Stuntz, 2002; White, 2004). Beyond that, the Bush administration authorized a range of tactics such as warrantless wiretapping of suspected terrorists, indefinite detention of suspected terrorists, increased use of "rendition" (the transfer of suspects to the custody of foreign governments to be tortured), and torture of suspected terrorists and "enemy combatants" by

American intelligence, law enforcement, and military personnel (Controversy Continues, 2005; Isikoff, 2008, 2009; Pious, 2006; Savage, 2009; Sinnar, 2003; Taylor & Thomas, 2009; U.S. Office of Legal Counsel, 2002a, 2002b, 2003, 2005). Additionally, a secret memo was written by the Justice Department's Office of Legal Counsel that suggested that U.S. military forces be allowed to conduct warrantless searches of homes and warrantless arrests of suspected terrorists in the United States (Meyer & Barnes, 2009; U.S. Office of Legal Counsel, 2001). The Bush administration also made information control a priority and undertook a massive censorship campaign that included the removal of information from government Web sites, strict restrictions on the dissemination of information about counterterrorism measures, and the issuance of broad directives encouraging federal officials to deny requests for information that included nonsensitive and declassified information (Bumiller, 2001; Clymer, 2003; Savage, 2006).

In summary, whereas *community policing* had been the buzz term among law enforcement officials during the 1990s, following the terrorist assault of September 11th, *homeland defense* became the new catchphrase. Prior to the attacks of September 11th, there had been a great deal of discussion and debate among national policy makers about how to secure the country against the threat of terrorist attacks, but relatively little action. After September 11th, the reverse was true: A great many actions were taken based on scant discussion or debate.

## ■ Combative Versus Cooperative Law Enforcement

Following the events of September 11th, law enforcement officials began using myriad tactics of dubious legality and showed little concern about alienating the people with whom they came into contact. For instance, in addition to the previously discussed roundup of more than 1000 people of Middle Eastern descent, the FBI sought to interview more than 5000 persons of Arab ancestry, focusing on immigrants from nations where Al Qaeda was active. The sheer scope of the undertaking led the FBI to request assistance from local law enforcement agencies throughout the country (Brill, 2002; Lengel, 2001). The official memorandum issued by the Office of the Deputy Attorney General (2001), titled "Guidelines for the Interviews Regarding International Terrorism," noted that interviewees were not criminal suspects and that interviews were to "be conducted on a consensual basis." Nonetheless, the guidelines also stated that agents "should feel free to use all appropriate means of encouraging an individual to cooperate" and that if any doubt existed as to whether an interviewee was a legal resident of the United States, immigration authorities should be contacted to ascertain whether the interviewee should be detained. As to the topics to be discussed, the guidelines dictated that interviewees be asked for information such as telephone numbers and residences they had used. In addition, they were to be asked about the reason they were in the United States, foreign travel, involvement with terrorist activity, and sympathy for terrorist groups. Clearly, having federal investigators and local police officers go to individuals' homes—many of whom had been selected via ethnic profiling—and question them about their knowledge of and involvement with terrorist organizations, and then checking their immigration status to ascertain whether they should be deported, is not the best means of fostering positive police–citizen relationships.

The fact that the nation's policy makers favored combative counterterrorism strategies over cooperative law enforcement efforts can be further illustrated by a cursory examination of post-September 11th disparities in federal funding for counterterrorism and community policing. Whereas in the late 1990s the COPS office was providing local law enforcement agencies with approximately $1 billion a year to hire new officers, a few years after the attacks of September 11th the COPS office was providing less than $200 million a year for hiring (DeSimone, 2003; Geraghty, 2003; Wade, 2002). Despite cuts in funding for the COPS office, the Bush administration would have liked to see even more funds shifted from community policing programs to counterterrorism. After President Bush signed a spending bill in February, 2003, White House officials began bemoaning the lack of funds for counterterrorism measures, the implication being that too much money had been earmarked for law enforcement expenditures unrelated to counterterrorism such as community policing programs. One White House official was quoted in the *New York Times* as having stated, "We wanted specific counterterrorism funding. . . . We weren't talking about community policing programs" (Shenon, 2003, p. A14).

At the same time federal officials were cutting funds for local law enforcement agencies, they began requesting that local law enforcement agencies implement new homeland security measures such as increased patrols of landmark buildings, public water supplies, and public transportation infrastructure (e.g., bridges, tunnels, airports)—activities that required local law enforcement agencies to divert funds from existing programs such as community policing activities (Geraghty, 2003; Haberman, 2001; Miller & Duggam, 2002; White, 2004). In the following years, federal funding for community policing was further decimated while funding for federal counterterrorism efforts ballooned. For example, in 2006 the operating budget of the FBI was almost $6 billion, a figure that represented an increase of more than 75% in the few years that had elapsed since the attacks of September 11th. As for community policing, in 2006 the COPS office provided no funds to local police departments for the hiring of new officers (U.S. Office of Management and Budget, 2007, pp. 170–177). It is important to clarify, however, that while there was enormous support for hard-line homeland security strategies in the wake of the September 11th attacks, there was also opposition to the new security measures.

Before the USA PATRIOT Act was passed by Congress, there was concern that the new powers granted to law enforcement and intelligence agencies could be abused. The concern was great enough that Congress included a sunset clause in the USA PATRIOT Act in order to assess the use of the new investigative powers and, if necessary, retract the powers. To provide a few specifics, the powers afforded to law enforcement and intelligence officials to share intercepted electronic information and wiretap-generated intelligence, the authority to seize unopened voice mail pursuant to a warrant, and the authority for nationwide execution of search warrants for electronic information were all considered to have severe potential for abuse and were thus included in the sunset clause (Doyle, 2005a, 2005b). As the Bush administration continued to advocate combative homeland security measures, criticism of the new measures intensified. Officials in a number of cities such as Boulder, Colorado; Santa Fe, New Mexico; and Amherst, Massachusetts, passed resolutions that called upon federal officials to be respectful of the basic rights of the citizenry while engaging in counterterrorism efforts and, in some cases, discouraged municipal employees from assisting with federal law enforcement tactics that might violate the civil rights of the citizenry. In addition, police officials in a number of cities such as Austin, Texas, and San Jose, California, objected to the mass interviewing of immigrants as part of the September 11th investigation. In a few instances, police officials did more than object. For example, police officials in Portland and Corvallis, Oregon, declined to aid federal agents with the interviewing of immigrants (Janofsky, 2002; Lengel, 2001; Miller & Duggan, 2002; Wilgoren, 2001).

Even within federal law enforcement and intelligence communities there were concerns that some of the new homeland security measures were too extreme. Case in point, in 2003 the Counterintelligence Field Activity agency within the U.S. Department of Defense launched an intelligence-gathering operation titled Threat And Local Observation Notice (TALON) designed to gather information about an array of activities such as public protests against corporations that could be linked to or indicate support for terrorist groups. Over time, Pentagon officials began questioning the legitimacy of collecting information about private citizens and organizations, especially when there were no clear links to terrorist activity, and the TALON operation was shut down (Isikoff, 2006; Pincus, 2007). To provide another example, a number of FBI officials questioned the wisdom of openly targeting immigrants as terrorist suspects, opposed the surveillance of mosques and other facilities where Muslims meet, and argued against the torture of suspected terrorists (Hosenball & Klaidman, 2002; Isikoff, 2009; Vest, 2005).

At issue was not only the question of whether post-September 11th homeland security tactics were constitutional or violated civil liberties, but whether the tactics were effective. Numerous incidents and studies have indicated that aggressive law enforcement tactics have little impact on crime and encourage public malice toward law enforcement. In fact, as discussed in Chapter 2, during the latter half of the 20th century, virtually every major urban riot in the United States was fueled by widespread hostility toward the police (Brown & Benedict, 2002;

Pelfrey, 1998; Uchida, 1997). Additionally, scholars have found that harsh counterterrorism tactics yield little or no increase in public safety and foster pervasive distrust of the government. Analyses of the Palmer Raids of the early 1900s, which involved mass arrests and deportations of immigrants, indicated the raids had no impact on public safety and generated widespread suspicion of the government (Braeman, 1964; Renshaw, 1968). To provide another example, based on an analysis of the Italian government's efforts to contain the Red Brigades during the 1970s and 1980s, Albini (2001) noted that "during the period that the Italian government employed its most repressive measures toward the brigades, Italy experienced one of the most violent waves of terrorist activity in its history" (p. 267). Building upon such findings, critics suggested that the implementation of aggressive homeland security tactics may backfire and generate widespread malice toward law enforcement (Brown, 2007; Stuntz, 2002).

Moreover, several scholars argued that many of the strategies used in the investigation of September 11th, such as the mass detentions and interviews of persons of Middle Eastern heritage and the surveillance of mosques, did little more than alienate Muslims and Arab immigrants. It has also been suggested that the use of security strategies that target Arab Americans (e.g., invasive airport security procedures that discourage Middle Easterners from flying, mass rejections of visa applications from persons in Middle Eastern nations) may contribute to increased anti-American sentiment among persons of Middle Eastern heritage, both in the United States and elsewhere in the world. Thus these strategies could discourage Muslim immigrants from engaging in social and political activities, which, in turn, could prevent the disclosure of information about anti-American activities, including terrorist activities (Carey, 2002; Lyons, 2002).

Despite calls from top-level federal officials to be assertive in the investigation of September 11th, many criminal justice officials clung to the central tenets of community policing. During the investigation of the events of September 11th, Michigan became one of the nation's battlegrounds in the dispute over whether combative or cooperative law enforcement tactics are best. Because Michigan has one of the largest Arab populations in the nation, the state was the focus of a substantial portion of the PENTTBOM investigation. Being concerned about public backlash to the targeting of Arabs by investigators, local and state leaders urged federal officials to be cautious about the investigative tactics they utilized and responded to federal requests for assistance in varying ways. For instance, when contacted by federal investigators about assistance with interviewing foreign students as part of the government's efforts to interview thousands of immigrants, officials at the University of Michigan and Michigan State University offered some cooperation but refused to have campus police officers conduct the interviews. Concerns about a lack of cooperation and public backlash to the interviews were great enough that instead of sending federal agents to knock on doors unannounced (an approach that had been used in some areas), U.S. Attorneys in

Michigan sent politely worded letters to interviewees asking that they contact the office and arrange a time and a place for the interviews to take place (Brill, 2002; Lengel, 2001; Peterson, 2001; Wilgoren, 2001).

To provide another example, consider the actions of the police in Dearborn, Michigan, a Detroit suburb wherein almost a third of the denizens are of Middle Eastern ancestry. During the investigation into the events of September 11th, several hundred Dearborn residents were sought by federal agents for questioning. Owing to the magnitude of the task, federal officials requested the assistance of the Dearborn police. Prior to the events of September 11th, the Dearborn Police Department had engaged in community-oriented efforts and worked to earn the trust of the local Arab community. In light of these efforts, the department was concerned about the ramifications of interviewing hundreds of local Arab residents to inquire about terrorist activity. In fact, after the attacks of September 11th, the police in Dearborn increased patrols in areas with concentrations of Arabs not due to concerns that terrorist cells were operating in the city but because of worries about Arabs being the targets of hate crimes carried out by people upset about the terrorist attacks. When the Dearborn police were contacted about the interviews by the Justice Department they agreed to assist but declined to conduct the interviews themselves. Police officers in Dearborn sat in during the interviews but acted as observers rather than interrogators. During the interviews, Dearborn police officers emphasized that the investigation was a federal initiative and monitored the behaviors of federal agents, asking the interviewees afterward if they had found the questioning offensive (Thacher, 2005).

In the years following the attacks of September 11th, many local law enforcement agencies did not make counterterrorism a priority and remained committed to the community policing philosophy. Hickman and Reaves' (2006) analysis of 2003 Law Enforcement Management and Administrative Statistics data showed that a far greater portion of the nation's police departments utilized full-time community policing officers (58%) and had agreements with local agencies or community groups that emphasized cooperative problem solving (60%) than had a written plan for responding to a terrorist attack (39%). Pelfrey's (2007) study of terrorism preparedness and prevention efforts among local and state law enforcement agencies in South Carolina showed that the largest police and sheriff departments had engaged in terrorism preparedness operations, but that the majority of agencies in the state had neither developed formal policies or procedures for responding to terrorist threats nor conducted any scenario-based training exercises in preparation for responding to a terrorist attack. Ortiz, Hendricks, and Sugie (2007) conducted a study of 16 police agencies operating in communities with sizeable Arab American populations and found that none of the agencies developed counterterrorism or intelligence-gathering units in response to the events of September 11th. Rather than adopting a combative style of law enforcement following the September 11th attacks, the majority of police agencies studied by Ortiz and colleagues engaged in community policing. Indeed, in response to September 11th, many of the agencies increased community policing programs, particularly in Arab American neighborhoods. As to counterterrorism operations, the only palpable change that occurred as a result of the September 11th attacks among the police departments was that the majority had increased their cooperation with federal agencies, most notably the FBI.

In sum, despite the outrage over the attacks of September 11th and the Bush administration's emphasis on aggressive homeland security tactics, many people opposed combative law enforcement tactics and remained committed to cooperative policing. A number of criminal justice scholars and practitioners argued that whereas highly aggressive public security tactics are likely to encourage hostility toward law enforcement and discourage cooperation with law enforcement, the use of amiable security tactics such as getting law enforcement officials involved with the communities they serve and seeking the cooperation of individuals and groups likely to have knowledge relevant to terrorist operations will generate favorable attitudes toward law enforcement and quality counterterrorist intelligence (Brown, 2007; Davies et al., 2005; Dyer, McCoy, Rodriguez, & Van Dyun, 2007; Gaylord, 2008; Lyons, 2002; Peed, 2008; Stuntz, 2002). As Carl Peed, the director of the COPS office, summed it up: "Community policing brings law enforcement agencies closer to their communities, and if you're close to your community, you are better able to gather intelligence on suspicious activities and individuals" (Community Policing More, 2002, p. 27). Among the multitude of arguments offered by the advocates of cooperative law enforcement is the underlying and unifying theme that respect for basic human dignity and limited government intrusion into the private domain are essential for the welfare of a democratic nation.

## ■ The Rift of September 11th: Effects of New Policies on Immigration Control

As discussed, the terrorist siege of September 11th sparked a rift in the field of public safety, with some individuals advocating stringent law enforcement tactics while others remained committed to the principles of community policing. Nowhere was this rift more evident or more contentious than in the arena of immigration policy. As the investigation into the attacks of September 11th progressed and it became clear that none of the hijackers were U.S. citizens, all 19 men being of Arab origin, immigrants came to be viewed with suspicion, and in the ensuing months and years, federal officials concentrated greater resources on immigration control.

To provide a few examples, in the aftermath of the September 11th attacks, a directive issued by Attorney General John Ashcroft authorized the indefinite detention of noncitizens; roughly 1000 Arab immigrants were detained during the initial stage of the PENTTBOM investigation; and the USA PATRIOT Act enhanced the authority of federal officials to investigate and detain noncitizens (Corn, 2002; Fine, 2003; National Commission on Terrorist Attacks Upon the United States, 2004). Of especial note was the haphazard process of arresting immigrants following the September 11th attacks. Based upon an examination of the PENTTBOM investigation, the Office of the Inspector General reported that if "agents searching for a particular person on a PENTTBOM lead arrived at a location and found a dozen individuals out of immigration status, each of them were considered to be arrested in connection with the PENTTBOM investigation." Owing to concerns that a terrorist might slip through the dragnet, field agents typically made no distinction "between the subjects of the lead and any other individuals encountered at the scene 'incidentally'" (Fine, 2003, p. 16).

In the years following the assault of September 11th, numerous efforts were made to define and treat immigration control as a matter of national security. In this new environment, federal immigration authorities increased their efforts to crack down on undocumented immigrants. For example, in 2006 and 2007, in what was dubbed Operation Return to Sender, Immigration and Customs Enforcement agents carried out a nationwide sweep of immigrants, resulting in more than 15,000 arrests. The goal was to apprehend criminal immigrants such as drug traffickers and drunk drivers; however, much like the law enforcement officials involved in the PENTTBOM investigation who made no distinction between the targeted immigrants and those encountered "incidentally," agents involved with Operation Return to Sender made no distinction between the criminal immigrants being sought and other immigrants unlawfully residing in the United States. As a result, roughly a third of the Operation Return to Sender arrestees were "collateral arrests"—arrests of undocumented immigrants not purposely targeted nor known to have committed any crimes other than having illegally entered the United States, and who, in numerous instances, were encountered by immigration officials only because they resided at locations previously occupied by the immigrants named in the arrest warrants. The most tragic consequence of the collateral arrests was the large number of children who were stranded following the detention of their parents and wound up in the care of a hodgepodge assortment of friends, family, and social service agencies (Armour & Leinwand, 2005; Associated Press, 2005; Smith, 2007; Spagat, 2007; Welch, 2006).

In addition to the merciless tactics by federal immigration officials after September 11th, efforts were made to have local law enforcement officials enforce federal law. Similar to the investigation into the September 11th attacks when federal officials called upon local law enforcement for assistance in interviewing thousands of immigrants, in an effort to crack down on illegal immigration, federal officials sought the assistance of local law enforcement. In the spring of 2002, while trying to devise new means of enhancing homeland security, Attorney General Ashcroft argued that local law enforcement officials should be authorized to enforce federal immigration law. The power of the Attorney General to authorize local law enforcement officials to enforce federal immigration law was granted in 1996 when President Clinton signed the Illegal Immigration Reform and Immigrant Responsibility Act, but the Clinton administration had not opted to put the measure into effect. In contrast, during President Bush's tenure, federal immigration authorities formulated agreements with local and state agencies that allowed the local and state agencies to help enforce federal immigration laws by means such as questioning stopped drivers about their nationality or immigration status and screening arrestees to ascertain whether they were legal residents of the United States (Axtman, 2002; Constable, 2008; Dorell & Welch, 2007; Garza, 2006; Ortiz et al., 2007; U.S. Immigration and Customs Enforcement, 2008; Wade, 2002).

In a few cases local law enforcement officials used extraordinarily pugilistic tactics. For instance, the Sheriff's Office in Bay County, Florida, devised an unorthodox means of identifying, capturing, and aiding in the deportation of undocumented immigrants: Deputies pulled into construction sites in force, arrested anyone who ran, and forwarded the names of arrestees suspected of illegally residing in the country to immigration officials (Nelson, 2007). As another example, in an effort to identify and apprehend undocumented immigrants, Maricopa County, Arizona, Sheriff Joe Arpaio used assertive means such as having deputies stop pedestrians and vehicles in Hispanic neighborhoods to question people about their nationality or immigration status and detain anyone whose answers or documents appeared suspicious. Sheriff Arpaio went so far as to have a special weapons and tactics (SWAT) team conduct a late night raid on the Mesa City Hall to determine whether the janitors were legal U.S. residents (Giblin, 2008; Hsu, 2008).

At the opposite end of the spectrum are government and law enforcement officials who opposed efforts to have local officers enforce federal immigration law. For instance, in 2003 New York City Mayor Michael Bloomberg signed an executive order that prohibited the police from arresting persons solely for violations of federal immigration law. From Oakland, California, to New Haven, Connecticut, city officials implemented policies that prohibit local law enforcement from making inquiries about immigration status. A small number of states (e.g., Alaska, Maine, and Oregon) also implemented policies discouraging state employees from inquiring about an individual's immigration status (Arnold, 2007; Constable, 2008; Fahim, 2007; McKinley, 2006). Consistent with the basic tenet of the community policing philosophy that good relations between the police and the public are essential for effective crime control, one reason that local agencies may refuse to help enforce federal immigration law is that certain segments of the population—in particular, immigrants and minorities often perceived to be immigrants—will be less likely to cooperate with the police, which, in turn, will reduce the efficacy of the police.

The idea is that if people are hesitant to contact the police for assistance and to provide the police with information about criminal activity, the police will be less able to handle crime. If there were few immigrants in the United States, then the reluctance of immigrants to interact with the police might not present an obstacle to effective crime control, but that is not the case. As of the early 2000s, roughly 1 out of every 10 people in the United States was born outside of the country (Camarota, 2002, 2005). Having local law enforcement aggressively enforce federal immigration law may cause not only undocumented immigrants to distrust the police but also documented immigrants and even native-born citizens. For instance, the aforementioned tactics used by deputies in Maricopa County, Arizona, irritated both Hispanic immigrants and native-born Hispanic citizens. The clash over whether local law enforcement should implement federal immigration law is but one example of the debate in the field of law enforcement between the advocates of combative tactics and the proponents of cooperative strategies. The rift of September 11th has ramifications for law enforcement agencies at all levels, of all sizes, and in every region of the nation.

## ■ Conclusion

When confronted with threats to public safety or the legitimacy of the government, the conventional government response has been to crack down on suspected troublemakers with hard-line tactics. In response to an increase in violence in Northern Ireland in the early 1920s, the British government authorized an array of oppressive tactics ranging from censorship of the press to the detention and interrogation of suspected troublemakers (Donohue, 2002). Similarly, in Italy during the 1970s and 1980s when the terrorist group the Red Brigades embarked upon an assortment of violent activities such as bombings and bank robberies, Italian authorities responded by arresting and imprisoning hundreds of suspected members (Albini, 2001; Simonsen & Spindlove, 2004). In the 1990s when the popularity of Falun Gong (a spiritual exercise) spread in China and its practitioners began to protest restrictions placed upon them by the Chinese government, China's leaders responded by arresting, interrogating, and imprisoning tens of thousands of Falun Gong practitioners (Lum, 2006). In brief, regardless of the character of the threat, the typical government response is to get tough with anyone and everyone suspected of being involved with the threat. The government of the United States is no exception.

From the crackdown on suspected communists during the Red Scare of 1919 to the internment of more than 100,000 Japanese immigrants and Japanese Americans in the 1940s following the attack on Pearl Harbor to the covert counterintelligence operations of the FBI during the 1950s, 1960s, and 1970s that targeted groups considered to be extremist and subversive, such as the Ku Klux Klan and civil rights activists, American authorities have time and again resorted to extreme tactics in response to threats against the social order. In other words, there was nothing unusual about the nature of the response by U.S. officials to the assault of September 11th. What was unusual about the response to the September 11th terrorist attacks was the intensity and scope.

For decades prior to the events of September 11th, American policy makers and criminal justice officials at all levels had made incremental yet consistent strides toward the creation of a law enforcement system consistent with the ideals of a modern democracy—a system that emphasizes human rights, civil liberties, due process of law, and cooperation between law enforcement officials and the people they serve. The response to the events of September 11th by the Bush administration, Congress, and federal law enforcement agencies marked a break from those ideals and practices. Many post-September 11th homeland security measures, ranging from the new powers afforded to law enforcement by the USA PATRIOT Act to the Presidential authorization of the indefinite detention of suspected terrorists, violated both the U.S. Constitution and the United Nations Universal Declaration of Human Rights. For years U.S. presidents and government officials had criticized nations such as China and Iraq for using brutal methods of control, yet in response to the attacks of September 11th, U.S. policy makers authorized the use of repressive security measures inclusive of mass arrests and the torture of suspected terrorists—practices no different than those U.S. officials have often condemned.

The other significant characteristic of the response to the events of September 11th was its scope. The response to the September 11th terrorist attacks affected an amalgam of local law enforcement agencies, state law enforcement agencies, federal law enforcement agencies, national intelligence agencies, military forces, and private enterprises. For instance, the creation of the Transportation Security Administration, a measure that in effect federalized airport security, was unprecedented. To provide another example, whereas prior to the events of September 11th the nation's law enforcement and intelligence agencies had operated independently of one another for decades, the USA PATRIOT Act included provisions designed to encourage the sharing of information between law enforcement and intelligence agencies. These provisions eroded constitutional safeguards previously considered essential to the welfare and progress of the American people. From the dissolution of the Customs Service, one of the nation's oldest law enforcement agencies, to the practice of granting local law enforcement the authority to enforce federal immigration law, the post-September 11th homeland defense measures had profound implications for an array of law enforcement and intelligence agencies.

It is important to clarify, however, that while post-September 11th legislation and federal policy changes had a substantial impact on the nation's federal law enforcement apparatus, the impact on local law enforcement is unclear. A number of scholars have argued that post-September 11th reforms contributed to a fundamental transformation of the nation's police forces. Kappeler and Miller (2006), for instance, suggested that post-September 11th efforts to get local law enforcement agencies involved in federal operations contributed to "the federalization of municipal police" (p. 562). Oliver (2007) argued that the events of September 11th "set in motion a new era of policing, namely the era of homeland security" (p. 43). Contrary to such sweeping generalizations, the fact is that many local law enforcement agencies actively opposed federal homeland security initiatives, especially measures that target immigrants. While some local and state law enforcement agencies have begun aiding in the enforcement of federal immigration law, such practices have by no means become the norm. Empirical studies of the impact of September 11th on local law enforcement agencies have shown that many local agencies made few if any substantive operational changes in response to the terrorist attacks and calls for enhanced homeland security measures from federal authorities (e.g., Pelfrey, 2007). The most blanketing change among local law enforcement agencies appears to be increased cooperation with the FBI's Joint Terrorism Task Forces (Barker & Fowler, 2008; Casey, 2004; Ortiz et al., 2007).

Although the nation's largest local law enforcement agencies, such as the New York City Police Department, Los Angeles Police Department, and the Chicago Police Department created or enhanced counterterrorism operations following the events of September 11th, the effect that such operations have on the daily activities of the agencies' officers is unclear. While the assault of September 11th was atrocious, it did not fundamentally alter life in the United States or the multitude of problems the police must contend with. Criminal activity such as domestic violence and drug trafficking continues to plague communities across the nation, and the police must continue their efforts to handle such problems. Similarly, noncriminal events such as storms that down electrical lines, parades that require streets to be cleared of parked and moving vehicles, and funeral processions that necessitate police escorts have continued unabated, and the police must help handle them. From mundane duties such as handling traffic accidents to disturbing tasks such as investigating child sexual abuse, the diversity of services that local law enforcement officials provide has remained largely unchanged since the events of September 11th, and the bulk of such services have nothing to do with organized terrorism.

Another simple fact is that the police, like every other public service agency, have limited funds. While Congress and the Bush administration substantially increased funding for federal law enforcement agencies following the siege of September 11th, little funding was provided to local law enforcement agencies to increase homeland security. As a result, local law enforcement agencies struggled to find money to pay for first-responder equipment (e.g., gas masks and protective suits), counterterrorism and emergency response training, overtime, and intelligence-gathering operations (James, 2003; Miller & Duggan, 2002; O'Hanlon & Weiss, 2004; Wilgoren, 2002). Pelfrey's (2007) analysis of counterterrorism activity among local law enforcement agencies in South Carolina showed "that the presence of funding (from the state, federal, or local levels) significantly predicts whether an agency has taken meaningful steps toward terrorism preparedness" (p. 318). In short, considering the multitude of services the police provide, the economic realities of municipal and state governments, and the empirical research on policing in post-September 11th America, there is little reason to believe that the events of September 11th led to a fundamental transformation of policing in the United States.

With respect to the future of policing in the United States and the question of whether the terrorist attacks of September 11th will have a lasting impact on the nature of American law enforcement, there are a couple of significant events that must be taken into consideration. As noted in the introduction to this essay, law enforcement agencies must adapt to social change. Law enforcement agencies must respond not only to changes in the nature of illegal activity but to economic and political changes as well. And in 2008 there were a couple significant societal changes to which the nation's law enforcement agencies had to adapt.

First, there was the global financial meltdown of 2008. The economic situation in the United States began to falter in 2007, and over the next year the situation went from grim to ghastly, with the economic havoc spreading from the United States to Asia and Europe. Owing to a combination of issues (e.g., subprime lending, deregulation of credit markets, quixotic banking and business management practices) what appears to have begun as a liquidity crisis metastasized into a cycle of failed loans and credit restrictions that resulted in enormous losses in the stock and housing markets, failures of well-established businesses such as Lehman Brothers and Circuit City, skyrocketing numbers of home foreclosures, and a substantial increase in unemployment (Gross, 2009; Jones, 2009; Slania, 2008; Stewart, 2007; Von Drehle & Curry, 2009; Zuckerman, 2007). As a result, tax revenues decreased and municipal, county, and state governments were forced to cut back on a plethora of public services including law enforcement, with numerous police agencies implementing hiring freezes and cutting overtime (Billups, 2009; Fausset & Riccardi, 2008; C. Johnson, 2009; Pfeifer, 2008; Schworm, 2008; Shister, 2009; Slack & Cramer, 2008; Willon, 2008). Faced with a decrease in funding yet consistent (if not increased) levels of crime and calls for service, it is doubtful that many local or state law enforcement agencies will make counterterrorist intelligence gathering and the surveillance of potential terrorist targets (e.g., bridges and public reservoirs) their top priority.

The second important social phenomenon was the election of 2008 and associated political power shifts. In November 2008, Democrats took control of the Office of the President, the House of Representatives, and the Senate (Herszenhorn & Hulse, 2008; Hulse, 2008; Schor & MacAskill, 2008; Simon, 2008). In the aftermath of the September 11th attacks, the Bush administration had the support of a conservative Republican-controlled Congress and, as evidenced by the rapid passage of the USA PATRIOT Act, had little trouble getting congressional authorization for intrusive law enforcement powers. And with the support of a Republican-controlled Congress and, concomitantly, a paucity of serious congressional inquiries into internal White House operations, the Bush administration had hardly any difficulties authorizing a variety of tactics of dubious legality such as the censorship of government documents and the torture of suspected terrorists. In contrast to President Bush, President Obama has expressed disapproval for such homeland security measures. Shortly after taking office, President Obama issued several executive orders and memorandums, which, among other things, dictated that the detention facility at Guantanamo Bay be closed, that U.S. intelligence agents cease the use of torture, that U.S. intelligence agencies shut down secret prisons, and that federal employees begin complying with requests for information. By the spring of 2009, the Justice Department had launched investigations into the torture of suspected terrorists and attempts to cover up such activities. In addition, whereas funding for the COPS office was dramatically reduced by the Bush administration, President Obama lauded and vowed to support the COPS office and community policing operations (Isikoff & Hosenball, 2009; K. Johnson, 2009; Lewis, 2009a, 2009b; Shane, Mazzetti, & Cooper, 2009; Spetalnick, 2009; Stolberg, 2009; U.S. Office of Legal Counsel, 2009; Warrick & DeYoung, 2009). In brief, there are numerous indicators that the nation's homeland defense agencies will be adopting a new operational philosophy and that the federal government will cease encouraging local law enforcement to use combative tactics.

While it is impossible to predict the future, there is little doubt that in much the same manner that the advocates of cooperative law enforcement strategies opposed the implementation of combative homeland security measures, the advocates of combative law enforcement strategies will oppose the revival of cooperative community-oriented law enforcement operations. In addition, it is essential to keep in mind that terrorist organizations such as Al Qaeda remain intact and continue to pose threats to the United States and its allies—a reality requiring the nation's law enforcement agencies to be vigilant in their efforts to deter terrorist strikes. Concisely stated, the political ideological pendulum has begun to swing away from combative law enforcement toward cooperative law enforcement, but the rift of September 11th has by no means been sealed. One can only hope that in the years to come America's leaders, law enforcement officials, and intelligence officials can find a way to secure the nation from the threat of organized terrorism, while also respecting and extending the basic human dignity that is fundamental to the ideals of an advanced democratic society.

## ■ Discussion Questions ▬▬▬▬▬▬▬▬▬▬▬▬▬▬▬▬▬▬▬

1. Discuss pre-September 11th examples of terrorism and their implications for policing policy in the United States.
2. How have the terrorist attacks of September 11, 2001, influenced policing policy in the United States and in other nations?
3. Describe the origins and implications of the PENTTBOM investigation.
4. Provide examples of combative and cooperative law enforcement.
5. How might recent developments in the economic and political sphere influence policing policy in the future?

## ■ References ▬▬▬▬▬▬▬▬▬▬▬▬▬▬▬▬▬▬▬▬▬▬▬▬▬▬

Albini, J. L. (2001). Dealing with the modern terrorist: The need for changes in strategies and tactics in the new war on terrorism. *Criminal Justice Policy Review, 12*(4), 255–281.

Armour, S., & Leinwand, D. (2005, November 18). 120 arrested on immigration violations at Wal-Mart site. *USA Today*, p. 4B.

Arnold, B. (2007, September 25). "Sanctuary" cities for illegals draw ire. *Christian Science Monitor*, 2.

Ashcroft, J. (2001, October 25). *Prepared remarks for the US Mayors Conference.* Washington, DC: U.S. Department of Justice. Retrieved February 5, 2010, from http://www.usdoj.gov/archive/ag/speeches/2001/agcrisisremarks10_25 .htm

Associated Press. (2005, November 19). 125 illegal workers arrested; Pennsylvania Wal-Mart center raid will bring deportations. *Houston Chronicle*, p. B2.

Axtman, K. (2002, August 19). Police can now be drafted to enforce immigration law. *Christian Science Monitor, 94*(186), 2.

Barker, B., & Fowler, S. (2008). The FBI joint terrorism task force officer. *FBI Law Enforcement Bulletin, 77*(11), 12–15.

Betts, R. K. (2004). The new politics of intelligence: Will reforms work this time? *Foreign Affairs, 83*(3), 2–8.

Billups, A. (2009, March 11). Schools cut budgets where it hurts children most: Little stimulus relief expected. *Washington Times*, p. A1.

Bodrero, D. D. (1999, March). Confronting terrorism on the state and local level. *FBI Law Enforcement Bulletin, 68*(3), 11–18.

Braeman, J. (1964). World War One and the crisis of American liberty. *American Quarterly, 16*(1), 104–112.

Brill, S. (2002, January 28). The FBI gets religion. *Newsweek, 139*(4), 32–33.

Brown, B. (2007). Community policing in post-September 11 America: A comment on the concept of community-oriented counterterrorism. *Police Practice and Research, 8*(3), 239–251.

Brown, B., & Benedict, W. R. (2002). Perceptions of the police: Past findings, methodological issues, conceptual issues, and policy implications. *Policing: An International Journal of Police Strategies and Management, 25*(3), 543–580.

Bulzomi, M. J. (2003, June). Foreign Intelligence Surveillance Act before and after the USA PATRIOT Act. *FBI Law Enforcement Bulletin, 72*(6), 25–31.

Bumiller, E. (2001, October 7). A nation challenged: Flow of information. *New York Times*, p. 1B.

Camarota, S. A. (2002, November). *Immigrants in the United States—2002: A snapshot of America's foreign-born population.* Washington, DC: Center for Immigration Studies.

Camarota, S. A. (2005, December). *Immigrants at mid-decade: A snapshot of America's foreign-born population in 2005.* Washington, DC: Center for Immigration Studies.

Carey, H. F. (2002). Immigrants, terrorism and counter-terrorism. *Peace Review, 14*(4), 395–402.

Casey, J. (2004, November). Managing joint terrorism task force resources. *FBI Law Enforcement Bulletin, 73*(11), 1–6.

Chalmers, D. (1990). The security state in America: Inside J. Edgar Hoover's FBI. *Reviews in American History, 18*(1), 118–123.

Clymer, A. (2003, January 3). Government openness at issue as Bush holds on to records. *New York Times*, p. A1.

Community Policing Consortium. (1994, August). *Understanding community policing: A framework for action* (NCJ 148457). Washington, DC: U.S. Department of Justice, Office of Justice Programs, Bureau of Justice Assistance.

Community policing more important today than it was before 9/11. (2002, January/February). *Sheriff, 54*(1), 26–28.

Constable, P. (2008, August 23). Many officials reluctant to help arrest immigrants. *Washington Post*, p. B1.

Controversy continues regarding detainees held by the CIA, renditions to other countries. (2005, July). *American Journal of International Law, 99*(3), 706–707.

Cordner, G. W. (1997) Community policing: Elements and effects. In R. G. Dunham & G. P. Alpert (Eds.), *Critical issues in policing: Contemporary readings* (3rd ed., pp. 451–468). Prospect Heights, IL: Waveland Press.

Corn, D. (2002, March/April). The fundamental John Ashcroft. *Mother Jones, 27*(2), 39–43.

Davies, H. J., Plotkin, M. R., Filler, J., Flynn, E. A., Foresman, G., Litzinger, J., et al. (2005, November). *Protecting your community from terrorism: Strategies for local law enforcement.* Washington, DC: U.S. Department of Justice, Office of Community Oriented Policing Services and Police Executive Research Forum.

DeSimone, D. C. (2003, April). Federal budget a mixed bag for state and local governments. *Government Finance Review, 19*(2), 66–69.

Donohue, L. K. (2002). Bias, national security, and military tribunals. *Criminology and Public Policy, 1*(3), 339–344.

Dorrell, O., & Welch, W. M. (2007, March 30). Local police confront illegal immigrants. *USA Today*, p. 1A.

Doyle, C. (1996, June 3). *Antiterrorism and Effective Death Penalty Act of 1996: A summary* (96-499A). Washington, DC: Library of Congress, Congressional Research Service.

Doyle, C. (2002, April 18). *The USA PATRIOT Act: A sketch* (RS21203). Washington, DC: Library of Congress, Congressional Research Service.

Doyle, C. (2005a, August 10). *USA PATRIOT Act Reauthorization in Brief* (RS22216). Washington, DC: Library of Congress, Congressional Research Service.

Doyle, C. (2005b, January 27). *USA PATRIOT Act Sunset: A sketch* (RS21704). Washington, DC: Library of Congress, Congressional Research Service.

Dreyfuss, R. (2000, September/October). The phantom menace. *Mother Jones, 25*(5), 40–91.

Dyer, C., McCoy, R. E., Rodriguez, J., & Van Dyun, D. N. (2007, December). Countering violent Islamic extremism: A community responsibility. *FBI Law Enforcement Bulletin, 76*(12), 3–9.

Eck, J. E., & Spelman, W. (1997). Problem-solving: Problem-oriented policing in Newport News. In R. G. Dunham & G. P. Alpert (Eds.), *Critical issues in policing: Contemporary readings* (3rd ed., pp. 489–503). Prospect Heights, IL: Waveland Press.

Fahim, K. (2007, April 29). Should immigration be a police issue? *New York Times*, p. C14.

Fausset, R., & Riccardi, N. (2008, October 19). States could face historic financial crisis: Shriveling tax revenue is forcing layoffs and program cuts. *Los Angeles Times*, p. A1.

Fine, G. (2003, June). *The September 11 detainees: A review of the treatment of aliens held on immigration charges in connection with the investigation of the September 11 attacks*. Washington, DC: U.S. Department of Justice, Office of the Inspector General.

Fogelson, R. M. (1968). From resentment to confrontation: The police, the negroes, and the outbreak of the nineteen-sixties riots. *Political Science Quarterly, 83*(2), 217–247.

Garza, C. L. (2006, October 23). HPD procedure shift draws concern among immigrants—Officers now must ask anyone arrested for proof of citizenship. *Houston Chronicle*, p. B1.

Gaylord, A. A. (2008, April). Community involvement: The ultimate force multiplier. *FBI Law Enforcement Bulletin, 77*(4), 16–17.

Geraghty, J. (2003, April 9). Democratic study indicates police strain under homeland securities duties. *States News Service*, p. 1008099u1552.

Giblin, P. (2008, October 18). Arizona sheriff conducts immigration raid at city hall, angering officials. *New York Times*, p. A10.

Goldstein, H. (1979) Improving policing: A problem-oriented approach. *Crime and Delinquency, 25*(2), 236–258.

Gross, D. (2009, March 9). Reigning in bubbles so they won't pop. *Newsweek, 153*(10), 44–46.

Haberman, C. (2001, October 27). Visions of a long struggle, strains on police and help for victims. *New York Times*, p. 1B.

Herszenhorn, D. M., & Hulse, C. (2008, November 6). Democrats in congress vowing to pursue aggressive agenda. *New York Times*, p. A1.

Hickman, M. J., & Reaves, B. A. (2003, January). *Local police departments, 2000* (NCJ 196002). Washington, DC: U.S. Department of Justice, Office of Justice Programs, Bureau of Justice Statistics.

Hickman, M. J., & Reaves, B. A. (2006, May). *Local police departments, 2003* (NCJ 210118). Washington, DC: U.S. Department of Justice, Office of Justice Programs, Bureau of Justice Statistics.

Hosenball, M., & Klaidman, D. (2002, June 17). A leap of faith. *Newsweek, 139*(24), 32.

Hsu, S. (2008, July 17). Arizona sheriff accused of racial profiling. *Washington Post*, p. A2.

Hulse, C. (2008, November 19). Democrats gain as Stevens loses his senate race. *New York Times*, p. A1.

Isikoff, M. (2006, January 30). The other big brother. *Newsweek, 147*(5), 32–34.

Isikoff, M. (2008, December 22). The fed who blew the whistle. *Newsweek, 152*(25), 40–48.

Isikoff, M. (2009, May 4). "We could have done this the right way": How Ali Soufan, an FBI agent, got Abu Zubaydah to talk without torture. *Newsweek, 153*(18), 18–21.

Isikoff, M., & Hosenball, M. (2009, May 11/May 18). Fresh questions about the CIA's interrogation tapes. *Newsweek, 153*(19/20), 6.

James, G. (2003, April 13). Is New Jersey ready for an attack? *New York Times*, p. NJ1.

Janofsky, M. (2002, December 23). Threats and responses: Civil liberties; cities wary of antiterror tactics pass civil liberties resolutions. *New York Times*, p. A1.

Johnson, C. (2009, February 8). Double blow for police: Less cash, more crime. *Washington Post*, p. A3.

Johnson, K. (2009, February 27). Stimulus allots for more cops. *USA Today*, p. 7A.

Jones, C. (2009, March 13). Sour U.S. economy has put 40% of Americans on edge. *USA Today*, p. 1A.

Kappeler, V. E., & Miller, K. S. (2006). Reinventing the police and society: Economic revolutions, dangerousness, and social control. In V. E. Kappeler (Ed.), *The police and society: Touchstone readings* (3rd ed., pp. 552–565). Long Grove, IL: Waveland Press.

Kelling, G. L., & Moore, M. H. (1988, November). *The evolving strategy of policing* (Perspectives on Policing, No. 4, NCJ 114213). Washington, DC: U.S. Department of Justice, Office of Justice Programs, National Institute of Justice.

Lengel, A. (2001, November 27). Arab men in Detroit to be asked to see U.S. attorney. *Washington Post,* p. A5.

Lewis, J. F. (1999, March). Fighting terrorism in the 21st century. *FBI Law Enforcement Bulletin, 68*(3), 3–10.

Lewis, N. (2009a, February 6). Recovery bill has $1 billion to hire more local police. *New York Times,* p. A16.

Lewis, N. (2009b, February 2). Justice department under Obama is preparing for doctrinal shift in policies of Bush years. *New York Times,* p. A14.

Lum, T. (2006, May 25). *China and Falun Gong* (RL33437). Washington, DC: Library of Congress, Congressional Research Service.

Lyons, W. (2002). Partnerships, information, and public safety: Community policing in a time of terror. *Policing: An International Journal of Police Strategies and Management, 25*(3), 530–542.

McKinley, J. (2006, November 12). Immigrant protection rules draw fire. *New York Times,* p. A22.

Meyer, J., & Barnes, J. (2009, March 3). Memos gave Bush overriding powers. *Los Angeles Times,* p. A1.

Miller, B., & Duggan, P. (2002, February 12). Security worries bring a costly shift in priorities. *Washington Post,* p. A3.

Moore, M. H. (1994). Research synthesis and policy implications. In D. P. Rosenbaum (Ed.), *The challenge of community policing: Testing the promises* (pp. 285–297). Thousand Oaks, CA: Sage Publications, Inc.

Mueller, R. S. (2001, December). Responding to terrorism. *FBI Law Enforcement Bulletin, 70*(12), 12–14.

Mufson, S. (2001, February 1). Overhaul of national security apparatus urged: Commission cites U.S. vulnerability. *Washington Post,* p. A2.

National Commission on Terrorist Attacks Upon the United States. (2004). *The 9/11 Commission report: Final report of the National Commission on Terrorist Attacks Upon the United States.* New York: W. W. Norton & Company.

Nelson, M. (2007, June 28). Migrant-chasing draws fire: Florida sheriff defends raids that target workers who run. *San Antonio Express-News,* p. 14A.

Office of the Deputy Attorney General. (2001, November 9). *Memorandum for all United States attorneys and all members of the anti-terrorism task forces: Guidelines for the interviews regarding international terrorism.* Washington, DC: Author. Retrieved February 5, 2010, from http://www.usdoj.gov/dag/reading room/terrorism2.htm

O'Hanlon, M., & Weiss, J. (2004, August 18). How police can intervene. *Washington Times*, p. A16.

Oliver, W. M. (2002, September/October). 9-11, federal crime control policy, and unintended consequences. *ACJS Today, 22*(3), 1–6.

Oliver, W. M. (2007). *Homeland security for policing.* Upper Saddle River, NJ: Pearson Prentice Hall.

Ortiz, C. W., Hendricks, N. J., & Sugie, N. F. (2007). Policing terrorism: The response of local police agencies to homeland security concerns. *Criminal Justice Studies, 20*(2), 91–109.

Peed, C. (2008, November). The community policing umbrella. *FBI Law Enforcement Bulletin, 77*(11), 22–24.

Pelfrey, W. V. (1998). Precipitating factors of paradigmatic shift in policing: The origin of the community policing era. In G. P. Alpert & A. Piquero (Eds.), *Community policing: Contemporary readings* (pp. 79–92). Prospect Heights, IL: Waveland Press.

Pelfrey, W. V. (2007). Local law enforcement terrorism prevention efforts: A state level case study. *Journal of Criminal Justice, 35,* 313–321.

Peterson, J. (2001, December 3). *U to support students in investigation* (statement from University of Michigan Office of Communications). Retrieved January 17, 2009, from http://www.ur.umich.edu/0102/Dec03_01/6.htm

Pfeifer, S. (2008, November 25). O.C. halts hiring, plans budget cuts. *Los Angeles Times*, p. B3.

Pincus, W. (2007, April 25). Pentagon to end TALON data-gathering program. *Washington Post*, p. A10.

Pious, R. M. (2006). *The war on terrorism and the rule of law.* Los Angeles, CA: Roxbury Publishing Company.

Renshaw, P. (1968). The IWW and the Red Scare 1917–24. *Journal of Contemporary History, 3*(4), 63–72.

Roth, J. A., Roehl, J., & Johnson, C. C. (2004). Trends in the adoption of community policing. In W. G. Skogan (Ed.), *Community policing: Can it work?* (pp. 3–29). Belmont, CA: Wadsworth/Thomson.

Savage, C. (2006, February 17). Activists assert secrecy is Cheney's hallmark; say he has led efforts to curtail flow of information. *Boston Globe*, p. A3.

Savage, D. G. (2009, January 16). Court calls warrantless wiretaps legal: The ruling allows eavesdropping on international calls, even when Americans may be involved. *Los Angeles Times*, p. A14.

Schor, E., & MacAskill, E. (2008, November 6). Obama triumph. *The Guardian*, p. 7.

Schworm, P. (2008, December 27). Boston could lose hundreds of teaching posts: School officials prepare for worst in preliminary talks on budget. *Boston Globe*, p. B4.

Shane, S., Mazzetti, M., & Cooper, H. (2009, January 23). Obama reverses key Bush policy but questions on detainees remain. *New York Times,* p. A16.

Shenon, P. (2003, February 26). Threats and responses: Domestic security; In reversal, White House concedes that counterterrorism budget is too meager. *New York Times,* p. A14.

Shister, G. (2009, February 27). Budget pinch costs jobs at D.A.'s office. *Philadelphia Inquirer,* p. B2.

Simon, R. (2008, November 5). Democrats make big gains on Capitol Hill: Senate edge grows but appears short of the crucial 60 votes. *Los Angeles Times,* p. A18.

Simonsen, C. E., & Spindlove, J. R. (2004). *Terrorism today: The past, the players, the future* (2nd ed.). Upper Saddle River, NJ: Pearson Prentice Hall.

Sinnar, S. (2003, April). Patriotic or unconstitutional? The mandatory detention of aliens under the USA PATRIOT Act. *Stanford Law Review, 55,* 1419–1456.

Slack, D., & Cramer, M. (2008, October 17). Menino freezes hiring by city, urges approval of developments to bring new jobs. *Boston Globe,* p. B1.

Slania, J. T. (2008, December 22). An economic avalanche. *Crain's Chicago Business, 31*(51), 15.

Smith, R. L. (2007, June 2). Latinos stay home while feds launch more raids: Arrests in Lake total 37. *Plain Dealer,* p. B1.

Spagat, E. (2007, April 6). "Collateral arrests": Crackdown on fugitives nets thousands of undocumented immigrants. *Brownsville Herald,* p. A1.

Spetalnick, M. (2009, January 23). Obama acts to burnish U.S. image abroad: President moves quickly to roll back several Bush policies, names troubleshooters. *Toronto Sun,* p. AA1.

Stewart, J. (2007, August 29). Does credit crunch signal deep woes in economy? *Wall Street Journal,* p. D3.

Stolberg, S. G. (2009, January 22). On first day Obama quickly sets a new tone. *New York Times,* p. A1.

Stuntz, W. J. (2002). Local policing after the terror. *Yale Law Review, 111,* 2137–2194.

Taylor, S., & Thomas, E. (2009, January 19). Obama's Cheney dilemma. *Newsweek, 153*(3), 20–26.

Thacher, D. (2005). The local role in homeland security. *Law & Society Review, 39*(3), 635–676.

Uchida, C. (1997) The development of the American police: An overview. In R. G. Dunham & G. P. Alpert (Eds.), *Critical issues in policing: Contemporary readings,* (3rd ed., pp. 18–35). Prospect Heights, IL: Waveland Press.

U.S. Department of Justice. (1994). *Violent Crime Control and Law Enforcement Act of 1994: Briefing book.* Washington, DC: Author.

U.S. Immigration and Customs Enforcement. (2008, August 18). Delegation of immigration authority section 287(g) Immigration and Nationality Act. Washington, DC: Author. Retrieved February 5, 2010, from http://www.ice.gov/partners/287g/Section287_g.htm

U.S. Office of Legal Counsel. (2001, October 23). *Memorandum for Alberto R. Gonzales, Counsel to the President, William J. Haynes, II, General Counsel, Department of Defense Re: Authority for Use of Military Force to Combat Terrorist Activities Within the United States*. Washington, DC: U.S. Department of Justice. Retrieved August 17, 2009 from http://www.justice.gov/opa/documents/memomilitaryforcecombatus10232001.pdf

U.S. Office of Legal Counsel. (2002a, June 27). *Memorandum for Daniel J. Bryant, Assistant Attorney General, Office of Legislative Affairs Re: Applicability of 18 U.S.C. § 4001(a) to Military Detention of United States Citizens*. Washington, DC: U.S. Department of Justice. Retrieved February 5, 2010, from http://www.usdoj.gov/opa/documents/memodetentionuscitizens06272002.pdf

U.S. Office of Legal Counsel. (2002b, March 13). *Memorandum for William J. Haynes, II, General Counsel, Department of Defense Re: The President's power as Commander in Chief to transfer captured terrorists to the control and custody of foreign nations*. Washington, DC: U.S. Department of Justice. Retrieved February 5, 2010, from http://www.usdoj.gov/opa/documents/memorandum03132002.pdf

U.S. Office of Legal Counsel. (2003, March 14). *Memorandum for William J. Haynes, II, General Counsel of the Department of Defense Re: Military Interrogation of Alien Unlawful Combatants Held Outside the United States*. Washington, DC: U.S. Department of Justice. Retrieved February 5, 2010, from http://www.aclu.org/pdfs/safefree/yoo_army_torture_memo.pdf

U.S. Office of Legal Counsel. (2005, May 10). *Memorandum for John A. Rizzo, Senior Deputy General Counsel, Central Intelligence Agency Re: Application of 18 U.S.C. §§ 2340-2340A to the Combined Use of Certain Techniques in the Interrogation of High Value al Qaeda Detainees*. Washington, DC: U.S. Department of Justice. Retrieved February 5, 2010, from http://www.aclu.org/safefree/general/olc_memos.html

U.S. Office of Legal Counsel. (2009, January 15). *Memorandum for the Files Re: Status of Certain OLC Opinions Issued in the Aftermath of the Terrorist Attacks of September 11, 2001*. Washington, DC: U.S. Department of Justice. Retrieved February 5, 2010, from http://www.usdoj.gov/opa/documents/memostatusolcopinions01152009.pdf

U.S. Office of Management and Budget. (2007). *Budget of the United States Government, Fiscal Year 2007*. Washington, DC: Author.

Vest, J. (2005, July). Pray and tell. *The American Prospect, 16*(7), 47–50.

Von Drehle, D., & Curry, M. (2009, March 9). House of Cards. *Time, 173*(9), 22–29.

Wade, B. (2002, June 1). Local law enforcement is getting robbed. *American City and County*, p. 1.

Wadman, R. C., & Allison, W. T. (2004). *To protect and to serve: A history of police in America*. Upper Saddle River, NJ: Pearson Prentice Hall.

Warrick, J., & DeYoung, K. (2009, January 23). Obama reverses Bush policies on detention and interrogation. *Washington Post*, p. A6.

Welch, W. M. (2006, June 15). More than 2,000 illegal immigrants nabbed in sweep. *USA Today*, p. 2A.

White, J. R. (2004). *Defending the homeland: Domestic intelligence, law enforcement, and security*. Belmont, CA: Wadsworth.

Whitehead, J. W., & Aden, S. H. (2002). Forfeiting "enduring freedom" for "homeland security": A constitutional analysis of the USA PATRIOT Act and the Justice Department's anti-terrorism initiatives. *American University Law Review, 51*, 1081–1133.

Wilgoren, J. (2001, December 1). A nation challenged: The interviews; University of Michigan won't cooperate in federal canvas. *New York Times*, p. B6.

Wilgoren, J. (2002, June 19). Traces of terror: Domestic security; At one of 1,000 front lines in U.S., local officials try to plan for war. *New York Times*, p. A19.

Williams, D. (1981). The Bureau of Investigation and its critics, 1919–1921: The origins of federal political surveillance. *Journal of American History, 68*(3), 560–579.

Williams, H., & Murphy, P. V. (1990, January). *The evolving strategy of policing: A minority view* (Perspectives on Policing, No. 13, NCJ 121019). Washington, DC: U.S. Department of Justice, Office of Justice Programs, National Institute of Justice.

Willon, P. (2008, November 25). Police hiring freeze on table. *Los Angeles Times*, p. B4.

Wong, K. C. (2006a). The making of the USA Patriot Act I: The legislative process and dynamics. *International Journal of the Sociology of Law, 34*(3), 179–219.

Wong, K. C. (2006b). The making of the USA PATRIOT Act II: Public sentiments, legislative climate, political gamesmanship, media patriotism. *International Journal of the Sociology of Law, 34*(2), 105–140.

Zuckerman, M. B. (2007, December 24). The credit crisis grows. *U.S. News and World Report, 142*(22), 68.

## ■ Court Cases Cited

*Katz v. United States*, 389 U.S. 347, 88 S. Ct. 507 (1967).

*Payton v. New York*, 445 U.S. 573, 100 S. Ct. 1371 (1980).

# Policies Promoting School–Police Partnerships

# CHAPTER 4

Nicole L. Bracy

## ■ Introduction

Over the past two decades, public schools in the United States have undergone significant physical and ideological changes with regard to discipline and security. The introduction of police officers as permanent fixtures in public schools across the country is one particularly notable change. In 1998 in New York City, amid concerns that security was not being properly implemented in city schools, Mayor Rudolph Giuliani transferred control of school safety from the New York City Board of Education to the New York Police Department (NYPD). In the years following this transfer, the number of police personnel in New York City schools increased 50%; by 2005, the school safety division employed 4600 unarmed security guards (called School Safety Agents) and over 200 armed police officers (Mukherjee, 2007). The NYPD also introduced roving metal detectors to all public middle and high schools in the city, and permanent metal detectors were in place in about one-fifth of the city's schools. When students at one Bronx high school protested police practices that caused lengthy delays as they tried to enter their school building each morning, administrators responded by promising to install more metal detectors to speed up lines (Fisher, 2005).

In major cities like New York, Washington DC, Baltimore, and Los Angeles that generally have big-city crime problems, it may not be surprising to find public schools that use extensive surveillance and policing. Increasingly, however, even rural and suburban public schools that have not historically experienced significant problems with school violence are rapidly beefing up school security (Simon, 2007). These changes have largely been in response to public perception of school violence as escalating and permeating, and the perceived inability of schools to prevent major incidents and adequately protect students and staff. Consider the tragic events at Columbine High School in 1999, for example. Following the deaths of 13 people at the hands of 2 students with firearms, Columbine High School administrators and local law enforcement experienced a significant amount of criticism for failing to prevent the tragedy and for an inadequate response once the violence erupted (for examples, see Janofsky, 2000; O'Driscoll, 2001). These criticisms persisted even though at the time of the shootings Columbine High School had security measures in place, including a school police officer and surveillance cameras. The message seems to be that never can *enough* be done to secure schools and keep students safe.

Extensive media coverage of the murders at Columbine High School and other similar incidents that followed in schools across the country (e.g., Santana High School in 2001, Red Lake High School in 2005, West Nickel Mines School in 2006) kept school violence at the forefront of the minds of the American public and further perpetuated fears about the ubiquitousness of school violence (Kupchik & Bracy, 2009a). Many Americans might be surprised to know that school crime and violence were actually declining during this time. The victimization rate of youth 12 to 18 years old declined between 1992 and 2005, both at school and away from school. The rate of all crimes against students at school was reduced by 40% during this time (from 144 per 1000 in 1992 to 57 per 1000 in 2005), and the rate of violent crimes against students was cut in half (from 48 per 1000 in 1992 to 24 per 1000 in 2005) (Dinkes, Cataldi, & Lin-Kelly, 2007).

Despite the decrease in school crime and violence since the early 1990s, public schools in the United States continue to implement security strategies borrowed from law enforcement and commonly used in jails and prisons. Public high schools are most likely to use these strategies, followed closely by middle schools; however, even public elementary schools are incorporating some of these security approaches. The National Center for Education Statistics' School Survey on Crime and Safety reports that in the 2005–2006 school year, 85% of public schools maintained locked or monitored doors, 43% used surveillance cameras, 23% randomly brought in drug-sniffing dogs for searches (60% of high schools did this), and 5% performed random metal detector checks on students (Dinkes, Cataldi, & Lin-Kelly, 2007).

In addition to the increased use of physical security measures, concerns about crime and violence in American public schools have also brought about an increase in security personnel used in schools, including security guards, nonteaching disciplinary staff (e.g., hall monitors, in-school suspension room supervisors), and lastly—the focus of this chapter—sworn, uniformed police officers called school resource officers (SROs). While the level of policing in New York City schools is an extreme example, schools across the country are forging partnerships with local law enforcement. Developed and funded through a series of federal initiatives, the widespread growth of school–police partnerships have made police presence in public schools today the norm, rather than the exception.

## ■ The School Resource Officer Program

The SRO program is a collaborative partnership between schools and law enforcement agencies that sends armed, uniformed police officers to work in public schools. An SRO is defined in the Omnibus Crime Control and Safe Streets Act of 1968 as

> a career law enforcement officer, with sworn authority, deployed in community-oriented policing, and assigned by the employing police department or agency to work in collaboration with schools and community organizations to:
>
> 1. Address crime and disorder problems, gangs, and drug activities affecting or occurring in or around an elementary or secondary school
>
> 2. Develop or expand crime prevention efforts for students
>
> 3. Educate likely school-age victims in crime prevention and safety
>
> 4. Develop or expand community justice initiatives for students
>
> 5. Train students in conflict resolution, restorative justice, and crime awareness
>
> 6. Assist in the identification of physical changes in the environment that may reduce crime in or around the school
>
> 7. Assist in developing school policy that addresses crime and recommend procedural changes. (Atkinson, 2002)

The first SRO program began in Flint, Michigan, in the 1950s, with goals of improving police–youth relations in that community (Burke, 2001). In 1968, the Fresno (California) Police Department developed a comparable program with similar goals in Fresno schools. Early Fresno SROs were not first responders; instead, their job was primarily to follow up on criminal investigations involving students. When incidents occurred on a school campus in Fresno, a patrol officer was still called in to conduct the initial investigation. By the 1980s, in response to increased demands, the role of the Fresno SRO broadened to include the first-responder role; these SROs were called juvenile tactical officers (West & Fries, 1995).

The SRO program represents a shift from a traditional policing model in schools to a community policing model. Where a traditional policing model is incident-driven and reactive, the community policing model is problem-oriented and proactive, as SROs work with educators, parents, and students to address concerns regarding the safety of the school community (Atkinson, 2002). For example, prior to having SROs in schools, administrators would have to dial 911 and wait for a patrol to respond if a crime occurred on a school campus. SRO presence in schools eliminates this step because an officer is already there and able to respond immediately. Additionally, like other community policing programs, the SRO program attempts to lower barriers between the police and the (school) community through positive, informal contacts, and, in the process, bolsters community partnerships with police agencies (McDevitt & Panniello, 2005).

The National Association of School Resource Officers (NASRO), the largest professional organization of SROs, has adopted a model called the Triad Concept to characterize the three roles of SROs on campus: law enforcement officer, teacher, and counselor (NASRO, n.d.). The most formal and central role of an SRO is that of law enforcement officer. SROs are in schools to prevent and deter crime, to respond to dangerous situations that may occur on campus, and to help administrators develop safety plans for schools. Secondly, in their capacity as teachers, SROs often spend time in classrooms teaching law-related education lessons and giving preventive talks to students on topics like drunk driving or drug use. In some SRO programs, officers are required to spend a specified number of hours per week in the classroom, while in others the teaching role of an SRO is on a less formal basis. Finally, in their role as mentors and as role models, SROs counsel students on crime and security issues (Lawrence, 2007, p. 208; Brown, 2006).

Despite a spattering of early programs, prior to the 1990s the practice of placing police officers in schools nationwide was still relatively rare (Brown, 2006; Girouard, 2001). Even as late as the 1996–1997 school year, 78% of all public schools—including 54% of public high schools—were without police presence (Kaufman et al., 2000). Today, however, school policing has been identified as the fastest growing area of law enforcement (Beger, 2002; Burke, 2001; NASRO, n.d.). Not all police that work in schools today are part of an SRO program, therefore it is difficult to know exactly how many U.S. schools have some form of police presence. NASRO boasts an estimated 9000 members nationwide (NASRO, n.d.). As of 1997, there were approximately 9400 police officers and nearly 3000 sheriffs' deputies in schools across the United States (Bureau of Justice Statistics, 1999). In 1999, 54% of public middle and high school students reported having security guards or assigned police officers in their schools. This number increased to 68% by 2005 (Dinkes, Cataldi, & Lin-Kelly, 2007). These figures suggest that school policing is widespread in the United States and is a practice that has no indication of waning.

## ■ A Federal Understanding of School Violence

The emergence and expansion of school–law enforcement partnerships was not happenstance and was not without significant support from the federal government. In the 1970s, the federal government identified school crime and violence as an area of concern and participated in assessing the scope of the school crime problem. The Safe Schools Study, conducted by the National Institute of Education (1978), was the first nationwide study of school crime (Lawrence, 2007). With $2.4 million in funding from Congress, the Safe School Study aimed to understand the extent and severity of school crime in America's schools and identify ways to prevent it. This large-scale study employed (1) mail surveys to principals in 4000 public schools, (2) intensive studies of nearly 650 public junior and senior high schools, including thousands of interviews of students and teachers, and (3) case studies of 10 schools with major disruption problems that had shown dramatic improvement in a short time (National Institute of Education, 1978).

The final report to Congress stated that 8% of all schools had serious problems with crime and that secondary schools were notably more troubled than primary schools. Theft and vandalism were among the most frequent crimes reported in schools; in any given month, 11% of students reported having something stolen, and 25% of schools reported experiencing vandalism. The study also concluded that hundreds of thousands of students and teachers were assaulted each year in American secondary schools, which reinforced the American public's belief that schools were indeed plagued with drug and crime problems (Gottfredson & Daiger, 1979).

The "Safe School Study Report" (National Institute of Education, 1978), presented to Congress in 1978, suggested two very different types of policy improvements that could reduce the "problem" of crime and disorder in schools. On one hand, the recommendations were student focused: decreasing the size of schools and number of students in classes, reducing students' sense of powerlessness and alienation, treating students fairly and equally, improving the consistency with which school rules are applied and enforced, and improving the relevance of the subject matter taught in schools to better suit students' interests and needs. On the other hand, the report recommended a crime control approach, noting that school administrators desired increased security technology and personnel to make their schools safer (National Institute of Education, 1978). Over time, we have seen the latter approach become the dominant strategy in American public schools.

## ■ Policies Supporting Placement of Police in Schools

Following the Safe Schools Study, a series of education reforms have incorporated initiatives designed to target school crime and violence. Many of these initiatives have fostered the growth of police presence in American schools by giving direction and financial backing to school–police partnerships. The next several sections describe various federal policies that have contributed to the buildup of police presence in schools by offering financial incentives and other encouragement to schools to partner with law enforcement agencies.

### National Education Goals

Much of the impetus for the school–police partnerships that are in place today resulted from a series of education reforms in the 1990s that identified school safety as a key objective. In 1990, the National Education Goals Panel was formed as an independent agency in the Executive Branch of the U.S. government. The panel's mission was to assess and report on state and national progress on achieving several identified education goals. The panel established National Education Goals (20 U.S.C. § 7102), which identified specific improvements in education to be accomplished by the year 2000; these included improvements in school readiness, graduation rates, teacher development, achievement in science and math, literacy, and partnerships with parents. The National Education Goals also proposed that by the year 2000, every American school would be "free of drugs and violence and the unauthorized presence of firearms and alcohol, and offer a disciplined environment that is conducive to learning" and that in pursuit of this goal "community-based teams should be organized to provide students and teachers with needed support" (sec. 102(7)(B)(vi)).

In 1994, President Clinton signed the Goals 2000: Educate America Act (Public Law 103-227), which codified these National Education Goals. The act grants money to states and communities to help them pursue the National Education Goals and help their schools reach higher levels of achievement. During the first year, Congress appropriated $105 million for these grants; by 1998, this appropriation increased to $466 million (Stedman & Riddle, 1998).

### Safe Schools Acts

As made clear in the National Education Goals, school safety had become a concern of the federal government, and promoting safer schools and the overall reduction of school crime and violence became agendas that were subsumed in federal education reforms. Setting forth specific plans for reform was the Improving America's Schools Act (IASA) of 1994, the eighth reauthorization of the Elementary and Secondary Education Act (ESEA) of 1965. The IASA, like its predecessors, targets resources to the most underperforming schools and children to raise all students in the United States to equally high-performing levels. The IASA proposed four key strategies: clearly defining high academic standards for all students, improving training for teachers in order to help students meet these high standards, granting schools more decision-making authority and holding them more responsible for results, and promoting partnerships among families, communities, and schools (Riley, 1999).

Unlike earlier versions of the ESEA, the IASA reforms of 1994 also specifically included measures to address school security in an effort to reach the National Education Goals. One of the most significant components of the IASA affecting change in school programs was the Safe and Drug-Free Schools and Communities Act (SDFSCA), which amended the 1986 Drug-Free Schools and Communities Act.

While the focus of the 1986 act was primarily on drug abuse prevention in schools, the 1994 SDFSCA added violence prevention to its main goals (Title IV, § 4111-4116, 20 U.S.C. 7111-7116). In 2001, the SDFSCA was reauthorized by Congress under Title IV Part A of the No Child Left Behind Act (NCLB).

The SDFSCA grants money to state and local educational agencies and governors for school-based programs designed to prevent youth drug use and violence. Examples include comprehensive drug prevention programs (including drug testing), school security activities, training and monitoring of school personnel, and conflict resolution programs. In 2008 alone, $295 million was provided for state grants under SDFSCA (McCallion, 2008). To qualify for these grants, schools must submit detailed drug and violence prevention plans demonstrating that they have strategies in place to prohibit the possession of weapons and to otherwise maintain safe, disciplined, and drug-free environments. They also must have a crisis management plan in place to respond to violent or traumatic events on school grounds. States receiving funding from SDFSCA must implement a uniform management information and reporting system (UMIRS) to collect and make publicly available detailed information about violence or drug incidents in schools that result in suspensions or expulsions (ESEA, Section 4112(c)(3)).

When violence prevention became an added focus of the SDFSCA in 1994, enhancing school security measures emerged as one way to address this goal. The SDFSCA specifies that up to one-third of each grant can be spent on security-related measures for schools (Stefkovich & Miller, 1999). These security-related measures could include installing metal detectors and security cameras, developing and implementing comprehensive school security plans, and hiring and training school security personnel (including SROs). For many schools, grants received through the SDFSCA supported them in building partnerships with law enforcement (Casella, 2001).

Concern about rising levels of school violence has led to broad political support for the SDFSCA (Reuter & Timpane, 2001). However, over time, there have been mounting criticisms and concerns about whether the SDFSCA is working effectively. A RAND report (Reuter & Timpane, 2001) analyzing the SDFSCA finds multiple weaknesses with regard to budgeting, implementation, and effectiveness. First, schools with the most financial need are receiving too little of the SDFSCA funding because monies are being dispersed across schools based on enrollment and population-based formulas instead of need. This lack of funding is contributing to uncertainty in many schools about whether specific programs will be continued from year to year. Second, the report finds that the overall program goals of the SDFSCA are vague; although the aspiration is to make schools safe and drug free, there is no clear articulation of how the grant program leads to achievement of this aspiration. This lack of specific measurable objectives, argues Reuter and Timpane (2001), presents a major obstacle to change in schools. Finally, there are limited empirical data to suggest that the programs funded by the SDFSCA are effective in reducing school violence and drug use. The grants provided to schools through the SDFSCA are not large enough to conduct evaluation studies of the programs implemented, so for the most part it is unclear if the programs are working. And because there is such limited research on school-based drug and violence prevention programs, schools lack the best practice models necessary for them to build their own effective drug and violence prevention programs.

### School Crime Reporting Laws

Following the national Safe Schools Act, many states implemented their own versions of these policies, which further required their school districts to follow both state and federal guidelines. For some states, this included enacting laws that required mandatory reporting of certain school-based offenses to local law enforcement (Office for Victims of Crime, 2002). In Delaware, for example, schools are required to report to the police any incidents in which a student or school employee has been the victim of a violent felony, assault in the third degree, or a sexual offense, and any additional incidents in which a school employee has been the victim of "offensive touching" or "terroristic threatening" (14 Del.C. §4112). Any school employee who fails to report such incidents can be penalized with a fine, and any school board member who hinders or delays the filing of a report can be charged with a misdemeanor.

School crime reporting laws further encourage (and require) schools to open the lines of communication with local law enforcement. Before these reporting requirements were in place, it was largely within the discretion of school administrators to decide how to handle crimes that occurred on campus, and there was no law enforcement oversight (Kipper, 1996). If a fight occurred at school, for example, and one of the students in the fight was injured, school administrators might decide to hand down school punishment to the students involved (e.g., suspension). Under Delaware's school crime reporting law, a fight with an injury is categorized as an Assault III and therefore must be reported to police. Various forms of school crime reporting laws exist in Alabama, Arkansas, Connecticut, Georgia, Illinois, Kansas, Louisiana, Mississippi, Missouri, Michigan, Nebraska, New Hampshire, Texas, and Virginia.

### COPS in Schools (CIS)

While the SDFSCA was an early source of funding for schools to hire police officers as part of comprehensive school safety plans, the COPS in Schools (CIS) program increased the incentive and opportunity for schools to hire SROs. CIS is a grant program that started in 1999 and is funded by the COPS office under the U.S. Department of Justice (see Chapter 3 for more about COPS). It encourages the use of community policing strategies to prevent school violence (Atkinson, 2002). CIS grants provide incentives to law enforcement agencies to partner with local schools by contributing up to $125,000 per newly hired SRO over a 3-year grant period to cover the officer's salary and benefit costs (Office of Community Oriented Policing Services, n.d.).

CIS funding has been responsible for spurring tremendous growth in the SRO program. As of 2005, CIS has awarded over $750 million to fund over 6500 SROs across the country (Office of Community Oriented Policing Services, 2005). This funding has increased both the numbers of SROs in primary and secondary schools as well as the amount of time these SROs dedicate to working in their assigned schools. CIS grants require that SROs spend a minimum of 75% of their time working in or around schools in addition to the amount of time they would have dedicated absent the CIS grant. For many schools that once only had part-time or shared police presence, CIS grants allowed them to have SROs on campus full time.

### Safe Schools/Healthy Students Initiative

The COPS program also provided an additional $10 million to hire SROs through the Safe Schools/Healthy Students (SS/HS) initiative, which is a joint program between the Department of Education and the Substance Abuse and Mental Health Services Administration (SAMHSA), an agency of the U.S. Department of Health and Human Services. The SS/HS initiative is a discretionary grant program that provides funding for programs that focus on healthy childhood development and the prevention of violence and alcohol and drug abuse. Grantees are required to establish partnerships with community-based organizations, including local law enforcement, public mental health agencies, and juvenile justice agencies/entities.

### Challenges and Successes in School Policing

Federal initiatives have clearly contributed to the rapid growth of school–police partnerships over the past two decades by identifying school safety and security as key objectives of schools and by funding programs that allow schools to hire police officers. For many schools and many law enforcement officers, these partnerships represent new and unfamiliar territory. The next sections discuss the challenges and successes that schools and law enforcement agencies face as they forge these new partnerships.

### Implementation Challenges

As federal grant programs encouraged schools and police agencies to partner with one another, there have inevitably been growing pains and other difficulties encountered by both organizations. The placement of police officers in public schools represents a blending of two organizations with different norms: Law enforcement is typically a closed organizational structure, whereas schools are traditionally open organizational structures (Brady, Balmer, & Phenix, 2007). For school personnel who have never worked so closely with police officers on a regular basis, and for police officers who have never worked exclusively in schools, these new arrangements require a period of adjustment as the organizations learn to work together.

One of the first challenges schools face when beginning their partnership with an SRO is setting expectations about the SRO's role in the school and coming to an agreement with the SRO about these expectations. It may take time for administrators to get used to collaborating with officers on issues (like physical security or how to deal with students who commit certain offenses) about which they were previously the sole decision makers. Furthermore, because SROs typically report to their police supervisors and not to the school's principal, there can be conflict over the hierarchy of authority in the school (Devine, 1996; Mukherjee, 2007).

SROs may also have varying expectations about the role they will play in the school community. The job of a law enforcement officer working in a school is quite different when compared to a law enforcement officer working "on the street" (Brown, 2006). In forging partnerships, SROs and schools have to decide how involved SROs should be in everyday school discipline matters like enforcing dress codes, checking hall passes, dealing with minor student misbehaviors, and addressing other noncriminal incidents that occur in school. Some SROs, for example, are hesitant to be involved in enforcing school rules (Travis & Coon, 2005) or mentoring students (Kupchik & Bracy, 2009b). NASRO recommends that to reduce implementation stumbling blocks, law enforcement agencies and school districts should devise written contracts detailing expectations for the partnership before SROs are placed in schools (Kamleiter, 2000).

### Effects on School Crime

It has been assumed by the CIS program, by the SRO program, and by the schools that participate in these programs, that placing police officers in schools creates safer school environments (for examples, see Finn, 2006; Kipper, 1996; Trump, 2002). However, the empirical data supporting this contention are limited, and research on whether school–police partnerships effectively reduce school crime is limited overall. A study by Ida Johnson (1999) of SRO effectiveness in one southern U.S. city finds that the number of intermediate and major offenses decreased a year after SROs were permanently assigned to the city's schools. However, this research was conducted during a time when school crime and violence were declining nationally and employs no comparison group; thus it should be interpreted with caution. Another study of the state of Virginia's SRO program finds that while illegal behavior persists in schools with SROs, the majority of students and staff feel safe at school (Schuiteman, 2001). However, Schuiteman's (2001) conclusion that "Virginia's state-assisted SRO programs are reducing school violence" (p. 28) is not supported by his limited data. More research on the effectiveness of SRO programs in schools is needed to determine what effect SRO involvement has on school safety and what type of involvement by an SRO is most effective.

### Effects on Schools and Students

In addition to understanding whether police–school partnerships improve levels of school crime and violence, it is important to assess the possible "side effects" of police presence in schools. Researchers that have studied schools with police officers suggest that there is potential for both positive and negative consequences for students and school administrators as a result of police–school partnerships.

From a school's standpoint, the presence of an SRO can present several benefits. First, the SRO is assumed to serve as a deterrent: Students are less likely to bring a weapon or drugs to school, or commit other crimes while at school, if they know that a police officer is present (Johnson, 1999). If indeed this is the case, schools may experience fewer incidents of crime and violence with an SRO in place. The presence of the SRO also benefits schools by providing an immediate response to crime or threats (Finn, 2006). As previously mentioned, prior to having SROs, school administrators would have to call the local police department in the case of a problem or threat and then wait for a response. Having a police officer on site allows for immediate action and increased daily support for school administrators and staff.

Whether or not school–police partnerships result in any actual reduction in school crime or violence, having SROs in schools may be effective in that they contribute to students feeling safer at school. It is widely agreed that schools should be safe havens for students. In 2006, President George W. Bush participated in a national conference on school safety in which he called on the National Sheriffs' Association to take the lead in identifying practices for ensuring the safety of students and schools. At this conference, the President declared, "Our school children should never fear their safety when they enter into a classroom" (National Sheriffs' Association, n.d.). Something as simple as *feeling* safe can be critical to student learning and academic performance; if students are fearful at school, they are more likely to skip school and to have problems concentrating while in class (Bowen & Bowen, 1999; Martin, Sadowski, Cotton, & McCarraher, 1996).

It is plausible that just having a police officer around can make students feel more at ease. The majority of the students interviewed in Johnson's (1999) study, for example, report that the presence of police in their schools provides them with a sense of security. Even in cases where students do feel threatened by a schoolmate, they may be less likely to take matters into their own hands, knowing that there is a police officer on campus. In a national assessment of 19 SRO programs, McDevitt and Panniello (2005) found that a large majority (87%) of students report feeling safe in school, but feel even safer after having had multiple conversations with SROs. Providing a sense of safety to students may be one important benefit of police–school partnerships.

McDevitt and Panniello (2005) also find that students who have a positive opinion of their SRO are more likely to feel comfortable reporting a crime and are more likely to report feeling safe at school. These results suggest that students' perceptions of their SRO could be important in determining how effective an SRO can be in a school. Student perceptions of their SROs can also serve as an indicator of how well SRO programs are meeting their community policing goals of building better relationships between youth and law enforcement. The few studies that have investigated students' perceptions of their SROs, however, have not found evidence that the SRO program is reaching these goals. A four-school study by Arrick Jackson (2002) finds that SRO presence in school does not change students' views of the police, impact student perceptions of offending, or make students more likely to think they will be caught should they misbehave.

A second study, conducted in the United Kingdom by Hopkins (1994), produces similar findings—students' overall views of the police are not influenced by having a police officer (called school liaison officers [SLOs] in the United Kingdom) on campus. Hopkins concludes that this is, in large part, because students view their schools' SLOs as atypical police officers (different from the police on the streets) and interpret their friendliness as divergent from attitudes of typical police officers. The findings are important considering that the goal of these programs is to improve relations between police and youth (Atkinson, 2002; Finn, 2006; Scott, 2001). They suggest that this goal may not be accomplished through the current model of school policing. There is need for further examination of how officers interact with students, how students perceive these interactions, and what consequence these factors have for rapport building and school crime prevention.

In addition to the SROs' apparent inability to improve students' views of police, critical examinations of law enforcement officers' work in schools suggest that police presence may actually escalate minor situations and alienate students (Beger, 2002). The New York Civil Liberties Union's account of police behavior in New York City public schools (Mukherjee, 2007) documents a prisonlike environment created by police presence, where students and staff are subject to abusive behavior at the hands of school police officers. If police strategies in schools are too aggressive, it could lead students to distrust police and lose confidence in their school administrators, which may increase delinquency rather than improve it (Hyman & Perone, 1998).

Having a police officer at school can also complicate the issue of students' rights in school. Traditionally, schools and law enforcement are held to different standards by law when it comes to conducting physical searches of students or questioning student suspects about a crime. Schools are given greater leeway with students when it comes to these matters as they are seen as acting *in loco parentis*, meaning "in place of the parents." Courts have granted school administrators this leeway because, in their parentlike capacity, they are the custodial guardians of students while at school and are thought to be acting in students' best interests (see *New Jersey v. T.L.O.,* for example). Police officers, by contrast, are held to much more stringent standards of behavior when they encounter youth on the street; citizens, including youth, have rights that protect their privacy, property, and civil liberties.

However, now that police are a fixture in many schools, concerns have been raised about the negative impact they are having on students' rights, particularly if SROs act in a capacity more similar to school administrators and if courts are supporting them in these actions (Beger, 2002, 2003). In *People v. Dilworth* (1996), for example, the Supreme Court upheld the search of a student by a school police officer, even though the search was conducted without probable cause. The court justified their decision by citing the similarity of the duties of the school police officer to school officials, who only need to meet the lesser standard of reasonable suspicion to search students. While the SRO program has been designed to improve safety for school staff and students, SROs must be particularly careful to protect students' rights as they go about their daily work in schools (Kamleiter, 2000).

An additional concern about police presence in schools is that disciplinary situations that in previous years would have been handled within the school by administrators may now be immediately taken under the control of the police (Devine, 1996). This means that in some cases students are arrested when they previously may have been suspended, expelled, or otherwise punished within the school. In a study comparing New York City schools that were designated as "impact" schools (and were consequently assigned elevated levels of police presence to promote safer school environments) to nonimpact schools, Brady, Balmer, and Phenix (2007) document a series of negative outcomes at the impact schools. A year and a half after being designated impact schools, suspensions and attendance rates in these schools worsened. While major crime rates decreased slightly in impact schools, they experienced a significant increase in noncriminal incidents reported to the police. This study shows that police presence in schools may have a net-widening effect, such that students who would normally receive a school-based sanction for their (noncriminal) misbehavior come under police attention and could be subject to a more severe sanction, such as arrest.

For many of the reasons that Brady, Balmer, and Phenix (2007) document, some critics of police–school partnerships express concerns that these programs contribute to a worsening of the school-to-prison pipeline. Combined with the school-crime reporting laws in many states—which compel schools to take serious action on specified school misbehaviors—as well as other federal, state, and local education and safety policies, police presence in schools can contribute to increasing numbers of students being funneled out of schools and into the criminal justice system. Students of color, low-income students, and students with disabilities are particularly at risk for these severe consequences (New York Civil Liberties Union, n.d.).

# ■ Conclusion

Over the past several decades, educators, legislators, parents, and community members have all expressed concerns about nationally rising rates of violence in schools, despite data showing that school violence has been declining (Hyman et al., 1996; Morrison, Furlong, & Morrison, 1994). Furthermore, while the most severe crime problems are concentrated in a small proportion of urban schools, school crime has increasingly come to be understood as a serious problem in all schools (Simon, 2007). In response to these fears, the federal government, along with state governments, has implemented a series of policies and programs, many of which promote school–police partnerships as an effective way to reduce crime and violence in schools.

Federal policies such as the Safe Schools Acts, the CIS program, and the SS/HS initiative have increased the number of police placed in schools over the past two decades by encouraging schools and law enforcement agencies to partner. These partnerships have been nurtured with significant financial support. The result is that today the majority of public middle and high schools have some form of police presence. The SRO program is the most common model of contemporary school policing and is a program that continues to expand.

While parents, educators, community members, and even, in some cases, students may see the hiring of school police officers as a proactive approach to school safety, there is little definitive evidence that police presence improves school safety. In fact, many examinations suggest negative consequences: Students may be more likely to be arrested by school police for minor incidents that once were handled entirely within school disciplinary systems, and the police presence in schools may contribute to the erosion of students' rights. Going forward, schools would be well advised to consider both the costs and benefits associated with the current policy of forging ever-stronger partnerships with law enforcement.

# ■ Discussion Questions

1. Discuss the trend of incorporating security strategies borrowed from criminal justice into schools. Is this a positive development?

2. Is the School Resource Officer program an example of a successful police policy? Identify the specific criteria that might help to answer this question.

3. Discuss the various policies that have supported the placement of police officers in schools.

4. Identify and discuss the challenges that have emerged in relation to police–school partnerships.

5. What can be done to address the concern that police–school partnerships are responsible for an increase in the number of young people criminalized in the United States?

## ■ References

Atkinson, A. J. (2002). Fostering school-law enforcement partnerships. In *Safe and secure: Guides to creating safer schools* (Guide 5). Portland, OR: Northwest Regional Educational Laboratory.

Beger, R. R. (2002). The expansion of police power in public schools and the vanishing rights of students. *Social Justice, 29*, 119–130.

Beger, R. R. (2003). "The worst of both worlds": School security and the disappearing Fourth Amendment rights of students. *Criminal Justice Review, 28*(2), 336–354.

Bowen, N. K., & Bowen, G. L. (1999). Effects of crime and violence in neighborhoods and schools on the school behavior and performance of adolescents. *Journal of Adolescent Research, 14*(3), 319–342.

Brady, K. P., Balmer, S., Phenix, D. (2007). School-police partnership effectiveness in urban schools: An analysis of New York City's impact schools initiatives. *Education and Urban Society, 39*, 455–478.

Brown, B. (2006). Understanding and assessing school police officers: A conceptual and methodological comment. *Journal of Criminal Justice 34*, 591–604.

Bureau of Justice Statistics. (1999). *Personnel increases in local police and sherrifs' departments* (Press release October 29, 1999). Retrieved July 1, 2007 from http://www.ojp.usdoj.gov/bjs/pub/press/sdlpd.pr

Burke, S. (2001). The advantages of a school resource officer. *Law and Order Magazine, 49*, 73–75.

Casella, R. (2001). *Being down: Challenging violence in urban schools.* New York: Teachers College Press.

Devine, J. (1996). *Maximum security: The culture of violence in inner-city schools.* Chicago: The University of Chicago Press.

Dinkes, R., Cataldi, E. F., & Lin-Kelly, W. (2007). *Indicators of school crime and safety: 2007* (NCES 2008-021/NCJ 219553). Washington, DC: National Center for Education Statistics, Institute of Education Sciences, U.S. Department of Education, and Bureau of Justice Statistics, Office of Justice Programs, U.S. Department of Justice.

Finn, P. (2006). School resource officer programs: Finding the funding, reaping the benefits. *FBI Law Enforcement Bulletin, 75*(8), 1–7.

Fisher, J. (2005, September 20). Students protest use of metal detectors in their Bronx school. *New York Times*, p. B4.

Girouard, C. (2001, March). *School resource officer training program* (FS 200105). Washington, DC: U.S. Department of Justice, Office of Justice Programs, Office of Juvenile Justice and Delinquency Prevention.

Gottfredson, G. D., & Daiger, D. C. (1979). *Disruption in six hundred schools.* Baltimore: The Johns Hopkins University.

Hopkins, N. (1994). School pupils' perceptions of the police that visit schools: Not all police are "pigs." *Journal of Community & Applied Psychology, 4,* 189–207.

Hyman, I. A., & Perone, D. C. (1998). The other side of school violence: Educator policies and practices that may contribute to student misbehavior. *Journal of School Psychology, 36*(1), 7–27.

Hyman, I. A., Weiler, E., Dahbany, A., Shanock, A., & Britton, G. (1996). Policy and practice in school discipline: Past, present and future. In R. Talley & G. Waltz (Eds.), *Safe Schools, Safe Students* (pp. 77–84). Washington DC: The National Education Goals Panel and the National Alliance of Pupil Services Organizations.

Jackson, A. (2002). Police-school resource officers' and students' perception of the police and offending. *Policing: An International Journal of Police Strategies & Management, 25*(3), 631–650.

Janofsky, M. (2000, April 21). A year after Littleton, services and lawsuits. *New York Times (Late Edition).* Retrieved October 4, 2007, from ProQuest Newspapers database.

Johnson, I. M. (1999). School violence: The effectiveness of a school resource officer program in a southern city. *Journal of Criminal Justice 27*(2), 173–192.

Kamleiter, D. J. (2000, Fall). *School-based policing and SROs* (Fact sheet No. 8). Portland, OR: National Resource Center for Safe Schools.

Kaufman, P., Chen, X., Choy, S. P., Ruddy, S. A., Miller, A. K., Fleury, J. K., et al. (2000). *Indicators of school crime and safety: 2000.* (NCES 2001-017/NCJ-184176). U.S. Departments of Education and Justice. Washington, D.C: U.S. Government Printing Office.

Kipper, B. (1996, June). Law enforcement's role in addressing school violence. *The Police Chief, 63,* 26–31.

Kupchik, A., & Bracy, N. (2009a). The news media on school violence: Constructing dangerousness and fueling fear. *Youth Violence and Juvenile Justice, 7*(2), 136–155.

Kupchik, A., & Bracy, N. (2009b). To protect, serve and mentor? Police officers in public schools. In T. Monahan & R. Torres (Eds.), *Schools Under Surveillance.* Piscataway, NJ: Rutgers University Press.

Lawrence, R. (2007). *School crime and juvenile justice* (2nd ed.). New York: Oxford University Press.

Martin, S. L., Sadowski, L. S., Cotton, N. U., & McCarraher, D. R. (1996). Response of African-American adolescents to gun carrying by school mates. *Journal of School Health, 66,* 23–27.

McCallion, G. (2008). *Safe and drug free schools and communities act: Program overview and reauthorization issues.* Washington, DC: Congressional Research Service, Library of Congress.

McDevitt, J., & Panniello, J. (2005). *National assessment of School Resource Officer Programs: Survey of students in three large new SRO programs* (209270). Retrieved February 6, 2010, from http://www.ncjrs.gov/pdffiles1/nij/grants/209270.pdf

Morrison, G. M., Furlong, M. L., & Morrison, R. L. (1994). School violence to school safety: Reframing the issue for school psychologists. *School Psychology Review, 23,* 236–256.

Mukherjee, E. (2007). *Criminalizing the classroom: The over-policing of New York City schools.* New York: New York Civil Liberties Union.

National Association of School Resource Officers. *Membership.* (n.d.). Retrieved January 16, 2009 from http://www.nasro.mobi/cms/index.php?option=com_content&view=article&id=55&Itemid=100

National Institute of Education. (1978). *Violent schools—safe schools: The safe school study report to the Congress—Executive summary.* Washington, DC: U.S. Department of Health, Education, and Welfare.

National Sheriffs' Association. (n.d.). *Keeping America's schools safe: Law enforcement promising practices series: School resource officers.* Retrieved February 6, 2010, from www.sheriffs.org/userfiles/file/SchoolSafetyBrochure.pdf

New York Civil Liberties Union. (n.d.). School to prison pipeline fact sheet. Retrieved February 6, 2010, from www.nyclu.org/files/stpp_fact_sheet.pdf

O'Driscoll, P. (2001, May 17). Columbine investigator: "Nobody acted" on clues. *USA Today.* Retrieved October 4, 2007, from ProQuest Newspapers database.

Office for Victims of Crime. (2002, January). *Reporting school violence: Legal series bulletin #2.* Washington, DC: Office of Justice Programs: U.S. Department of Justice.

Office of Community Oriented Policing Services. (n.d.) *About COPS funding: CIS.* Retrieved February 6, 2010, from http://www.cops.usdoj.gov/Default.asp?Item=54

Office of Community Oriented Policing Services (2005). *COPS in Schools: The COPS commitment to school safety.* U.S. Department of Justice. Retrieved February 6, 2010, from http://www.cops.usdoj.gov/files/RIC/Publications/e09042494.PDF

Reuter, P. H., & Timpane, P. M. (2001). *Options for restructuring the Safe and Drug-Free Schools and Communities Act.* Santa Monica, CA: RAND, MR-1328-EDU.

Riley, R. W. (1999). Statement given before the U.S. Senate Committee on Health, Education, Labor, and Pensions on the Reauthorization of the Elementary and Secondary Education Act of 1965. Retrieved February 6, 2010, from http://www.ed.gov/Speeches/02-1999/990209.html

Schuiteman, J. G. (2001). *Second annual evaluation of DCJS funded School Resource Officer Program. Report of the Department of Criminal Justice Services, Fiscal year 1999–2000*. Richmond, VA: Virginia State Department of Criminal Justice Services.

Scott, M. (2001, September). Special report I: School enforcement, school liaison. *Law and Order Magazine, 49,* 68–70.

Simon, J. (2007). *Governing through crime: How the war on crime transformed American democracy and created a culture of fear.* New York: Oxford University Press.

Stedman, J. B., & Riddle, W. C. (1998). *Goals 2000: Educate America Act implementation status and issues.* Congressional Research Service (CRS 95-502): Library of Congress.

Stefkovich, J. A., & Miller, J. A. (1999). Law enforcement officers in public schools: Student citizens in safe havens? *Brigham Young University Education and Law Journal, 1999*(1), 25–68.

Travis, L. F., & Coon, J. K. (2005). *The role of law enforcement in public school safety: A national survey* (211676). Rockville, MD: National Criminal Justice Reference Service.

Trump, K. S. (2002). *NASRO school resource officer survey, 2002: Final report on the 2nd annual survey of school based police officers.* St. Anthony, FL: National Association of School-Based Police Officers.

West, M. L., & Fries, J. M. (1995). Campus-based police/probation teams—Making schools safer. *Corrections Today, 57*(5), 144–146.

## ■ Court Cases Cited

*New Jersey v. T.L.O.*, 469 U.S. 325 (1985).
*People v. Dilworth*, 661 N.E.2d 310 (1996).

# Procedural Fairness, Criminal Justice Policy, and the Courts

# CHAPTER
# 5

David B. Rottman

## ■ Introduction

Policy-making by and for the courts has been guided by a succession of theories adopted from the worlds of academia and management consultancy. Over the last century, key concepts from scientific management, open systems theory, contingency theory, and total quality management each enjoyed a heyday. Much was accomplished through policies guided by perspectives on what makes organizations and institutions succeed.

A new direction is needed nonetheless. The policies of the past overemphasized some aspects of what courts do to the exclusion of other aspects arguably more important to achieving the goals of the court system. Court policies animated by earlier theories sought to maximize timeliness of case disposition and efficiency in court operations by consolidating courts and centralizing and professionalizing court management. These court reforms improved the public's image of the courts.

Procedural fairness, a field of inquiry within social psychology, offers the court community a new direction for devising policies that maximize the likelihood that defendants and offenders will voluntarily comply with court orders. Compliance means that defendants appear in court on the date specified in a summons, pay their fines, keep the conditions of protection orders, complete community sentences, abide by treatment contracts, and refrain from reoffending. Procedural fairness offers the only viable guide to policy-making if compliance is what we seek to maximize. Moreover, practicing procedural fairness will build public trust in and support for the courts.

Procedural fairness is already guiding policy-making for all of the courts in one state (California, as dicussed later) and for individual trial courts around the country. Its mark also can be found in practices used throughout the criminal justice system, especially in policing and alternatives to incarceration (Rottman, 2007). Yet it is for the courts that procedural fairness can make the strongest contribution toward increasing the effectiveness of policy-making. From the founding of this country, the judiciary has been called the "least dangerous" branch. Former U.S. Supreme Court Justice Felix Frankfurter offers one of several eloquent elaborations of this point: "The Court's authority—possessed of neither the purse nor the sword—ultimately rests on sustained public confidence in its moral sanction" (*Baker v. Carr*, 1962).

The courts must by their actions generate a belief that court decisions should be adhered to even if the case outcome is not favorable. In short, they must be viewed by individual defendants as possessing legitimacy if they are to administer undesirable outcomes and still be obeyed.

The influence of procedural fairness in criminal justice policy-making is growing, but it remains a new concept for many. Therefore, some clarifications at this stage are appropriate.

First, procedural fairness is subjective in nature. It is a layperson's way of evaluating how he or she is being treated, not the evaluation of someone schooled in the law. People examine how they are being treated in the courtroom and courthouse to assess the fairness of what they are experiencing. Research concludes that all of us share a virtually identical set of criteria to use in deciding when we are being treated fairly (Tyler & Huo, 2002). Second, procedural fairness is distinct from due process considerations, the legal system's rules and procedures for ensuring fairness. A particular set of court procedures and rules designed to ensure due process may or may not be viewed by the public as evidence that they are being treated fairly. Finally, it should be noted that this chapter focuses on the state courts where interest in procedural fairness-based policy-making is greatest. In 2006, roughly 102 million new trial court cases and 283,000 new appellate court cases were filed in the state courts (LaFountain et al., 2008). By contrast, federal courts only received 320,000 new trial court filings and 67,000 new appeals (Administrative Office of the U. S. Courts, 2007).

## ■ What Is Procedural Fairness?

The field of procedural fairness emerged from decades of theory and research dedicated to isolating the relative importance of factors that can explain when people will be satisfied with the decisions of judges and other authority figures. As defined by Tom Tyler (2001),

> The procedural justice argument is that, on the general level, the key concerns that people have about the police and the courts center around whether these authorities treat people fairly, recognize citizen rights, treat people with dignity, and care about people's concerns. (p. 216)

The importance of procedural fairness in explaining public satisfaction with outcomes "is one of the most robust findings in the justice literature" (Brockner et al., 2001, p. 301). Procedural fairness is present when people perceive that they are experiencing (Tyler, 2004, pp. 443–447):

- *Respect:* Being treated with dignity and having one's rights respected
- *Neutrality:* Believing decision makers are honest and impartial, and that their decisions are based on facts
- *Participation:* Having an opportunity to express one's viewpoint to the decision maker
- *Trustworthiness:* Perceiving decision makers as benevolent, caring, motivated to treat individuals fairly, and sincerely concerned about the individual

One indication of the primacy of procedural fairness in policy-making is that, when it is taken into consideration, factors such as race, ethnicity, income level, and gender are no longer important for explaining whether people have trust and confidence in the courts. People share a common set of criteria for deciding what is fair—the criteria used to assess how they are being treated. If African Americans are less confident in the courts, research tells us it is because they perceive less procedural fairness.

Two research studies capture the profound impact that acting in a procedurally fair manner can have on compliance with decisions by criminal justice authorities. Both studies dealt with offenses and offenders generally regarded as among the most intractable in terms of being able to change their behavior.

### Domestic Violence

A study of recidivism rates among domestic violence offenders in Milwaukee, Wisconsin, suggested that the optimal policy for police officers responding to a complaint was to arrest the suspect. State legislators across the country responded by passing mandatory arrest laws in domestic violence cases. Subsequent research in six other cities failed to offer consistent support for the logic behind those laws.

This inconsistency was puzzling. The research followed a rigorous experimental design. Calls for help were randomly assigned to three kinds of police response: (1) warning with no arrest, (2) arrest followed by brief detention, and (3) arrest with a longer detention. The research data were reanalyzed to make sense of the implications for police policy. On the one hand, a policy of mandatory arrests follows the deterrence model. Arrest is a "cost" to the abuser that outweighs whatever "benefits" they might gain from subsequent spousal abuse. On the other hand, procedural fairness might explain why some abusers are less likely than others to reoffend. Being treated in a procedurally fair manner by the police will engender a sense that the abuser should not reoffend. The reanalysis of the experimental data concluded:

> When police acted in a procedurally fair manner when arresting assault suspects, the rate of subsequent domestic violence was significantly lower than when they did not. Moreover, suspects who were arrested and perceived that they were treated in a procedurally fair manner had subsequent assault rates that were as low as those suspects given a more favorable outcome (warned and released without arrest). (Paternoster, Bachman, Brame, & Sherman, 1997, p. 163)

Fair treatment proved more important than deterrence in predicting who would reoffend.

### Drunk Drivers

The data from another experimental study regarding convicted drunk drivers in Australia tested the impact that perceptions of fair treatment had on recidivism. Once again, procedural fairness was the strongest explanation of why some offenders continued to drive and drink and others desisted from doing so. The key factor examined was that of "legitimacy," the feeling on the part of the offender that, whatever the outcome in a particular case, one should comply with the sanctions imposed and change behavior. Tom Tyler (2008) describes the conclusions from the research as follows:

> Adults who were arrested for driving while drunk had their case disposed through different legal procedures, including traditional courts. After their case was disposed each person was interviewed. As expected, the fairness of the legal procedure was related to the legitimacy of the legal system. Two years later, those involved were reinterviewed and it was found their views about the legitimacy of the law were related to their initial perceptions of the fairness of their cases. Peoples' obedience to the law was then tracked for the two years following this second interview, and it was found that people who experienced their hearing as fairer, and therefore viewed the law as more legitimate two years later, reoffended at around 25% the rate of those who viewed the law as less legitimate during the two years following their second interview. In other words, the reduction in reoffending caused by experiencing a hearing as fairer extended to at least four years after the hearing. (p. 27)

How does the explanatory power of procedural fairness stack up against that provided by the perceived favorability and fairness of outcomes? A number of research studies explicitly make that comparison. Residents of Los Angeles and Oakland, California, who recently had a personal experience with the courts were the subject of one such study. A telephone survey collected information from a random sample of relevant individuals in the two cities. The results are presented in **Table 5–1** in the form of two multiple regression equations that tell us the relative importance of people's perceptions of procedural fairness, distributive fairness, and outcome favorability. A host of other factors are included, ranging from a person's political ideology, income, education level, and race or ethnicity (for technical reasons, whites are not listed in the table; the results tell us if African Americans and Latinos respond differently than whites).

TABLE 5-1 The Influence of Procedural Justice

| | Willingness to accept the decision | Evaluation of the courts and the law |
|---|---|---|
| **Experience-based judgements** | | |
| Procedural justice | .68*** | .36*** |
| Distributive justice | .20** | .15* |
| Outcome favorability | −.11* | −.11* |
| **Background factors** | | |
| Ideology | .08 | .07 |
| Age | −.06 | .02 |
| Education | −.12 | .05 |
| Income | .13* | .07 |
| Gender | .02 | .00 |
| African-American | −.03 | −.17^ |
| Hispanic | −.10 | .07 |
| City of residence | −.06 | .04 |
| Was contact voluntary | −.04 | .02 |
| **Adjusted R²** | | |

^p<.10; *p<.05; **p<.01; ***p<.001.

Source: Reproduced from T. R. Tyler, *Court Review* 1/2 (2008): 26–31.

In Table 5–1, each column is a predictive equation. The column on the left predicts a survey respondent's willingness to accept the decision in the case. The righthand column predicts the respondent's overall evaluation of the courts and the law. The table provides "beta weights," a standardized measure of how well each factor (e.g., procedural fairness, age, race or ethnicity) predicts the perception shown at the top of a column. The higher the beta weight, the more important a factor is as a predictor.

The results are straightforward. Procedural fairness trumps all other factors by a large margin. Both perceptions of outcome fairness (distributive justice) and outcome favorability contribute to the ability to predict decision acceptance, but these factors are far less influential than perceived procedural fairness. Only procedural fairness is important, to a statistically significant degree, in predicting evaluations of the courts and the law. Socioeconomic and demographic background factors are rarely of any importance whatsoever as predictors. The two multivariate equations can explain a large proportion (over two-thirds) of the variation in how willing people are to accept the judge's decision. There is less success for evaluations of the courts and the law (21%). The bottom line: People who believe they were treated fairly are likely to accept court decisions.

Such persuasive research findings are one reason that judges are attracted to policies based on procedural fairness. Policies that promote procedural fairness increase decision acceptance among individuals brought before the courts and thus reduce recidivism. The growing acceptance of the procedural fairness perspective is evidenced by a white paper adopted by the American Judges Association in 2007, "Procedural Fairness: A Key Ingredient in Public Satisfaction" (Burke & Leben, 2007). In 2008, the judges who cowrote the white paper made presentations to state court judges in Arkansas, Kansas, Maryland, Minnesota, New Hampshire, and Virginia, as well as to all federal bankruptcy judges (Hon. Steve Leben, personal communication, March 18, 2009). A variety of articles expanding upon the procedural fairness perspective were also included in the *Court Review* special issue (Leben & Tomkins, 2008).

These testimonies to procedural fairness are dwarfed by comprehensive applications of procedural fairness to redesign the trial court of Hennepin County, Minnesota, District Court (Minneapolis) and the entire Judicial Branch of California. Those applications, and several others across the country, are highlighted later in the chapter.

Judicial interest in procedural fairness comes despite its truly revolutionary nature. The dominant American approach to criminal justice policy-making is guided by deterrence theory. The underlying psychological theory is that people comply when the costs of doing so are less than the benefits of disobeying the law. Instead, procedural fairness focuses on what judges can do to inculcate a sense that they and other decision makers in the criminal justice system have legitimacy. As Tom Tyler (2008) explains:

> The important role played by legitimacy in shaping people's law-related behavior indicates the possibility of creating a law abiding society in which citizens have the internal values that lead to voluntary deference to the law and the decisions of legal authorities. The ultimate test of legitimacy is that whatever the cost imposed in a particular case, the defendant or offender still believes that the decision-maker had the right to impose it. (p. 833)

In fact, the available research evidence provides more support for advocates of procedural fairness than for advocates of a deterrence approach: "Despite the centrality of deterrence theory to many criminal justice policies, there is a surprising lack of research on deterrence in operation" (Walker, 2006, p. 444). Moreover, the evidence that does support deterrence as the basis for policy appears to show a stronger deterrent effect on law-abiding citizens than on offenders (Walker, 2006, p. 448).

The assembled research evidence for the primacy of procedural fairness is counterintuitive to many people, including judges. After all, since the majority of defendants in criminal proceedings are found guilty or plead guilty, most emerge as "losers" in their interaction with the courts. Yet people who are found guilty and sentenced to prison can still be satisfied with and comply with the orders of the court (see Hollander-Blumoff & Tyler, 2008).

Distributive justice is important, but less so than procedural fairness. A large number of research studies find that once procedural fairness has been taken into account when predicting people's reactions to court decisions, race and other demographic, economic, and social characteristics tend to become statistically insignificant. They have no predictive force (cf. Olson & Huth, 1998; Rottman, Hansen, Mott, & Grimes, 2003; Rottman, 2005; Tyler, 2001; Tyler & Huo, 2002). If African Americans and Latinos are less satisfied than whites with the courts, it is because they perceive less procedural fairness in the courthouse.

## ■ How Court Policy Is Made

Policy-making for the courts differs in some important respects from what has been characterized for other components of the justice system. First, it was only in the 1970s that judges and court managers assumed the primary responsibility for setting the direction for state court policy-making. Until then, lawyers, particularly the American Bar Association (ABA), took the lead. Starting in the early 20th century, reformers focused on policies that would consolidate the many and varied courts in a specific locality that often had overlapping jurisdiction to hear cases; out of them would be forged a single trial court to hear all manner of cases. Management responsibilities were to be assumed by a team consisting of a presiding judge and a professional trial court administrator. The ABA promulgated standards for court organization and structure. The same court reform agenda sought to build a statewide management structure for the courts that would wean them from dependence on local governments.

Second, in many states by the 1970s a true judicial branch had been formed with the capacity to develop and implement policies for a state court system. Before then, state supreme courts adjudicated cases at the appellate level; their authority to influence the operations or organization of trial courts was varied. Things became very different in the 1980s and 1990s. State supreme courts today, which can be described as "the boards of directors of state court systems," did not exist in all but a few states before the 1970s (Shepard, 2006). Today's chief justices spend 50% of their time on court management issues (Feldman & Smith, 2001).

Third, the federal government was less influential and supportive of policy-making for the courts than it was for the police, prosecutors, and corrections. There are exceptions. The U.S. Department of Justice was an important ally at critical moments in the history of the ABA-led policies for court reform through funding provided by the Law Enforcement Assistance Administration (LEAA). Policies supported in this way included steps toward statewide funding, establishing a state-level planning function for the courts, public opinion surveys, jury research, and judicial education. LEAA also funded key national conferences, such as the 1978 gathering State Courts: A Blueprint for the Future, that encouraged national discussion on the direction of state court reform. More enduringly, LEAA funds helped build an infrastructure of court-specific organizations to support the work of the state courts, including the National Center for State Courts and the National Judicial College. More recently (post-1990), the Department of Justice provided funding and resources for the rapid expansion of problem-solving courts (to be discussed later).

Fourth, the constitutions of many states created a significant impediment to making policy for trial courts. Trial courts in these states are in the unique situation of often having no managerial control over the staff on which they rely. Elected clerks of courts or clerks appointed by local government are responsible for maintaining court records and for hiring and supervising court staff (Aikman, 2007, pp. 101–109). Research in the mid-1990s found that the lack of flexibility judges and court administrators had over staffing and records could nullify the advantages associated with court consolidation (Rottman & Hewitt, 1996).

Fifth, and finally, policy-making in the courts, to a degree unmatched in the other parts of the criminal justice system, has relied upon public opinion surveys. This trend follows from an emphasis on treating public trust and confidence as a key criterion against which success is measured, one that stems from the court's reliance on the executive branch for enforcement of its orders and on the legislative branch for its funding. Since the late 1970s, 33 states have commissioned one or more opinion surveys as part of their planning efforts, most since the mid-1990s (Rottman et al., 2003). In contrast, it is only recently that executive branch agencies have integrated public opinion surveys as a management tool (Poister & Thomas, 2007).

## ■ A Balance Sheet: What Did Traditional Policy-Making Accomplish?

Past eras of court reform, guided by theories stressing maximizing instrumental factors such as timeliness, cost, and efficiency, accomplished a great deal. In 1950, there were 826 trial courts in California. Today, 58 trial courts—1 per county—hear all manners of cases (Sipes, 2002, p. 119). Management theories drawn from the business field provided the blueprint for court reform by (1) simplifying trial court structure through consolidation, (2) centralizing management, (3) replacing local court funding with state funding under a centralized budget, and (4) centralizing rule making.

By the 1970s, a more flexible approach to reform emerged, one that sought to optimize court performance by matching a court's organization with the broader socio-political context in which it operated. The inspiration came from new developments in the sociology and social psychology of organizations. Contingency theory views organizations as open systems responding to their unique specific environments. This theory was translated by judges, court administrators, and consultants into a reform program seeking decentralized coordination that encouraged innovation.

Subsequent theory-driven influences on court reform included total quality management (Osborne & Gaebler, 1992), expressed as court performance standards adopted for both trial and appellate courts by national court leadership organizations in the 1990s. The results of that era are best represented by the work of a Commission on Trial Court Performance and the 22 standards they developed under five headings:

- Access to justice
- Expedition and timeliness
- Equality, fairness, and integrity
- Independence and accountability
- Public trust and confidence

The choice of areas for which courts should set standards expanded the range of topics requiring policy-making by the courts. Further, by developing empirical measures of performance for each standard and creating a Web site to monitor implementation, the courts gained a way to assess the extent to which they were accomplishing the objectives of those policies (National Center for State Courts, 2005). These changes represent a decisive break from the era of policy-making for the courts dominated by the bar rather than the bench.

## ■ Procedural Fairness in Practice: Policies and Practices

### Policies for Individual Judges and Court Staff

The key policy insight derived from procedural fairness is that the reactions of defendants to their experience in court are shaped most powerfully by how fairly they perceive they were treated by the judge. This holds true for defendants in traffic and misdemeanor cases (Tyler, 1984) and for those facing felony charges (Casper, Tyler, & Fisher, 1988).

What kinds of policies emerge from a procedural fairness perspective? The American Judges Association's "White Paper on Procedural Fairness" offers these recommendations, among many others, for trial judges to follow (Burke & Leben, 2007):

- As a matter of practice, explain in understandable language, what is about to go on to litigants, witnesses, and jurors.
- Put something on the bench as a mental reminder that patience is a virtue not always easily practiced.

- Enlist the academic community. Professors who specialize in communications and nonverbal behavior can offer great insight.

Another recommendation is bold given the role of the judge as a constitutional officer who is officially accountable only to a disciplinary board or, in 39 states, to the electorate at fixed intervals:

> Arrange to have yourself videotaped, particularly when you preside over heavy calendars. Ideally review the tape with a professional or colleagues, but . . . you can still learn a lot about how you are perceived by the people before you. (Burke & Leben, 2007, p. 18)

Taken together, these recommendations call for a paradigm shift in how judges define their role and focus their energies in the courtroom. Indeed, adhering to procedural fairness during the sentencing phase is a recognized "evidence-based" practice for reducing recidivism (Warren, 2007, pp. 40–42).

One reason for terming this change a paradigm shift is that judges evaluate court decisions more in terms of what they see as the fairness of outcomes (distributive fairness) rather than the quality of treatment that litigants receive. While the public pays the greatest amount of attention to procedural fairness, judges place the greatest emphasis on their views of distributive fairness (Heuer, 2005; Rottman, 2005; Sivasubramaniam & Heuer, 2008). As a result, judges and the public may often be talking at cross-purposes when discussing the priorities for court policy-making. The findings of public opinion surveys and exit surveys of defendants as they leave the courthouse offer a tool to help judges understand the extent to which their own view of the world differs from the view of the people who appear before them in court and who generally live in their community.

### Policies for Entire Trial Courts

Presiding judges, the entire bench, and court administrators can devise policies and procedures that will support the efforts of individual judges to adopt procedural fairness. Equally important, the courthouse can be redesigned to express the concern for and reality of fair treatment that defendants and others will experience. Areas for policy application include the physical layout of the building, signage, staff training, and help desks to assist people in navigating the often confusing court system.

Policy recommendations for judicial and court administrators in the American Judges Association's "White Paper on Procedural Fairness" include (Burke & Leben, 2007, p. 19):

- Conduct courtwide training so that all employees understand the important role they play in providing procedural fairness. How litigants are treated by court employees from the moment they enter the courthouse door—or the moment they encounter security personnel at a metal detector—sets the tone.

- Make it a major project . . . to analyze the tone of public interaction that is set in your courthouse. Does it convey respect for the people who, often in stress, come there? Could it be improved? Many courthouses have child-care facilities, adequate handicapped-accessible areas (now required by the Americans with Disabilities Act), and domestic-violence waiting rooms. . . . Involve all stakeholders (judges, staff, attorneys, litigants, and the general public) in this process.

- Treat employees fairly. If court employees do not feel that they are fairly treated in their jobs by court leaders, it is unlikely that they will treat the public any better.

The last recommendation is important. Judges are managers as well as adjudicators. Leaders in any organizational setting who are perceived as behaving in a procedurally fair manner strengthen the sense of group belongingness in organizations, which, in turn, increases the level of cooperation (De Cremer & van Knippenberg, 2002, p. 864). Procedural fairness applies wherever there is a superior-to-subordinate relationship. Judges should bear in mind that the elements of procedural fairness will govern the results whenever they interact with and set policies for their courtroom staff and court staff generally. This applies to counter clerks, bailiffs, cleaners, and all other employees who make the courthouse function. In addition, judges should monitor the degree to which probation officers are adhering to the principles of procedural fairness. The officers, and thus the court, will be more effective at reducing recidivism if they are practicing procedural fairness (Taxman & Thanner, 2003).

The application of the procedural fairness perspective to an entire trial court is most comprehensively demonstrated in Minneapolis, Minnesota (Hennepin County District Court). That court (in March 2009) was served by 62 judges, 16 referees, and 550 staff. Through the encouragement of its then presiding judge, Kevin Burke, in 2002 the court partnered with Tom Tyler and Larry Heuer, two of the leading scholars in the field of procedural fairness, to develop instruments that measure procedural fairness as perceived by defendants and others who appear in court. As litigants, attorneys, jurors, and others leave the courtroom, they are handed short surveys to complete on the spot. The questions measure the degree to which individuals felt they had been treated fairly. Examples of defendant exit questions (to which they were asked to agree or disagree) are (Podkopacz, 2005):

- The judicial officer gave reasons for his or her decision.
- The judicial officer made sure I understood the decision.
- The judicial officer seemed to be a caring person.
- The judicial officer treated me with respect.
- The judicial officer listened carefully to what I (or my lawyer) had to say.

- I understand what is required of me in order to comply with the judicial officer's decision.

The results of these surveys are tabulated both for individual judges and for the court as a whole. At the court level, initiatives are developed to encourage and facilitate behavior by judges and court staff that are consistent with the criteria of procedural fairness.

A hallmark of the Hennepin approach is an emphasis on ensuring that defendants understand what they are supposed to do when they leave the courtroom. One study looked at whether defendants left the courtroom knowing the conditions of their probation or bail orders. When they did not, remedial steps were taken. Another hallmark of the Hennepin approach is making judges aware of the impact their nonverbal communication has on perceptions of procedural fairness. This led to a 2001 study by an independent researcher that first conducted in-court observations of nonverbal behavior by trial judges. The study continued with a survey on sentiments toward nonverbal communication given to those who were observed. The result found "almost all of the judges observed used nonverbal behaviors . . . that are considered to be ineffective and in need of improvement. About one-third of the judges used these ineffective behaviors frequently" (Porter, 2001, p. 4). Examples included failure to make eye contact, the use of an exasperated tone of voice, sighing audibly, and body position.

### Policies for Designing New Court Forums

Procedural fairness has especially strong implications for the design of forums that relax the imperatives of the adversarial process. In recent years, Tom Tyler (2008) has turned his attention to demonstrating the capacity of procedural fairness to design new institutional forums. He notes in particular that:

> Legal institutions are designed based upon the assumption that behavior is shaped by the risk of sanctioning. As a result, there is a fundamental misalignment of the organization, in this case the legal system, and models of motivation, leading the system to be less efficient and effective than might potentially be the case. (p. 873)

Recognition of the importance of motivation, even if not explicitly guided by procedural fairness theory and research, has led to the emergence of problem-solving courts.

In response to the unique difficulties presented by defendants with substance abuse problems, some judges concluded that these "new" kinds of cases were not susceptible to the traditional adversarial approach that guided the state courts from their inception. Attentiveness to an individual's needs and circumstances seemed necessary, along the lines pioneered by juvenile courts. In these nontraditional processes, people are more important than cases, needs-based considerations are balanced against rights-based considerations, the process is more collaborative than adversarial, and the attention paid to interpretation and application of social science knowledge rivals that given to interpretation and application of the law (Rottman & Casey, 1999, p. 140).

The first drug court opened in 1989. The rapid proliferation of problem-solving courts (really specialized dockets)—including mental health, domestic violence, truancy, and community courts—is the most radical change to the structure of the state courts over the last two decades. It is estimated that 3200 such court dockets exist today (Huddleston, Marlowe, & Casebolt, 2008).

Research suggests that these new court forums do indeed promote a sense of procedural fairness. Defendants in the Brooklyn, New York, Red Hook Community Justice Center, for example, reported experiencing more procedural fairness than did defendants in control groups drawn from the regular courts. Perceptions of the judges and bailiffs they encountered were more positive as well (Frazer, 2007). Although the planning process for Red Hook was not explicitly guided by the precepts of procedural fairness, the research found that the design nonetheless promoted fair treatment. For instance, "in the community court, the judge spoke directly to the defendant in 45 percent of the observed appearances, while in the traditional court this occurred in only 19 percent of appearances" (Frazer, 2007, p. 37).

A similar "boost" to procedural fairness occurs in restorative justice programs. Supporting evidence for this assertion includes a meta-analysis of victim–offender mediation and family conferencing programs. Mediation and conferencing elicited positive responses from both offenders and victims. Compared to their counterparts experiencing traditional court adjudication, those in mediation or conferencing reported being treated more fairly, were more satisfied with how their case was handled, felt that their individual voices were heard, and were more satisfied with the case outcome (Poulson, 2003).

Good feelings are important, but do they promote greater compliance? The answer is yes based on a drug court evaluation that randomly assigned defendants to a drug treatment court (DTC). Higher perceptions of procedural fairness were indeed linked to better outcomes: "More specifically [the study] suggests that the DTC program, especially the judicial hearings, contributes to an offender's perception of fairness and due process, thereby increasing his or her willingness to fulfill his or her part of the negotiated DTC agreement" (Gottfredson, Kearley, Najaka, & Rocha, 2007, p. 28). Such research findings are causing existing problem-solving courts to reassess their design and operations to better leverage their ability to generate a sense of fair treatment.

Procedural fairness is integral to the current High Performing Courts project at the National Center for State Courts (Ostrom, Ostrom, Hanson, & Kleiman 2007).[1] "CourTools," a new system for measuring court performance, has as its first measure "Access and Fairness." This measure is evaluated through five questions, designed to be completed by all those who appeared before a judicial officer on the day of the survey. Defendants and others leaving the courthouse are asked to rate the following statements from "strongly disagree" to "strongly agree" on a five-point scale:

- The way my case was handled was fair.

- The judge listened to my side of the story before he or she made a decision.

- The judge had the information necessary to make good decisions about my case.

- I was treated the same as everyone else.

- As I leave the court, I know what to do next about my case.

To be high performing, a court must demonstrate that defendants and others report experiencing fair treatment as judged by the elements of procedural fairness.

## ■ Procedural Fairness and Statewide Court Policy-Making

The influence of procedural fairness is nowhere more powerful than in California, where it is being implemented statewide. The judicial branch of the state of California is massive. Its trial courts receive 9 million new case filings annually, include 1900 judges and over 21,000 employees, and incorporate 461 court facilities. The branch's annual budget is $3.8 billion. It is the most self-governed court system in the United States. A Judicial Council sets policies for the branch using 6-year strategic plans, 3-year operation plans, and annual action plans. The Council consists of 28 members.[2] The policy-making process is notable for its built-in capacity for follow-through over the long term. **Figure 5–1** describes the stages to the process and identifies the groups contributing to the various plans.

The pursuit of procedural fairness is an explicit goal of the Operational Plan for California's Judicial Branch, 2008–2011, adopted by the state's Judicial Council, and is defined as follows:

> Ensure that all court users are treated with dignity, respect, and concern for their rights and cultural backgrounds, without bias or appearance of bias, and are given an opportunity to be heard. (Denton, 2008, p. 44)

Throughout the branch, task forces and committees are reviewing and revising policies to enhance the likelihood that the elements of procedural fairness are evident in the way people experience the state's trial and appellate courts. On a still more ambitious level, the operational plan seeks to increase the trust and confidence of Californians who do not have direct experience with the courts.

The adoption of procedural fairness as a primary goal of the policy-making process and as an instrument for achieving other goals emerged from research launched by the Judicial Council. The council sponsored a public opinion survey in 2005 that was discussed extensively within the branch. The report empha-

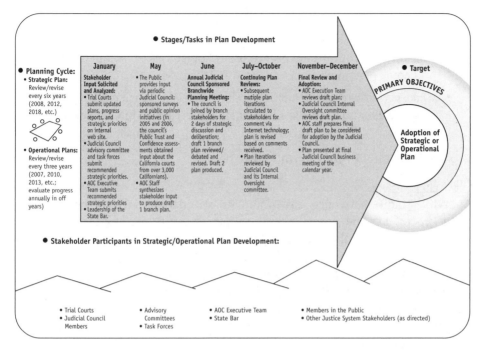

**Figure 5–1**  California's Judicial Branch: Summary of Strategic/Operational Planning Timeline and Processes
**Source:** Courtesy of the Administrative Office of the Courts, Judicial Council of California.

sized the critical role of perceptions of procedural fairness in establishing trust and confidence in the courts (Rottman, 2005). In 2006, the themes from that report were explored in greater depth through a program of focus group research (Wooden & Doble, 2006). One subset of the 24 focus groups consisted of court participants who had recently been involved in the kinds of cases found by the survey to be associated with the lowest levels of perceived procedural fairness: family, small claims, and traffic. Other focus groups brought together judges and court administrators to explore procedural justice issues through the lenses of their experiences.

In 2007, the California courts embarked on a 3-year procedural fairness initiative closely tied to their strategic planning process. "Work to achieve procedural fairness in all types of cases" is a goal of the 2006–2012 plan. Seventeen committees and other advisory groups were charged with identifying ways in which court rules and procedures could be changed to promote procedural fairness.

California's procedural fairness initiative will (Denton, 2008, p. 44):

- Identify procedural fairness best practices and model programs
- Study and evaluate efforts that have the potential to achieve procedural fairness for court users

- Develop procedural fairness guidelines, tools, and resources for judicial officers and judicial branch personnel
- Recommend educational programs and objectives to help judicial officers and court personnel achieve procedural fairness
- Make periodic recommendations to the Judicial Council regarding a variety of strategies and means to help courts achieve procedural fairness

## ■ The Public and Procedural Fairness

### Public Input

While traditional court reform and its considerable improvements to the court process left the general public unmoved by its success, the new court reforms consistent with procedural fairness expectations do appear to have an impact. For example, the Red Hook Criminal Justice Center, established in one of Brooklyn's most disadvantaged communities, boosted the public's positive view toward the courts from 57% to 78% after 2 years of operation (Frazer, 2006 p. 5).

A 2000 telephone survey of the American public also measured support for these nontraditional court processes (Rottman, et al., 2003). The preamble to the set of questions read:

> Some people think that courts should stick to their traditional role of looking at the facts in a specific case and then applying the law. Other people think that it is now necessary for the courts to go beyond that role and try to solve the problems that bring people into court. I am going to read you a few statements about the role of the court. Do you strongly agree, somewhat agree, somewhat disagree, or strongly disagree that courts . . .
>
> 1. Should hire drug treatment counselors and social workers as court staff members?
> 2. Should order a person to go back to court and talk to the judge about their progress in a treatment program?
> 3. Should take responsibility for making sure local agencies provide help to people with drug abuse and/or alcohol problems?
> 4. Should consider what psychologists and medical doctors know about the causes of emotional problems when making decisions about people in court cases?

Survey respondents reacted highly positively to all four changes to the traditional role of courts. Black respondents tend to be the most supportive of change, followed by Latinos. Whites are distinctly less enthusiastic in their support of new roles for judges and courts. The magnitude of the differences in levels of support is evident in the proportion of "strongly agree" responses to the four statements.

The lowest level of support, across all three groups, is given to the hiring of treatment counselors and social workers. This change to traditional processes, however, also marks the sharpest racial and ethnic difference. Whites are less supportive than are minority group members, although a majority still agree with the changes. One interpretation of the findings is that these nontraditional court processes fit more closely with the public's expectations for procedural fairness than traditional courts. Also, the population segments least supportive of the traditional courts are the same ones that express the strongest support for courts adopting a different approach to case adjudication.

## The Court's Two Publics

Trial court judges and courts have two publics. One public consists of the 50% of all adults who have had one or more direct experiences with the courts as a litigant, defendant, juror, or witness. Members of that public remember the details of their encounter decades later, even if the stakes in their case were low and the time involved short. A negative or positive experience will linger for many years in a person's mind. It becomes the person's point of reference when expressing their views on the judiciary or court system.

A national jury expert, Tom Munsterman, looking back on his career, testified to the durability of jury service:

> What amazes me is when I ask someone if they've been on a jury, and they say, "Yea, it was 10 years ago, and on the first day this happened, and on the second day that happened . . . and then this and this and this." It's an incredible, indelible experience. It stays in the memory for years and years, even down to what "he said" and "she said" and the details of deliberations. (Wolf, 2008, p. 384)

For defendants, this effect may be still more intense. Writing about the Australian recidivism study of drunk driving, Tom Tyler observed, "It is striking that people's experiences in a courtroom or at a conference with legal authorities, something that lasts at best a few hours, can be strongly affecting their behavior several years later" (Tyler et al., 2008, p. 27).

Procedural fairness offers a template on how to increase the proportion of people entering courtrooms who will leave satisfied with their day in court. Procedural fairness teaches court reformers that 50% of litigants can lose their case and still leave feeling that they did receive their day in court and believe the outcome to be fair.

This public (the one with court experience) is expanding rapidly. In recent decades, jury service has expanded from 6% to 25% of adults in response to a reduction in exemptions from jury service and reform of jury source lists. About 1.5 million Americans annually are impaneled as jurors (nearly 1% of the adult population) and another 32 million receive a jury summons (Mize, Hannaford-Agor, & Waters, 2007, p. 7).

The "other" public lacks direct experience on which to base opinions about courts. Instead, their perceptions of the courtroom experience are shaped by perceived sentencing leniency and popular perceptions regarding the antics of fictional judges on television, the movies, and "reality" TV judges. A lack of experience makes individuals' political orientation a significant predictor of their beliefs about the judiciary and court system (Olson & Huth, 1998, p. 227). Overall, their frame of reference about the courts is national, not local. In contrast, people with direct court experience are little influenced by factors over which the judiciary has no control such as political considerations.

## ■ Conclusion

Previous eras of policy-making made America's trial courts more efficient organizations, offering enhanced customer service. Yet something was missing. Court policies need to connect with the core concerns of respect, neutrality, participation, and trustworthiness—principles that encourage people to support and comply with court decisions. Adhering to procedural fairness throughout the court system is a program for reform capable of addressing the problems judges face in the 21st century.

It will be a radical program in that it downplays, but does not ignore, traditional concerns of court policy-making such as deterrence and efficiency. Instead, the focus is on securing voluntary compliance with court decisions. As Tyler (2008) notes, "social motivations are important because they are more powerful and more likely to produce changes in cooperative behavior than are instrumental motivations. Hence, social motivations are both more powerful and less costly than are incentives and sanctions" (p.874).

In this way, procedural fairness offers the courts a policy-driven reform program that strengthens the connection between courts and the public. The promise for the 21st century is that courts will become organized in a manner that generates satisfaction, trust, and compliance with court orders. Procedural fairness applies to all organizations, but it has particular relevance to judges and court administrators because it so clearly influences the effectiveness of court decisions. Protection orders are more likely to be followed, civil litigants are more likely to pay damages, and probationers are more likely to desist from crime.

Procedural fairness can even guide the judiciary branch as it fends off efforts to politicize its work. Judges should respond with arguments that demonstrate how courts embody the elements of procedural fairness and how those individuals and groups attacking the courts would harm those same elements.

## ■ Endnotes

1. The High Performing Courts initiative and CourTools reflect the influence of another theoretical approach to understanding and improving the performance of organizations: organizational culture. As described by Quinn (1988), "Simply put, culture is the set of values and assumptions that underlie the statement,

'This is how we do things around here'" (p. 66). Culture at the organizational level, like information processing at the individual level, tends to take on moral overtones. While cultures can vary dramatically, they share the common characteristic of providing integration of effort in one direction while often sealing off the possibility of moving in another direction.

2. The membership includes the chief justice, 14 judges appointed by the chief justice (1 associate Supreme Court justice, 3 justices of the intermediate appellate court, and 10 trial judges), 4 attorneys appointed by the Board of Governors of the State Bar, 1 member from each house of the legislature, and 7 advisory members representing the California Judges Association and agencies of the office of the state court administrator.

## ■ Discussion Questions

1. Why is procedural fairness an important policy consideration for criminal courts when carrying out their responsibilities?

2. Using examples, describe the difference between due process and procedural fairness.

3. Is procedural fairness an important feature of criminal justice policy today? Can it be applied to criminal justice decisions and processes outside the court system?

4. Compare and contrast traditional court policy-making with policies and practices guided by procedural justice.

5. Drawing on research and case studies, describe some of the benefits of criminal justice policies anchored by procedural fairness considerations.

## ■ References

Administrative Office of the U.S. Courts. (2007). *Judicial Business 2007* (p. 13, 20). Retrieved February 7, 2010, from http://www.uscourts.gov/judbus2007/2007judicial%20business.pdf

Aikman, A. (2007). *The art and practice of court administration.* Boca Raton, FL: Auerbach Publications.

Brockner, J., Ackerman, G., Greenberg, J., Gelfand, M., Francesco, A. M., Chen, Z., et al. (2001). Culture and procedural justice: The influence of power distance on reaction to voice. *Journal of Experimental Social Psychology, 37,* 300–315.

Burke, K., & Leben, S. (2007) Procedural fairness: A key ingredient in public satisfaction. *Court Review, 44,* 4–25 [American Judges Association White Paper].

Casper, J., Tyler, T. R., & Fisher, B. (1988). Procedural justice in felony cases. *Law & Society Review, 22,* 483–507.

Denton, D. (2008). Procedural fairness in the California courts. *Court Review, 44,* 44–54.

De Cremer, D., & van Knippenberg, D. (2002). How do leaders promote cooperation? The effects of charisma and procedural fairness. *Journal of Applied Psychology, 87,* 858–866.

Feldman, H., & Smith, C. E. (2001). Burdens of the bench: State supreme courts "non-judicial tasks." *Judicature, 84,* 304–309.

Frazer, M. S. (2006). *The Impact of the Community Court Model on Defendant Perceptions of Fairness: A Case Study at the Red Hook Community Justice Center.* New York: Center for Court Innovation.

Frazer, M. S. (2007). Examining defendant perceptions of fairness in the courtroom. *Judicature, 91,* 36–37.

Gottfredson, D., Kearley, B., Najaka, S., & Rocha, C. (2007). How drug treatment courts work: An analysis of mediators. *Journal of Research in Crime and Delinquency, 44,* 3–35.

Heuer, L. (2005). What's just about the criminal justice system? A psychological perspective. *Journal of Law and Policy: Criminal Law and Procedural Justice, 13,* 209–227.

Hollander-Blumoff, R., & Tyler, T. R. (2008). Procedural justice in negotiation: Procedural fairness, outcome acceptance, and integrative potential. *Law & Social Inquiry, 33,* 473–500.

Huddleston, W., Marlowe, D., & Casebolt, R. (2008). *Painting the current picture: A national report card on drug courts and other problem-solving court programs* (vol. II, no. 1). Alexandria, VA: National Drug Court Institute.

Lafountain, R., Schauffler, R., Strickland, S., Raftery, W., Bromage, C., Lee, C., et al. (2008). *Examining the work of state courts, 2007: A national perspective from the Court Statistics Project* (National Center for State Courts 2008). Retrieved February 7, 2010, from http://www.ncsconline.org/D_Research/csp/2007B_files/ EWSC-2007-v21-online.pdf

Leben, S., & Tomkins, A. (Eds.). (2008). Procedural Fairness [Special issue]. *Court Review: The Journal of the American Judges Association, 44.* Retrieved February 7, 2010, from http://aja.ncsc.dni.us/courtrv/cr44-1/CR44-1-2.pdf

Mize, G., Hannaford-Agor, P., & Waters, N. (2007). *The state-of-the-states survey of jury improvement efforts: A compendium report 7.* February 7, 2010, from http://www.ncsconline.org/D_Research/cjs/pdf/SOSCompendiumFinal. pdf

National Center for State Courts. (2005). *Trial court performance standards & measurement system.* Retrieved February 7, 2010, from http://www.ncsconline. org/D_Research/TCPS/index.html

Olson, S., & Huth, D. (1998). Explaining public attitudes toward local courts. *Justice System Journal, 20,* 41–61.

Osborne, D., & Gaebler, T. (1992). *Reinventing government: How the entrepreneurial spirit is transforming the public sector.* Reading, MA: Addison Wesley.

Ostrom, B., Ostrom, C., Hanson, R., & Kleiman, M. (2007). *Trial Courts as Organizations.* Philadelphia, PA: Temple University Press.

Paternoster, R., Bachman, R., Brame, R., & Sherman, L. (1997). Do fair procedures matter? The effect of procedural justice on spouse assault. *Law & Society Review, 31,* 163–204.

Podkopacz, M. (2005). *Fourth judicial district of the state of Minnesota report on the judicial development survey.* Retrieved February 7, 2010, from http://www.mncourts.gov/district/4/?page=545

Poister, T., & Thomas, J. C. (2007). The wisdom of crowds: Learning from administrators' perceptions of citizen perceptions. *Public Administration Review, 67,* 279–289.

Porter, L. (2001). *Nonverbal communication in courtrooms at the Hennepin County Government Center: A report on observations of Fourth Judicial District judges in March and April 2001.* Hennepin County District Court.

Poulson, B. (2003). A third voice: A review of empirical research on the psychological outcomes of restorative justice. *Utah Law Review, 1,* 167–203.

Quinn, R. (1988). *Beyond rational management.* San Francisco: Jossey-Bass.

Rottman, D. (2005). *Trust and confidence in the California courts: A survey of the public and attorneys, part I: Findings and recommendations.* San Francisco: Judicial Council of California/Administrative Office of the Courts.

Rottman, D. (2007). Adhere to procedural fairness in the justice system. *Criminology & Public Policy, 6,* 835–842.

Rottman, D. & Casey, P. (1999, July). Therapeutic jurisprudence and the emergence of problem-solving courts, *National Institute of Justice Journal, 240,* 12–19.

Rottman, D., Hansen, R., Mott, N., & Grimes, L. (2003). *Perceptions of the courts in your community: The influence of experience, race, and ethnicity.* Williamsburg, VA: National Center for State Courts.

Rottman, D. & Hewitt, W. (1996). *Trial court structure and performance: A contemporary reappraisal.* Williamsburg, VA: National Center for State Courts.

Shepard, R. T. (2006). The new role of state supreme courts as engines of court reform. *New York University Law Review, 81,* 1535–1552.

Sipes, L. (2002). *Committed to justice: The rise of judicial administration in California.* San Francisco: Administrative Office of the California Courts.

Sivasubramaniam, D., & Heuer, L. (2008). Decision makers and decision recipients: Understanding disparities in the meaning of fairness. *Court Review, 44,* 62–70.

Taxman, F., & Thanner, M. (2003). Probation from a therapeutic perspective: Results from the field. *Contemporary Issues in Law, 7,* 39–63.

Tyler, T. R. (1984). The role of perceived injustice in defendants' evaluations of their courtroom experience. *The Law & Society Review, 18,* 51–74.

Tyler, T. R. (2001). Public trust and confidence in legal authorities: What do majority and minority group members want for the law and legal institutions? *Behavioral Sciences & the Law, 19,* 215–235.

Tyler, T. R. (2004). Procedural justice. In A. Sarat (Ed.), *The Blackwell Companion to Law and Society.* London, U.K.: Blackwell Publishing.

Tyler, T. R. (2008). Psychology and institutional design. *Review of Law and Economics, 4,* 801–887.

Tyler, T. R. & Huo, Y. (2002). *Trust in the law: Encouraging public cooperation with the police and courts.* New York: Russell Sage Foundation.

Walker, S. (2006). Too many sticks, not enough carrots: Limits and new opportunities in American crime policy. *University of St. Thomas Law Journal, 3,* 430–461.

Warren, R. W. (2007). *Evidence-based practices to reduce recidivism: Implications for state judiciaries.* Washington, DC: Crime and Justice Institute and the National Institute of Corrections.

Wolf, R. A. (2008). Interview: G. Thomas Munsterman: Guilty (of a career dedicated to improving the jury system). *Journal of Court Innovation, 1,* 372–386.

Wooden, R., & Doble, J. (2006). *Trust and confidence in the California courts phase II: Focus groups and interviews.* San Francisco: California Administrative Office of the Courts.

■ **Court Cases Cited**

*Baker v. Carr*, 369 U.S. 186 (1962).

# Criminal Justice Policy and Problem-Solving Courts

## CHAPTER 6

Susan T. Krumholz

## ■ Introduction

The courts have historically been the location of final adjudication for the criminal justice system. State trial courts may be courts of limited jurisdiction or courts of general jurisdiction. Courts of limited jurisdiction, often called district courts, are the court of first resort, hearing misdemeanor cases or cases that may only result in relatively short sentences. Courts of general jurisdiction, generally referred to as superior courts,[1] may rule on any state criminal matter. In either case, these trial courts render decisions of guilt or innocence, but more commonly accept pleas and issue consequences such as fines, jail time, or probation. Historically, courts have been limited to choosing fines and/or incarceration, restricting the availability of alternative sentences. This chapter looks at one model for integrating treatment alternatives into the sentencing process: specialized or—as they are often called—problem-solving courts. Specialized courts have been around since the first juvenile courts in the 1890s, but any reference to "problem solving" does not appear until a century later, in the 1980s. We will study what this shift means for the work of the courts by exploring the theory behind these courts and by examining two unique manifestations: drug courts and domestic violence courts.

The policy challenges are three-tiered. The first tier includes issues that are common to criminal justice programs. In this tier, policy makers may consider, "Can we save money and improve the efficiency of the criminal courts?" The second tier speaks to the role of the courts in managing social harms, addressing such questions as, "Can we respond to the needs of victims and offenders while also respecting their rights?" The third tier of policy issues might be referred to as the micro issues, those that explore what makes a program effective. Here policy makers may, for instance, consider, "What are the best treatment options? And how should the program be staffed?"

After examining the workings of these courts and some of the current research into their efficacy, perhaps we will be in a better position to determine whether they provide alternatives to incarceration that can reduce recidivism, thereby reducing the monetary, social, and individual cost of crime.

## ■ Background of Problem-Solving Courts

Rapid growth in the number of individuals being processed through the criminal courts during the 1980s and 1990s shifted the focus to the criminal courts and made reforms to the processing of crime more than desirable. Several models for reforming the law to include a response to criminal transgressions emerged. Some focused exclusively on models of lawyering.[2] Other reform models, including restorative justice,[3] therapeutic jurisprudence, and problem-solving courts, had broader goals, and thus have had a wider influence on the criminal justice system. While restorative justice proposes a system that would effectively remove many matters from the jurisdiction of courts and place it within the community, therapeutic jurisprudence (TJ) advocates contend that TJ is compatible with our current system of justice and is indeed a direct precursor to the problem-solving courts we see today (for example, see Daicoff & Wexler, 2003).

Therapeutic jurisprudence emerged in the early 1990s as a promising approach to addressing issues of civil commitment and other aspects of mental health law. Slobogin (1995) has defined TJ as "the use of social science to study the extent to which a legal rule or practice promotes the psychological and physical well-being of the people it affects" (p. 196). Once rooted in mental health law, TJ has extended its reach to an ever-increasing number of legal arenas. According to Daicoff and Wexler, TJ has significant implications for understanding the nature and scope of the impact of legal rules, processes, and practices on the well-being, both physically and psychologically, of all persons involved in the legal system of our society. As they state, "Therapeutic jurisprudence recognizes that legal rules, procedures, and actors are social forces that intentionally or unintentionally often produce therapeutic or anti-therapeutic consequences" (2003, p. 561). Among the concerns raised about the veracity and viability of TJ, Roderick and Krumholz (2006) ask, "Can the law be therapeutic, how do we determine what is therapeutic and anti-therapeutic, who defines therapeutic/anti-therapeutic, and how does one empirically test these constructs?" (p. 206).

The proliferation and expansion of TJ has resulted in claims of profound changes in how legal scholars, practitioners, social scientists, and even lawmakers and judges have attempted to understand broad topics primarily, but not exclusively, within the field of criminal justice. Such claims include TJ's evolution into a number of specific areas of practical application such as therapeutic lawyering and judging and problem-solving courts. Daicoff and Wexler (2003) claim that by applying therapeutic models to legal problems, courts could "fashion remedies and issue opinions that lessen contentiousness and promote harmony and dialogue" (p. 571). While harmony and dialogue may be admirable goals, are they compatible with the goals of criminal control? Despite these and other concerns, TJ continues to be closely linked with problem-solving courts.

A description of problem-solving courts, taken from the U.S. Department of Justice's report "Stewards of the American Dream" (2007), is representative of the extraordinarily broad mandate given to this new model of adjudication:

> Problem-solving courts are designed to treat offenders while, at the same time, considering the harm to victims and the community. These courts work with other criminal justice institutions, and across disciplines, such as health and social services, to address underlying issues that contribute to criminal behavior and to design appropriate interventions. (p. 90)

Dorf and Fagan (2003) ask not only whether we want these courts to take on every issue confronting the criminal justice system, but also how we decide which problems these courts are best suited to solve. To date we have seen the development of mental health courts, drug courts, community courts, family courts, juvenile drug courts, and domestic violence courts. What is the common bond among these diverse courts? What impact do these courts have on the traditional goals of the criminal justice system? Is it the role of the courts to judge or to repair? And what role should the courts play in social change?

The problem-solving courts that have emerged to date are primarily aimed at treatment first, reserving punishment for those individuals who fail to comply. They are "team-oriented, multi-disciplinary, and non-adversarial" (Nolan, 2003, p. 1543). A 2007 report prepared for the Bureau of Justice Assistance by the Center for Court Innovation created a list of underlying principles, including (1) enhanced information, "gathered with the assistance of technology and shared in accordance with confidentiality laws"; (2) community engagement; (3) collaboration; (4) individualized justice; (5) accountability; and (6) measuring, processing, and disseminating outcomes (Wolf, 2007).

Not all view the move toward problem-solving justice favorably. Judge Morris Hoffman (2002) describes the problem-solving courts as follows:

> Defendants are "clients"; judges are a bizarre amalgam of untrained psychiatrists, parental figures, storytellers, and confessors; sentencing decisions are made off-the-record by a therapeutic team . . . and court proceedings are unabashed theater. (p. 2066)

This quote dramatically illustrates what some argue is an inherent conflict that exists within courts that seek to dispense justice while also solving the problems of individuals and society. Judge Hoffman voices the concerns of those who fear that in an attempt to solve problems, adjudicators lose their objectivity and defendants lose the presumption of innocence and their right to counsel. For example, the

Web site of the Minnesota Judicial Branch (www.courts.state.mn.us/?page=626) describes a court in which the attorneys and other system participants partner "to develop a strategy that will *pressure an offender into completing a treatment program* and abstaining from repeating the behaviors that brought them to court" (emphasis added). Hoffman's fears are not abated by representations such as this, which suggest that the court may be comfortable compromising an offender's rights for the sake of treatment.

McCoy (2003) argues that the move toward judicial problem solving is merely an act of recovering some of the discretion judges had prior to the rapid rise of determinate sentencing that began in the 1970s. Additionally, McCoy suggests that we have only to look at juvenile courts, which have long wedded therapeutic goals with those of justice, to see that the two can work together.

In trying to argue in favor of problem-solving courts, Berman (2004) makes what is perhaps the best argument against them. As he points out, "more than nine out of ten cases in state courts are resolved . . . by plea bargain" and "the most common form of punishment in state courts is . . . probation," often including mandated treatment or reparation (p. 1318). If it is true that most defendants are already receiving mandated treatment, then perhaps problem-solving courts do not represent the extreme paradigm shift that some suggest. Do these courts really represent such a radical departure from traditional courts? And can they further the overarching goal of reducing recidivism? For now let us turn to an exploration of two of the most visible examples—drug courts and domestic violence courts. We will return to a discussion of the policy implications of problem-solving courts later in the chapter.

## ■ Drug Courts

### Background

Drug courts are recent innovations in a two-century old court system. The first drug court was created less then 20 years ago in Miami, Florida. For close to 20 years in the United States, there has been a trend toward guiding nonviolent drug offenders into treatment rather than incarceration. Drug courts are designed to provide treatment to participants who admit to having an addiction.

Drug treatment courts have seen rapid expansion and are considered "one of the fastest growing programs designed to reduce drug abuse and criminality in nonviolent offenders in the nation" (Carey, Pukstas, Waller, Mackin, & Finigan, 2008, p. 1). As of March 2008, there were 1853 adult and juvenile drug courts operating in all 50 states, the District of Columbia, Northern Mariana Islands, Puerto Rico, and Guam (Bureau of Justice Assistance, 2008). According to Dorf and Fagan (2003), the rapid rise in drug courts can be traced to four converging problems within the justice system: (1) the growth of drug cases coming before the courts as a result of the escalating "war on drugs"; (2) the public perception that punishment of drug crimes amounted to no more then a "revolving door"; (3) growing judicial resentment of the restrictive nature of mandatory sentencing laws; and (4) increasing reliance on the courts to solve social problems. Drug courts, which "emphasized both the individual responsibility of drug addicts and the disease model of addiction" (p. 1501), promised to address all these problems. Nolan (2003) asserts the following:

> In particular, drug courts offer drug offenders, as an alternative to the normal adjudication process, an intensive court-based treatment program. Participants or "clients" (as they are typically called in drug courts) return regularly to the courtroom, where they directly interact with the judge. They also submit to regular urinalysis tests, receive acupuncture treatment, and participate in a variety of individual and group counseling sessions, including Alcoholics Anonymous (AA) and Narcotics Anonymous (NA) twelve-step programs. (p. 1547)

In other words, the courts use their coercive authority to mandate treatment. Successful completion comes with a promise that charges will be dropped. Failure results in sentencing. The hope of drug court proponents is that they will increase access to effective treatment and decrease recidivism. It is presumed that a decrease in recidivism would also decrease costs, both in time and dollars. However, as noted by Nolan (2003), "It's a process advertised to take one year, although it often lasts much longer" (p. 1548). Extending the length of an offender's involvement with the drug court increases the anticipated costs.

In a typical drug court program, participants are under the strict supervision of the courts. Once an individual is referred to the drug court, a group—typically comprised of a probation officer, a prosecutor, a defense attorney, a judge, and one or more service providers—will convene a meeting to determine the appropriate treatment response. Individual courts tend to have a specific protocol, including the level of supervision and steps or phases that measure the individual's progress, and an expected schedule of progress meetings. It is not unusual to have a specific judge, a specific member of the probation department staff, and a specific prosecutor assigned to the drug court. Along with representatives from the various treatment providers, they form a working group that convenes regularly to monitor the progress of the participants. While progress meetings develop with distinct characteristics from jurisdiction to jurisdiction, they generally include drug testing results reported by probation, a status report from the treatment agencies, and either an admonition (often accompanied by sanctions) or positive recognition from the judge.

Early in this period of rapid expansion, the National Association of Drug Court Professionals offered a model of standardization. In "Defining Drug Courts: The Key Components" (1997), the organization set out 10 "key components" that all drug courts should share as performance goals or benchmarks.

Included in what are defined as essential elements of an effective drug court are such items as:

- Integrating treatment services with other aspects of case processing
- Employing a nonadversarial approach
- Assisting participants in accessing rehabilitation services
- Regular monitoring and continuing interaction with the judge
- Building and nurturing community partnerships to assure that holistic services designed to treat and rehabilitate are provided to participants

These key components, widely adopted by drug courts, are actually a loosely defined set of guidelines allowing for a great deal of discretion by individual courts. From the suggestion that courts use "nonadversarial" processes to the recommendation that drug testing be "frequent," practices in different courts vary widely. Indeed, the drafters of the "Key Components" intended that they serve as "a practical, yet flexible framework" (National Association of Drug Court Professionals, 1997) for emerging courts. These variations are seen as a strength, and as Fox and Wolf (2004) note, they will enable researchers to test whether different implementation decisions lead to different outcomes or costs, which in turn might lead to programmatic changes. For example, one court might include job skills training while another does not; one judge might rely strictly upon positive reinforcement while another may not. If we can evaluate the different outcomes brought about by these variations, we might learn that job skills training has a significant effect on participant success and thus encourage others to follow suit.

### Research on Drug Courts

Extensive research is available on drug courts, which is especially noteworthy considering that they have only been in existence for two decades. Since the late 1990s, any federal money granted for the creation or continuation of a drug court has the stipulation that an evaluation be conducted. Given the expansion of such courts, there are literally hundreds of evaluations. According to the Office of National Drug Control Policy, a "decade of research indicates that drug court reduces crime by lowering re-arrest and conviction rates, improving substance abuse treatment outcomes, and reuniting families, and also produces measurable cost benefits" (n.d., p. 1). But the evidence appears to be more ambivalent than that statement might suggest. The following examples are designed to illustrate the evaluation findings.

Early research on drug courts collected data on rates of treatment progress indicated by such measures as retention in treatment, number of treatment sessions actually attended (versus those participants were expected to attend), and status at the end of 1 year. These measures were considered the best available, as not enough time had passed to accurately assess rates of recidivism. One such study of drug courts in Clark County, Nevada, and Multnomah County, Oregon, found that participants attended about 70% of the expected treatment sessions. At the end of 1 year, 51% were in good standing, but only 2% had graduated (Goldkamp, White, & Robinson, 2002). As suggested here, and by Nolan (2003), studies indicate that although most participants are in some degree of compliance with their treatment protocol, few graduate in the prescribed year. Extending the duration of contact with the court and with the treatment provider results in an increase in services and time, and can be expected to increase the cost per participant. It remains uncertain whether the longer-than-anticipated court involvement plays a role in the rate of subsequent offending.

A good example of a study that explores rates of recidivism is a 2003 evaluation of multiple drug courts in New York State (NYS). For purposes of this study, recidivism was defined as a new arrest leading to conviction 1 year postarrest and 3 years postarrest. Researchers found that recidivism rates more than 1 year after an initial arrest ranged from 12% to 30% in six NYS drug courts. By comparison, for individuals with similar charges who were not participating in the drug court, recidivism rates ranged from 23% to 37%. Lower rates of recidivism are expected during that first year, as many of the drug court participants are in residence or under strict supervision during that period. At 3 years postarrest, rates of recidivism ranged from 29% to 56% for drug courts, as compared with rates of 40% to 65% for non-drug courts. Even after 3 years, the drug court appears to have a significant impact, lowering the likelihood of repeat offending by up to 45%. The reduction in rates of recidivism is even more striking when counting only those who have "graduated" or satisfactorily completed the requirements imposed by the court. In the six courts, an average of only 9% of drug court graduates were counted among the recidivists (Rempel et al., 2003, p. 273–278). The researchers note that "while drug court graduates are far less likely than comparison defendants to recidivate, drug court failures are just as or more likely than comparison defendants to recidivate" (p. 277). These findings emphasize the importance of program retention, while noting that more needs to be learned to understand the factors that determine retention.

The result of the New York State study is fairly representative of findings to date. Overall, there does appear to be a slight reduction in the rate of recidivism for those who have participated in drug court as compared to those similarly charged who have not. But that reduction varies significantly and, as with program retention, studies have only recently begun to isolate the program factors that impact this variation.

Another important policy consideration is whether drug courts save money. According to the research, the savings to governments vary widely. A study of five Indiana drug courts, for example, found a range of $314 to $7040 in per person cost savings (NPC Research, 2007). None of these studies indicate that drug court cost per participant is higher than the cost of incarceration, though it should be noted that if drug courts have resulted in significant savings to the states, at least some of that saving is attributable to the fact that most of these courts have been supported by federal grant money. There is concern that the loss of such "soft money" will reduce the number of states willing to support drug courts. As stated by Fox and Wolf (2004),

> Drug courts will not survive for long unless they are institutionalized. Federal funding for drug courts will not last forever. And those states fortunate enough to receive federal grants must deal with the reality that grants usually terminate after a few years, and states will have to cover costs to keep drug courts running. (p. 2)

In a move designed to support the growth and institutionalization of drug courts, recent research has moved from simply examining cost savings and recidivism to addressing more in-depth policy issues. Carey et al. explore the intricacies of court functioning by operationalizing the 10 key components discussed earlier and then comparing these factors across 18 courts. For example, the third key component recommends that "eligible participants are identified early and promptly placed in the drug court program." Among the relevant practices, they examine whether each court uses a central intake for referral (100% say yes), whether the participant enters the drug court 20 days or less after arrest (61.1% say yes), and whether the case load of the court is fewer than 150 (58.8% say yes) (2008, p. 30). They then test those practices for both cost and graduation rates. While their conclusions are extensive, they do find that courts with the most positive outcomes in both cost savings and graduation set high standards for the participation of criminal justice personnel and impose strict guidelines for treatment. Overall they identify 27 practices as "promising . . . in that they are related to avoided costs due to lower recidivism" (p. 88).

Another exploration of the future of drug courts can be found in Aubrey Fox and Robert Wolf's (2004) review of four state's efforts to institutionalize their drug courts. They ask whether drug courts in the future will resemble those that exist today and whether they should remain as separate courts or as alternative approaches available to any judge. While they do not offer opinions on these questions, they do examine the organizational arrangements of the courts, including the role of leadership and the best practices for managing the courts and assuring accountability. They conclude that "there is no magic wand" (p. 47) and that the goal should be to find what works for each jurisdiction—to balance the "ease of implementation with the need to protect the integrity of an innovative practice" (p. 48).

The successes and rapid expansion of drug courts have encouraged a growth of specialized courts in a multitude of areas, including courts that only hear cases involving businesses, courts that offer alternatives to incarceration to teenage misdemeanor offenders, and courts designed to address the effects of long-term deinstitutionalization. Domestic violence courts hold particular promise as a locus of specialization, as more traditional courts have been frustrated by the victim's unwillingness to cooperate and by the high likelihood of persistent offending. Can a specialized domestic violence court better serve the victim, the offender, and the community?

## ■ Domestic Violence Courts

### Background

Domestic violence courts are a recent addition to the criminal justice repertoire, with the first of them appearing in 1993 in Miami, Florida. As with drug courts, the impetus leading to their creation was a rapid increase in the volume of domestic violence cases, along with few available resources. The courts were feeling the pressure of the volume of cases, and judges were often dissatisfied with their lack of options and apparent inability to effect a change. The need for specialization was also felt within the community that advocated for battered women. Studies indicated that the lack of education of court personnel resulted in victim blaming, which often left the victim feeling vulnerable or even revictimized by the process. In response to these concerns, by 1998 there were reportedly close to 200 courts that identified themselves as focusing most or all of their attention on matters of family violence.

The court is tested by what might be viewed as the conflicting needs of the state—speed, effectiveness, and expediency—and the goals of the battered women's movement to empower and give the right to be heard to every individual being battered. The battered women's movement, begun in earnest in the late 1960s, has focused on promoting change. Changes sought include giving voice to women who have been battered by making them active participants in shaping a remedy, effecting a change both in society's attitudes and in behaviors toward violence against women, and moving beyond patriarchy or a model of top-down power and control. There has been a good deal of concern in the movement that the legal system itself reinforces the existing power hierarchy.

Despite the concerns, there are high hopes for the effectiveness of these new courts. The unique challenge faced by domestic violence courts is to increase efficiency while addressing the needs of both offenders (participants) and victims. Supporters suggest that creating dedicated teams of judges and prosecutors will lead to more knowledgeable personnel who can use their expertise to attend to the needs of victims even while processing the offender through the court.

Mirchandani (2005) examined a domestic violence court in Salt Lake County, Utah.[4] Four years after its inception in 1997, she observed that the technocratic goals of the court existed side by side with the goals of social change promoted by the battered women's movement. As an example, she notes, "Domestic violence court attorneys who spend most of their time either defending or prosecuting domestic violence offenders quickly understand offender behavior as fundamentally about power and control" (2005, p. 401). It is her conclusion that the efficiency of the court actually promotes change. She saw this in the judges' adoption of the rhetoric of the battered women's movement when addressing both victim and offender. She also observed that the court is more "porous," enabling the community to engage and learn from the process.

Her findings were replicated in a study of courts in Lexington County, South Carolina. In this observational study, researchers concluded that the process was collaborative, that victims were given a chance to speak "in nearly every case," and that options for responding to the defendant were discussed in an open and transparent manner. On the other hand, the observers found that judges rarely asked defendants why they had acted as they had or what might be done to reduce the possibility of future violence, thereby missing a clear opportunity to provide the offender with a stake in the case outcome (Gover, MacDonald, Alpert, & Geary, 2004).

Some research has measured qualitative outcomes such as recidivism and case pursuit (whether the case goes forward or is dropped). An evaluation of the Miami court found a lower rate of recidivism toward the same victim than was found in the control group. The chance of having a misdemeanor dismissed was 37% lower for cases processed in the domestic violence court (Goldkamp, Weiland, Collins, & White, 1998). In Lexington County, where researchers conducted an 18-month follow-up, it was found that both overall arrests and arrests for domestic violence were lower for offenders who had participated in the specialized court (Grover et al., 2004).

But the participants do not always agree with these conclusions of researchers. In a study of Florida's domestic violence courts, victims reported a "general lack of assistance" with the court process that was associated with a frequent turnover in staff. Victims also reported that they did not feel sufficiently protected or feel that their need for privacy was respected (Coulter, Alexander, & Harrison, 2005).

Among activists and other stakeholders, there are also concerns that these courts will be viewed as panaceas, even though there are numerous issues yet to be resolved. Dedicated teams run the risk of being marginalized, much as we see when examining domestic violence units within police departments.[5] Even the consistency—often identified as an advantage—can be a disadvantage when the judge or prosecutor assigned to the court is not an enthusiastic participant. And the possibility of resistance is high when the court needs to rely upon an extensive network, including court staff, prosecutors and defense lawyers, community programs, and law enforcement and judges, any of whom might not understand or be sympathetic to a victim's actions. Nor can one ignore the very real prospect of professional burnout, a hazard experienced by other trauma workers.

Dorf and Fagan (2003) note an additional concern: "Domestic violence courts . . . were established before there was any known effective treatment for batterers" (p. 25). Indeed, the research on the effectiveness of programs for batterers continues to indicate inconsistent and not very optimistic results. Dropout rates are high. For those who do not drop out, recidivism in the form of re-arrest is reduced as compared with those not completing a program. But reports by partners of escalating verbal abuse suggest that batterers have simply discovered less overt means of control. And in many locations where batterer programs do not provide adequate coverage, batterers are referred instead to anger management programs, though there is no evidence that these programs address the batterer's therapeutic needs. Does the court system's reliance upon batterer treatment programs that have not been proven very effective, or on anger management programs that are deemed unsuitable to address battering, create an insurmountable hurdle? Alternately, perhaps the successes of the domestic violence courts, however modest, can be a building block for more intervention that is both efficient and effective.

### Domestic Violence and a Coordinated Community Response

An interesting thing happened to domestic violence courts along the way. They became the central focus of a new movement toward coordinated community response to intimate partner violence (IPV), umbrella organizations that bring together domestic violence courts, prosecutors, and community advocates to provide coordinated services. Most initial efforts to address the criminal justice response to domestic abuse were aimed at the actions of police. However, if police make arrests but the court system fails to follow through, that effort may simply aggravate the situation. Models were developed to bring together court and community actors—specifically prosecutors, victim-witness advocates, and community-based services for battered women—in an effort to encourage prosecution. Early examples include the Domestic Abuse Intervention Project in Minnesota and the Alexandria Domestic Violence Intervention Project in Virginia. These projects offered new promise by creating a dialogue between community advocates and prosecutors, and effecting ongoing victim support.

This model was all but institutionalized in 1994 with the passage of the Violence Against Women Act (VAWA). STOP (Services-Training-Officers-Prosecutors), a component of VAWA, required that states applying for grants encourage collaboration between law enforcement, prosecution, and victim services. In the late 1990s, the U.S. Department of Justice designed the Judicial Oversight Demonstration (JOD) and selected three jurisdictions to serve as sites for criminal justice and community-based agency collaboration. Based upon earlier findings that success or failure of such programs was largely determined by the method of coordination, JOD placed "special focus on the role of the court, specifically the judge," to direct the collaboration and to bring accountability to the process (Harrell, Visher, Newmark, & Yahner, 2009). The model employed was not very different from other problem-solving courts, though the circle was drawn wider. After all, responses to intimate personal violence require services for victims as well as offenders; victims are conspicuously absent in drug court. Components included "coordination of court and community agencies," "specialized law enforcement, prosecution, and court procedures," "specialized probation and batterer intervention services," and "enhancement of victim services" (Visher, Newmark, & Harrell, 2008).

As with previous studies of domestic violence courts, the evaluations of JOD sites were encouraging, but largely inconclusive. Victims reported increased involvement with service agencies, greater satisfaction with the process, and a decreased level of fear than did a comparison group of victims. And while JOD increased offender compliance during their period of involvement with the court, it did not change their expectations of negative consequences for their behaviors in the future. Regarding recidivism, in two of the JOD courts, victims reported lower levels of revictimization than the comparison victims, while in the third court there was no significant difference. Based upon offender self-reporting, there was no difference between the JOD offender and the comparison offender in either frequency or severity of violence.

The evaluations of the JOD projects find that the most successful programs have a diverse group of justice practitioners and community agencies. Building a well-managed group, where all partners play a role in articulating polices and practices, creates the most promising collaboration for addressing domestic violence.

## ■ Discussion of the Policy Questions

As discussed earlier, the three-tiered policy analysis refers to (1) systemic issues regarding the overall efficiency of these new courts, (2) societal/legal issues about the proper role of the court in addressing social problems, and (3) micro-issues concerning best practices. Efficiency, or cost saving, and best practices are addressed in the literature we have been reviewing and, as such, will be discussed here summarily. Some of the policy questions concerning the stance of the courtroom participants are presented in greater depth.

### Systemic: Do Problem-Solving Courts Reduce Costs?

The first policy concern with any new endeavor is whether it increases efficiency. In this instance, increased efficiency would mean a decrease in the time and money spent on processing each individual as well as a decrease in recidivism so that the individual would not cost the system in the future. The studies of drug courts have tested these questions by providing data on rates of successful program completion, rates of recidivism after program completion, and costs associated with participation in the program versus traditional processing and punishment. Even in the domestic violence courts there is some evidence that future arrests are reduced through program participation. Although the positive outcomes reflected in these studies are modest, they do represent an improvement in court efficiency.

## Societal: Can Courts Solve Problems?

Courts have historically been the venue that provided problem resolution through adversarial litigation. Can they now become what is, in effect, a social service provider? How will this new responsibility impact the courts' longstanding job of conflict resolution? Can the courts retain the fundamental goals of due process and proportionality when faced with the prospect of overseeing treatment? These questions are largely ignored in the literature, perhaps because they are difficult to quantify. Because they are both interesting and important considerations, a few of the more salient issues will be elaborated upon.

### Jurisdiction

Jurisdiction refers to a court's authority—that is, which cases it can consider. Judge Hoffman warns that these new courts may result in even more people under the authority of the legal system. Commenting on the drug court over which he presides, he observes, "There was massive net widening in Denver. What was happening was the usual constraints on arrest discretion, on prosecutorial charging discretion, were gone because police and prosecutors were not out searching for criminals, they were out trolling for clients" (2002, p. 1796). Net widening is the idea that, rather than reducing participation in the criminal justice system, problem-solving courts will actually increase the participation, perhaps by extending the reach of the criminal justice system to include less serious offenders.

On the other hand, drug courts, especially those funded by the 1994 Crime Act, may process only nonviolent offenders (Gebelein, 2000, p. 6). An offender who is charged with drug possession and assault is not eligible to enter that drug court. The case remains in the traditional court setting where available services are finite and the court is overwhelmed with cases. We do not know that this offender would be any less amenable to rehabilitation, but he or she is not likely to be offered much beyond Alcoholics Anonymous or Narcotics Anonymous. This narrow construction of eligibility serves a gatekeeping function, allowing access to only a certain type of offender. Can we support a legal system that favors some with attention, therapy, and services that are not available to others?

### Adversarialism and Due Process

Our court system is based upon a concept of adversarialism that posits that we find the truth by having two sides argue before a deciding body. In this model, it is the role of the prosecutor and defense attorney to argue and the role of a judge to oversee the process. Problem-solving courts change all of that. The lawyers become a "team" trying to cure a problem; the judge becomes the caring parent, stern but supportive. This shift raises a variety of concerns for the judges, the attorneys, and the offenders, only a few of which are raised here.

The judge needs to be an active participant in the court. For some, who are more accustomed to the role of referee, this new role is demanding. And what if there are not enough interested judges? Without a highly motivated judge, the drug court approach simply does not work well (Gebelein, 2000). To effectively participate, judges must agree to be further educated. Clearly not all judges will desire this new role with the attendant responsibilities.

Prosecutors are not usually in the business of advocating for offenders, and moving to effective advocacy could be challenging. There is additional concern that prosecutors will use these courts as places to "dump" weak cases, cases that would not otherwise proceed to prosecution. This again raises the possibility of net widening but also challenges the role of defense counsel. Defense counsel is required by legal ethics to provide clients with a vigorous defense. Problem-solving courts are premised on the defendants entering a plea of guilty. According to Quinn (2000–2001), "without knowing with certainty all that a treatment modality and regime of sanctioning entails, the attorney may not be able to assist meaningfully a . . . client considering a guilty plea" (p. 55). After the plea is entered, the role of defense counsel in problem-solving courts is unclear. The conditions of participation are prescribed, the consequences for offenders are largely predetermined, and counsel is often not provided with progress reports until they enter the courtroom; all are factors that make it difficult to mount an effective defense. The underlying policy question here is whether offenders are being required to compromise their due process rights in exchange for needed treatment.

### Provision of Services

At the core of problem-solving courts is the agreement to offer a remedy to a problem—e.g., to cure a drug addict of addiction. Once the court assumes the responsibility of providing a remedy, it likewise assumes an obligation to find services that are available and effective. The lack of available services has been a problem for those seeking drug treatment, where extensive waiting lists have been the rule. And as was discussed previously, there continues to be serious concerns about the effectiveness of batterer treatment programs. These examples merely illustrate the problem, but the question is much broader—what standard of treatment is the court obligated to provide? One side of the argument suggests that any services are likely to address the problem better than no services, and perhaps this is true. But is that standard satisfactory?

## Best Practice

Given what we have learned about drug courts and domestic violence courts, there is emerging evidence that some models and practices produce better results. Outstanding practices include having strong and clear team leadership, creating an extensive network of support by involving a diverse group in the work of the team, and having well-articulated rules and regulations for participants, especially offenders. Effective leadership is important to the individual court team, but it is also important to establish state-level management. Central administration has access to resources and oversight not present at the local level. For example, in New York State the Office of Court Drug Treatment Programs has created a management information system and assessment tools to be used by all drug courts in the state (Fox & Wolf, 2004).

A National Institute of Justice special report entitled "Drug Courts: The Second Decade" describes some of the recommendations arising from a review of drug court research. Included among these recommendations are two that affirm the importance of providing clear guidance to program participants. They suggest that courts adhere "to a balanced system of sanctions and rewards that can be applied consistently and appropriately in response to participant behaviors" and "carefully defin[e] the conditions that must be met in each step of a treatment phase before a participant can move on to the next one" (2006, p. 2). Clarifying required action makes accountability possible, and accountability is an essential component of problem solving, whether we are talking about restorative justice (Zehr, 2002), therapeutic jurisprudence (Winick & Wexler, 2003), or problem-solving courts (Casey & Rottman, 2003).

## ■ Conclusion

Problem-solving courts have been participants in the criminal justice system for barely 20 years. As with other innovations such as community policing or intensive probation, these courts will continue to take shape in the coming years. By studying the policy concerns that are now facing the courts, we can play a role in determining the form and function that the mature courts will play in the criminal justice system of the future.

## ■ Endnotes

1. One notable exception is New York State, where the court of general jurisdiction is the supreme court.
2. Other movements were introduced at the same time, primarily aimed at altering the experience of the lawyer rather than that of the client (recognizing that one necessarily impacts the other). These movements include collaborative lawyering, holistic lawyering, and finding spirituality in lawyering. For a good discussion about the overall motivations for these movements, see McArdle (2004).
3. For more information of restorative justice, see Restorative Justice Online (2009), available at www.restorativejustice.org.
4. Simply as a matter of interest for students of criminology, it should be noted that most studies of drug courts are looking at recidivism and costs, and rely principally upon quantitative evaluations to make these determinations, while studies of domestic violence courts are generally concerned with the perceptions of the participants and rely more heavily on qualitative data to evaluate the outcomes.

5. See, for example, Krumholz, S. T. (2002). Policing domestic violence. In D. M. Robbins (Ed.), *Policing and crime prevention*, ed. Upper Saddle River, NJ: Prentice Hall.

## ■ Discussion Questions ▬▬▬▬▬▬▬▬▬▬▬▬▬▬▬▬▬▬

1. Discuss the impetus behind the creation and use of problem-solving courts in the criminal justice system.
2. Compare and contrast problem-solving courts with traditional criminal courts.
3. Provide and discuss examples of successful problem-solving courts? What factors are important to consider when making this determination?
4. What policy issues must court officials consider before developing and implementing problem-solving courts?
5. How can research help to determine whether problem-solving courts present a viable alternative to traditional criminal courts?

## ■ References ▬▬▬▬▬▬▬▬▬▬▬▬▬▬▬▬▬▬▬▬▬▬

Berman, G. (2004). Redefining criminal courts: Problem-solving and the meaning of justice. *The American Criminal Law Review, 41*(3), 1313–1319.

Bureau of Justice Assistance Drug Court Clearinghouse Project. (2008). *Summary of drug court activity by state and county* (data file; Justice Programs Office). Retrieved March 18, 2009, from http://spa.american.edu/justice

Carey, S. M., Pukstas, K., Waller, M. S., Mackin, R. J., & Finigan, M. W. (2008). *Drug courts and state mandated drug treatment programs: Outcomes, costs, and consequences* (No. NCJ 223975). Washington, DC: U.S. Department of Justice, National Institute of Justice.

Casey, P. M., & Rottman, D. B. (2003). *Problem-solving courts: Models and trends*. Retrieved February 14, 2010, from National Center for State Courts Web site: http://www.ncsconline.org/D_Research/publications.html

Coulter, M. L., Alexander, A., & Harrison, V. (2005). Specialized domestic violence courts: Improvement for women victims? *Women & Criminal Justice, 16*(3), 91–106.

Daicoff, S., & Wexler, B. (2003). Therapeutic jurisprudence. In A. Goldstein (Ed.), *Handbook of Psychology: Forensic Psychology, 11*, 561–580.

Dorf, M. C., & Fagan, J. (2003). Problem-solving courts: From innovation to institutionalization. *American Criminal Justice Review, 40*(14), 1501–1512.

Fox, A., & Wolf, R. V. (2004). *The future of drug courts: How states are mainstreaming the drug court model* (Think Piece). New York: The Center for Court Innovation.

Gebelein, R. S. (2000). *Rebirth of rehabilitation: Promise and perils of drug courts* (Research in Brief No. NCJ 211081). Washington, DC: U.S. Department of Justice, National Institute of Justice.

Goldkamp, J. S., Weiland, D., Collins, M., & White, M. D. (1998). The role of drug and alcohol abuse in domestic violence and its treatment: Dade County's domestic violence court experiment. In *Legal interventions in family violence: Research finding and police implications.* Washington, DC: National Institute of Justice, Office of Justice Programs.

Goldkamp, J. S., White, M. D., & Robinson, J. B. (2002). *Retrospective evaluation of two pioneering drug courts: Phase I findings from Clark County, Nevada and Multnomah County, Oregon* (No. 197055). Washington, DC: U.S. Department of Justice.

Gover, A. R., MacDonald, J. M., Alpert, G. P., & Geary, I. A. (2004). *Lexington County Domestic Violence Court: A partnership and evaluation* (No. NCJ 204023). Washington, DC: U.S. Department of Justice, National Institute of Justice.

Harrell, A. V., Visher, C. A., Newmark, L. C., & Yahner, J. (2009). *The judicial oversight demonstration: Culminating report on the evaluation* (No. NCJ 224201). Washington, DC: U.S. Department of Justice, National Institute of Justice.

Hoffman, M. B. (2002). Therapeutic jurisprudence, neo-rehabilitationism, and judicial collectivism: The least dangerous branch becomes most dangerous. *Fordham Urban Law Journal, XXIX*(5), 2063–2098.

McArdle, E. (January 11, 2004). From Ballistic to Holistic. Retrieved February 14, 2010, from *Boston Globe* Web site: http://www.boston.com/news/globe/magazine/articles/2004/01/11/from_ballistic_to_holistic

McCoy, C. (2003). The politics of problem-solving: An overview of the origins and development of therapeutic courts. *The American Criminal Law Review, 40*(4), 1513–1534.

Minnesota Judicial Branch. (2009). Problem-solving courts. Retrieved February 14, 2010, from http://www.courts.state.mn.us/?page=626

Mirchandani, R. (2005). What's so special about specialized courts? The state and social change in Salt Lake City's domestic violence court. *Law & Society Review, 39*(2), 379–417.

National Association of Drug Court Professionals. (1997). *Defining drug courts: The key components* (Bureau of Justice Assistance). Retrieved February 14, 2010, from http://www.ojp.usdoj.gov/BJA/grant/DrugCourts/DefiningDC.pdf

National Institute of Justice. (2006). *Drug courts: The second decade* (Special Report No. NCJ 211081). Washington, DC: U.S. Department of Justice, Office of Justice Programs.

Nolan, J. L. J. (2003). Redefining criminal courts: Problem-solving and the meaning of justice. *The American Criminal Justice Review, 40*(14), 1541–1565.

NPC Research. (2007). *Indiana drug courts: A summary of evaluation findings in five adult programs*. Retrieved February 14, 2010, from http://www.in.gov/judiciary/pscourts/docs/eval-summary.pdf

Office of National Drug Control Policy. (n.d.). Drug courts. Retrieved February 14, 2010, from http://www.whitehousedrugpolicy.gov/enforce/DrugCourt.html

Quinn, M. C. (2000–2001). Whose team am I on anyway? Musings of a public defender about drug treatment court practice. *NYU Review of Law & Social Change, 26*(1–2), 37–75.

Rempel, M., Fox-Kralstein, D., Cissner, A., Cohen, R., Labriola, M., Farole, D., et al. (2003). *The New York State drug court evaluation: Policies, participants and impacts* (New York: Center for Court Innovation). Retrieved February 14, 2010, from http://www.courts.state.ny.us/whatsnew/pdf/NYSAdultDrugCourtEvaluation.pdf

Roderick, D., & Krumholz, S. T. (2006). Much ado about nothing? A critical examination of therapeutic jurisprudence. *Southern New England Roundtable Symposium Law Journal, 1*, 201–223.

Slobogin, C. (1995). Therapeutic jurisprudence: Five dilemmas to ponder. *Psychology, Public Policy & Law, 1*, 193.

U.S. Department of Justice. (2007). *Stewards of the American Dream: FY 2007–FY 2012 Strategic Plan* (p. 90). Retrieved February 14, 2010, from http://www.justice.gov/jmd/mps/strategic2007-2012/goal3.pdf

Visher, C. A., Newmark, L. C., & Harrell, A. V. (2008). *The evaluation of the judicial oversight demonstration: Findings and lessons on implementation* (No. NCJ 219077). Washington, DC: U.S. Department of Justice, Office of Justice Programs.

Winick, B. J., & Wexler, D. B. (Eds.). (2003). *Judging in a therapeutic key: Therapeutic jurisprudence and the courts*. Durham, NC: Carolina Academic Press.

Wolf, R. V. (2007). *Principles of problem-solving justice*. New York: Center for Court Innovation. Retrieved March 11, 2010, from http://www.courtinnovation.org/_uploads/documents/Principles.pdf

Zehr, H. (2002). *The little book of restorative justice*. Intercourse, PA: Good Books.

# U.S. Corrections Policy Since the 1970s

# CHAPTER

# 7

Steven E. Barkan

## ■ Introduction

In February 2009, a panel of three federal judges ordered California to reduce the size of its prison population by one-third—representing a decrease of 55,000 inmates—by 2012 to relieve drastic overcrowding in the state's prisons. At the time of the ruling, more than 150,000 inmates were being held in California's 33 prisons, which were built to hold only 84,000 inmates. Noting the severe overcrowding in the prisons and the lack of adequate medical and mental health care for so many inmates, the judges concluded that California was violating the inmates' Eighth Amendment protection against cruel and unusual punishment, as dozens of prisoners had died during the previous decade from suicide or preventable illness. The judges approximated that the state would save between $800 million and $900 million annually by reducing its prison population and advised that these savings could be used to enhance services for offenders given nonprison sentences and for others released from prison on parole (Moore, 2009a). They gave the state time to negotiate a settlement with inmates' attorneys before the ruling would take effect.

Just days before the judges issued this ruling, a California state court upheld a life sentence for a defendant convicted of possessing 0.03 gram of methamphetamine. It was the defendant's third felony conviction, and he had been sentenced to life under California's "three-strikes" law mandating sentences of 25 years to life for offenders convicted of a third felony. Critics blamed the law for contributing to the state's prison overcrowding crisis (Moore, 2009b).

These developments came on the heels of the nation's economic downturn that became a financial crisis beginning in 2008. To try to deal with their economic problems, many states took a hard look at the high cost of their prisons and began to institute various kinds of changes to reduce their prisons' populations. According to The Sentencing Project, 17 states in 2008 enacted various reforms to achieve this goal, including: (1) reducing the number of parolees returned to prison for technical violations of the rules for their parole (for example, missing a drug test); (2) expanding nonincarceration alternatives, such as home confinement, for certain nonviolent offenders; and (3) expanding the use of drug courts, which require substance abuse counseling and other services in lieu of imprisonment for offenders convicted of drug offenses (King, 2009).

All these initiatives are a response to the huge increase in incarceration that has been the hallmark of U.S. corrections policy since the 1970s. How and why did California and most other states end up with prisons overflowing with inmates and billions of dollars in annual expenses? What have been the economic and noneconomic costs of the incarceration boom of the last few decades? What impact has the incarceration boom had on crime rates?

In trying to answer these questions, this chapter takes a critical stance that can be summarized as follows: First, the incarceration boom has cost the nation tens of billions of dollars that have been diverted from other governmental needs, including various programs and policies that would almost certainly offer a more cost-effective strategy for crime reduction. Second, rising incarceration is not primarily the result of an increase in crime; rather, its origins lie largely in inflammatory pronouncements by politicians and like coverage by the news media, both of which led to changes in sentencing policy that increased incarceration. Third, the huge increase in incarceration has had only a minimal impact, if that, on crime rates. And fourth, it has had several negative, "collateral" consequences that have increased racial inequality and may exacerbate the crime problem in years to come. These are certainly harsh charges, but each reflects a wide variety of research studies by criminologists and criminal justice scholars.

## ■ The Incarceration Boom

From the 1930s until the early 1970s, the United States had a fairly stable incarceration rate (defined as the number of prisoners per 100,000 members of the U.S. population). The rate varied only slightly from year to year and averaged about 110 inmates per 100,000 people for any given year. This rate was in the "high middle" of Western nations, meaning it was higher than the rates in many nations but lower than those in some other nations (Tonry, 2008b). During the 1960s, the U.S. incarceration rate fell from 117 at the start of the decade to 97 at the end of the decade even though reported crime increased throughout the decade. The stable and then declining incarceration rate during the pre-1970s period reflected an American commitment to what has been called the "rehabilitation ideal," the belief that many criminal offenders can be reformed with appropriate correctional practice, including vocational training for inmates and nonincarceration alternatives such as probation for first-time and nonviolent offenders and others viewed as capable of being rehabilitated (Cullen, 2007). Scholars even wrote that the general stability of the incarceration rate was the almost inevitable outcome of certain social processes contributing to social integration (Blumstein & Cohen, 1973).

Beginning in the early 1970s and through the 2000s, the number of prisoners and thus the incarceration rate soared, from about 196,000 inmates and a rate of 93 in 1972 to about 1.6 million inmates and a rate of 506 in 2007 (West & Sabol, 2008) (see **Figure 7–1**). The incarceration rate thus increased by about 700% during this 35-year period. The number of jail inmates also surged, more than quadrupling from about 183,000 in 1980 to almost 781,000 in 2007. Adding the numbers of prison and jail inmates together, almost 2.4 million offenders are now under local, state, or federal custody. Not surprisingly, the cost of corrections also soared, rising from $6.9 billion for the state, federal, and local governments in 1980 to almost $69 billion in 2006 (see **Figure 7–2**), an increase greatly outstripping inflation.

Today more than 1% of American adults are behind bars, and more than 3% are under correctional control (in prison or jail, or on probation or parole) (Warren, 2009). The U.S. incarceration rate for prison and jail inmates combined, 762 per 100,000 residents in 2007, is the highest in the world and by far the highest of all Western nations (see **Figure 7–3**); the U.S. rate is about eight times higher

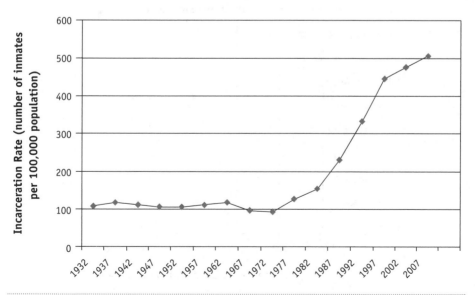

**Figure 7–1** Incarceration Rates, 1932–2007 (number of prisoners per 100,000 population)
**Source:** Adapted from Pastore, Ann L. and Kathleen Maguire, eds. *Sourcebook of Criminal Justice Statistics,* table 6.28.2007 [Online]. Available: http://www.albany.edu/sourcebook [accessed February 2010].

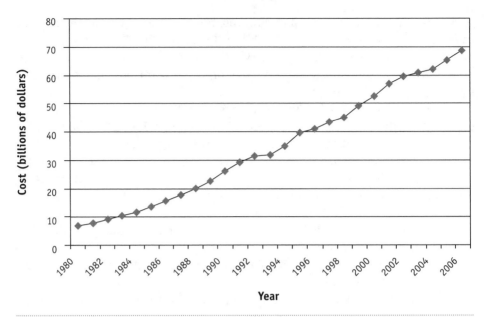

**Figure 7–2** The Cost of Corrections for State, Federal, and Local Governments, 1980–2006 (in billions of dollars)
**Source:** Adapted from Pastore, Ann L. and Kathleen Maguire, eds. *Sourcebook of Criminal Justice Statistics*, table 1.2.2006 [Online]. Available: http://www.albany.edu/sourcebook [accessed February 2010].

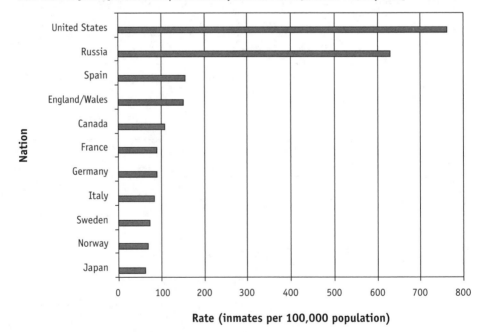

**Figure 7–3** International Incarceration Rates (number of inmates per 100,000 population)
**Source:** Adapted from King's College London International Centre for Prison Studies, *World Prison Brief.*
Available: http://www.kcl.ac.uk/depsta/law/research/icps/worldbrief/.

than all of western Europe combined (Tonry, 2008b). The number of prisoners is so great that some 700,000 are now being released annually back to their communities after serving their sentences. About two-thirds of these ex-inmates can be expected to be rearrested within 3 years and more than half can be expected to return to prison for committing new offenses and/or violating terms of parole (Jacobson, 2005).

Many scholars have criticized the surge in incarceration, often called the "incarceration boom" (Jacobson, 2005) or "imprisonment binge" (Austin & Irwin, 2001), since the 1970s. One critic says the United States has experienced a reckless "race to incarcerate" (Mauer, 2006), while another says the United States is "addicted to incarceration" (Pratt, 2008). These and other scholars identify many problems stemming from the incarceration boom that later sections of this chapter will discuss. Before turning to these problems, we review the evidence on the reasons for the incarceration boom, then address whether this boom has substantially reduced the crime rate in a cost-effective manner.

## ■ Why Has the Incarceration Boom Occurred?

Recall that the U.S. incarceration rate was fairly stable from the 1930s to the early 1970s and then began to soar through the first decade of the new century. This boom resulted from a change in penal practices, as the rehabilitation ideal gave way to a "get-tough" approach involving, among other things, mandatory imprisonment, mandatory minimum sentences, and longer prison terms for offenders who went to prison (Pratt, 2008; Raphael, 2009; Spelman, 2009). But why did the get-tough approach underlying the incarceration boom arise in the first place, and why has incarceration continued to increase every year since the boom started?

Criminologists have discussed several possible reasons for the change in penal practices underlying the onset and continuation of the incarceration boom, including (1) a possible rise in the crime rate, (2) a possible rise in public concern and fear about crime, and (3) alarmist statements by politicians and other governmental officials about crime. Each explanation will be considered in turn.

### A Rise in the Crime Rate?

For many Americans, it would make sense to think that a steep rise in crime must have initiated the incarceration boom and prompted its continuation. Why else would incarceration have begun to rise and continued to rise? However, data on U.S. crime rate trends provide only mixed or perhaps no support for the assumption that rising crime inevitably produces rising incarceration (Beckett & Sasson, 2004; Western, 2006), even if it is true that state crime rates predict state incarceration rates (Spelman, 2009).

For example, although incarceration has risen every year since the early 1970s, reported crime (homicide, rape, robbery, aggravated assault, burglary, larceny, motor vehicle theft, and arson, as measured by the FBI from police reports) has *not* steadily risen during this time. It rose during the 1960s and 1970s, fell in the early 1980s, rose during the late 1980s and early 1990s, and then fell sharply after the early 1990s before leveling off during the past few years. Thus, incarceration fell during the 1960s even though reported crime rose; incarceration rose during the 1970s as reported crime rose; incarceration rose during the early 1980s even though reported crime fell; incarceration rose during the late 1980s and early 1990s as reported crime rose; and incarceration continued to rise after the early 1990s even though reported crime fell sharply.

As should be apparent from these patterns, no consistent linkage emerges between rising crime and rising incarceration, leading Raphael (2009) to conclude that "increases in crime cannot explain a substantial portion of incarceration growth" (p. 91). However, there is some evidence that the U.S. crime rise of the 1970s did help produce the incarceration boom that began that decade, even if crime rates were not always related to incarceration rates during the other periods just examined. Western (2006) found that states (primarily Southern states) with the highest increases in their homicide rates between 1965 and 1980 tended to have the highest increases in their incarceration rates between 1980 and 2000, and he notes that reported violent crime of all types tripled between 1965 and 1980. Thus, he says, "the large growth in crime rates predated the explosion of the penal population" (p. 48), and he argues that the rise in violent crime set a political context for incarceration to surge: "The growth in crime contributed to new feelings of vulnerability among the affluent and created a political opening for a change in crime policy that ultimately increased the incarceration rate" (p. 48). For the post-1980 period, however, Western agrees that the growth of incarceration was not due to growth in crime rates, because, as we have seen, incarceration continued to grow even though crime dropped during the early 1980s and then again after the early 1990s.

Comparative evidence reinforces the idea that rising crime does not have to produce rising incarceration. Although crime rates rose in most other Western nations from the 1960s through the early to mid-1990s, the incarceration rates in several nations did not rise, and in a few nations these rates even fell (Lappi-Seppälä, 2008). As Tonry (2007) concludes from this evidence, "Any assumption or hypothesis, therefore, that there is a simple, common, or invariant relationship between crime patterns that befall a country and the number of people it confines is wrong. Faced with similar crime trends, different countries react in different ways" (p. 3). In the United States, the fact that incarceration fell during the 1960s even though reported crime rose again indicates that rising incarceration is not an inevitable consequence of rising crime and that rising crime does not inevitably produce rising incarceration.

Looking at the entire evidence on crime and incarceration trends, Raymond Michalowski (2006) offers a conclusion with which most criminologists would probably agree: "the research is fairly clear that changes in penal practices are not a direct reflection of increases in rates of crime" (p. 189). Beckett and Sasson (2004) similarly conclude, "The expansion of prisons and jails is not a result of a worsening crime problem. . . . The massive expansion of the criminal justice system has not been primarily a consequence of rising levels of crime" (pp. 17–18). Spelman (2009), who found state crime rates linked to their incarceration rates, nonetheless concluded that "nothing was inevitable about prison buildup in the United States," as crime could have been "dealt with through alternative means" (p. 73). What matters more than changes in crime rates, then, are changes in how a nation chooses to respond to whatever crime it has (Raphael & Stoll, 2009). In the United States, the continually rising incarceration rate since the early 1970s cannot be said to reflect rising crime rates.

### A Rise in Public Concern and Fear About Crime?

Even if rising incarceration in the United States cannot be linked to rising crime rates, it may still have stemmed from growing public concern and fear about crime. Perhaps lawmakers initiated the penal practices that produced greater incarceration because the public has become more and more concerned about crime and thus increasingly punitive toward criminal offenders since the 1970s. Beckett and Sasson (2004) summarize this argument as follows:

> One version of the conventional wisdom on the expansion of the penal system goes like this: Whatever crime rates may be doing, the American public has become more fearful of crime and increasingly punitive in its policy preferences. Although these sentiments are not necessarily a response to trends in criminal activity, they are widespread and are the driving force behind the war on crime and drugs. (p. 103)

However, they then conclude that evidence "casts serious doubt on the claim that the war on crime is primarily a response to the spread and dominance of such sentiments" (pp. 103–104).

Several kinds of evidence led them to this conclusion. For example, fear of crime did rise from about 32% in 1968 to 45% by 1975, but it then leveled off before falling after the mid-1990s. (Fear was measured by the standard polling question, "Is there any area right around here—that is, within a mile—where you would be afraid to walk alone at night?") Although the initial phase of the incarceration boom coincided with the rise in fear of crime before 1975, the boom continued long after fear of crime stabilized. Perhaps more telling, another measure of concern about crime, the percentage of people who say they feel unsafe in their homes at night, remained stable through the 1970s at less than 20% before declining to 10% by 1989 and staying at that level through 2000. The incarceration boom thus began after 1973 even though public concern about crime as reflected in this measure did not rise, and the boom accelerated even though the percentage of people feeling unsafe did not change and then even declined.

There is one "kernel of truth" (Bohm & Walker, 2006) in the conventional wisdom that the public became more punitive and that the incarceration boom reflected this increase in punitiveness. Two common measures of punitiveness are support for the death penalty and the belief that courts in one's local area are not harsh enough. Support for the death penalty rose fairly steadily from about 42% in 1968 to 80% by 1989 before declining after the mid-1990s as evidence grew of mistakes in the convictions of capital defendants (Unnever & Cullen, 2005). Similarly, the belief that local courts are not harsh enough also rose, climbing from 49% in 1965 to almost 75% in 1969 to more than 80% beginning in 1976, where it held through the mid-1990s until declining to 64% by 2006 (Maguire & Pastore, 2009). Although punitiveness as reflected by both measures did increase in the 1970s as the incarceration boom began, the boom continued after the mid-1990s even though punitiveness as reflected by both measures declined. Because incarceration continued to rise even after punitiveness declined, the relationship between punitiveness and incarceration remains unclear, and the exact role played by the rising punitiveness of the 1970s and 1980s in the onset of the incarceration boom also remains unclear.

### Public Officials, the News Media, and the Incarceration Boom

We have seen that the onset of the incarceration boom cannot be attributed to a rising crime rate and that it also probably cannot be attributed, at least to any strong degree, to changes in public concern about crime or in public punitiveness. Why, then, did the incarceration boom occur?

Many scholars trace the boom to alarmist statements by public officials and corresponding coverage by the news media. In their view, public officials began in the 1960s and 1970s to make crime a problem by emphasizing the danger of violent crime and by talking about violent crime in racially coded language suggesting that African Americans were becoming more violent and represented a growing threat to whites. The news media readily reported these officials' statements and speeches, and they also gave disproportionate attention to violent crime generally and particularly to violent crime committed by African Americans against whites. To the extent the public did become more concerned about crime, this rise in concern stemmed from the alarmist comments from public officials and from the inflammatory coverage by the news media (Beckett & Sasson, 2004; Chambliss, 1999; Gottschalk, 2006; Michalowski, 2006; Pratt, 2008; Simon, 2007; Western, 2006). In the 1980s, attention paid by public officials and the media to crack cocaine and other drugs was again racially tinged, and it again inflamed public opinion and prompted a harsh response by the criminal justice system (Mauer, 2006; Provine, 2007).

The alarmist statements by public officials in the 1960s and 1970s originated with Republican Party officials, who attempted to exploit the fears of whites across the nation in the aftermath of the 1960s urban riots and especially of Southern whites angered by the success of the civil rights movement in their region. These officials blamed the civil rights movement for undermining law and order and for contributing to a mindset that fostered criminal behavior. This threat to law and order was a key issue in Republican Barry Goldwater's 1964 presidential campaign, and he declared in his acceptance speech at the Republican Party convention that "history shows us that nothing prepares the way for tyranny more than the failure of public officials to keep the streets safe from bullies and marauders" (quoted in Beckett & Sasson, 2004, p. 49).

Within a few years, statements by Republican officials about crime took on a racial tone. Under the leadership of Richard Nixon during his 1968 presidential campaign, the Republican Party began its so-called Southern strategy by appealing to Southern white Democrats angered by the help that Democratic President Lyndon B. Johnson had given to the cause of civil rights. A major part of this appeal involved inflammatory statements that implicitly linked violent crime to African Americans. The Republican Party later used such statements to help win the votes of white Catholics and blue-collar workers in the North. As John Ehrlichmann, Special Counsel to President Nixon, later wrote of the 1968 campaign, "We'll go after the racists. That subliminal appeal to the anti-African-American voter was always present in Nixon's statements and speeches" (quoted in Beckett & Sasson, 2004, p. 54). Ronald Reagan echoed the Republican Party's emphasis on law and order when he successfully ran for president in 1980. Perhaps not surprisingly, research shows that the highest increases in incarceration rates between 1980 and 2000 tended to occur in states with Republican governors (Western, 2006).

As this brief historical survey should indicate, public officials' statements and media coverage were the cause, and not the result, of public concern about crime. Beckett and Sasson (2004) summarize this argument as follows:

> Crime-related issues rise to the top of the popular agenda in response to political and media activity around crime—not the other way around. By focusing on violent crime perpetrated by racial minorities, . . . politicians and the news media have amplified and intensified popular fear and punitiveness. . . . Americans have become most alarmed about crime and drugs on those occasions when national political leaders and, by extension, the mass media have spotlighted these issues. (pp. 104, 128)

By extension, the onset and continuation of the get-tough approach and incarceration boom are best also seen as the result of officials' statements and media coverage about crime, with the racial underpinnings of these statements and coverage playing no small role in why the get-tough approach proved so popular.

## ■ The Incarceration Boom and the Crime Rate

Regardless of why the incarceration boom occurred, has it lowered the crime rate? Before considering the answer, it is helpful to reframe the question. Rather than asking simply whether the incarceration boom has lowered the crime rate, we need to ask whether it has lowered the crime rate significantly and cost-effectively. Because mass incarceration is so expensive, we need to justify the use of tax dollars, especially during the recent dire economic period, to keep so many people behind bars. If mass incarceration has lowered the crime rate but only to a small degree, and if this small crime-reduction impact has cost billions of dollars that might produce larger crime reduction if spent in other areas (for example, on parenting training), then our tax dollars have not been spent wisely all these years. On the other hand, if mass incarceration has produced great reductions in the crime rate, then our tax dollars have perhaps been spent wisely.

The following discussion begins with the two principal goals of the get-tough approach, deterrence and incapacitation, and examines several issues that indicate why deterrence and incapacitation may be difficult for mass incarceration to achieve.

### Deterrence

One goal of the get-tough approach and corresponding incarceration boom has been to send a signal to potential criminal offenders that they will be punished with imprisonment (instead of just a fine or probation) and, once imprisoned, with a longer prison term than would have been expected before the 1970s when the rehabilitation ideal prevailed. Thus, as noted earlier, the get-tough approach has involved the use of mandatory minimum prison terms and longer terms than before for many types of crimes, including terms of 25 years to life for someone convicted of a third (or in some states, second) felony (i.e., the "three-strikes" laws) (Austin et al., 2007; Pratt, 2008). A primary stated or implicit goal of sentencing reforms like these has been to discourage potential criminals from committing a crime for fear that they will be locked up for a lengthy period.

If potential offenders are so discouraged, then sentencing and incarceration have a deterrent impact. This impact is called *general deterrence*. The belief that harsher sentencing may deter potential offenders rests on the assumption that potential offenders act rationally by carefully weighing the possible benefits and costs of their actions before deciding whether to commit a crime. However, although there is certainly good reason to believe that the threat of incarceration should deter much crime for this reason, most studies conclude that the general deterrent effect of incarceration for most types of crime is, in fact, very small or even nonexistent (Tonry, 2008a). Several reasons appear to explain these results.

First, although many people, criminals and noncriminals alike, do act rationally, studies of criminals find that these individuals often do *not* carefully weigh the possible benefits and costs of their potential offending before they decide to commit a crime. Why do they not weigh the benefits and costs? Consider violent crime. Much violent crime, particularly homicide and assault, is committed for intense, emotional reasons—e.g., anger or jealousy. When people act on emotion, they ordinarily do not take the time to think about the consequences of their actions and instead lash out because they simply want to hurt the person who angered them. Violent crime of this type cannot be expected to be deterred by harsher sanctions; offenders are acting too emotionally to consider the sanctions.

Second, studies find that robbers, burglars, and other offenders who plan their crimes without undue emotion naturally do so in ways that minimize their chances of getting caught and thus do not think they will be caught (Decker, Wright, & Logie, 1993; Tunnell 1996). Other offenders have a more fatalistic attitude and think they will eventually get caught and commit their crimes with this expectation. And some simply do not think about getting caught one way or the other and just commit their crimes impulsively (Robinson & Darley, 2004). For any of these reasons, many (and perhaps most) offenders are unlikely to be deterred by the prospect of a harsh sentence, either because they do not think they will be caught, because they think they will be caught but resign themselves to this fate, or because they just do not think about what might happen at all.

Third, many offenders are using alcohol and/or drugs at the time of their offense. About one-third to one-half of prisoners report being under the influence of alcohol and/or drugs when they committed the offense that led to their incarceration (Beck, 2000; Mumola & Karberg, 2006). In this condition, it is doubtful that they could have carefully weighed the benefits and costs of their behavior, and in particular their chances of being arrested and receiving a harsh sentence, before committing their crime. These offenders, too, are unlikely to be deterred by the prospect of a harsh sentence.

Fourth, a major goal of increasing the severity of sentences for certain crimes has been to deter potential offenders from committing these crimes. Yet most offenders have little, if any, advance knowledge of the average or maximum sentence for the crimes they commit (Kleck, Sever, Li, & Gertz, 2005; Western, 2006). If they do not know the penalties for a crime they are considering committing, they are unlikely to be deterred by these penalties

Finally, the threat of imprisonment may become less stigmatizing as so many people in a community go to prison; it may seem common and akin to a rite of passage. If the stigma of incarceration declines, any deterrent effect that incarceration might have had also declines (Kovandzic & Vieraitis, 2006). Ironically, then, the incarceration boom may have reduced any deterrent effect of imprisonment, which was likely to be low to begin with.

For all these reasons, then, it would be somewhat surprising if the prospect of going to prison for a lengthy period had a significant general deterrent effect on potential offenders, especially in an era of unprecedented mass imprisonment, and a recent review of the evidence concludes that sentence severity has no such deterrent effect (Doob & Webster, 2003). We will return to this issue in discussing the evidence of the effect of the incarceration boom on the crime rate.

### Incapacitation

*Incapacitation* is another goal of the get-tough approach and refers to the inability of individuals behind bars to commit crimes against members of the public. When scholars study the crime-reduction impact of the incarceration boom, they usually assume that any impact they find is due more to incapacitation than to general deterrence. At first glance, it makes sense to think that many crimes must be prevented if hundreds of thousands of offenders are incarcerated and that the crime-reduction impact of mass incarceration must be fairly large. However, once again there are several reasons for believing that this impact must be fairly small.

One problem is that in any given year, most crimes do *not* result in anyone being imprisoned. For example, in 2004, the latest year for which data were available at the time of this writing, about 20.6 million felonies (i.e., homicide, rape, aggravated assault, robbery, burglary, larceny, and motor vehicle theft) occurred according to the National Crime Victimization Survey. However, only about 300,000 individuals, amounting to only 1.5% of the 20.6 million felonies, were sentenced to prison or jail in 2004 for committing one of these felonies (Barkan, 2009). Even if we grant that many of these individuals probably committed more than one crime in 2004 (the typical annual number of offenses per criminal is very difficult to determine but probably is in the 15–25 range; see Piquero & Blumstein, 2007), it is clear that the vast majority of serious crimes do not result in anyone being incarcerated within a given year. If so, the incapacitation effect of incarceration must be fairly small relative to the amount of crime committed by nonincarcerated offenders.

In another problem, a large incapacitation effect is most likely to occur if high-rate or chronic offenders are imprisoned (*selective incapacitation*), but it is difficult in practice to identify such offenders (Auerhahn, 2006). The incarceration boom thus has followed a policy of *gross incapacitation*—that is, the mass imprisonment of offenders without regard to their rates of offending or likelihood of future offending. However, the highest-rate offenders are generally already incarcerated because they commit so many offenses that the "odds" catch up to them, and they are eventually arrested and imprisoned. Thus, as the incarceration boom proceeded, lower-rate offenders were increasingly the ones being imprisoned, and the boom has had diminishing returns in its crime-reduction impact. Samuel Walker (2006) summarizes this problem as follows:

> As we lock up more people, we quickly skim off the really high-rate offenders and begin incarcerating more of the less serious offenders. Because they average far fewer crimes per year, . . . we get progressively lower returns in crime reduction. (p. 143)

A third problem involves the fact that some crime, especially drug crime, is committed by groups of offenders. If one member of the group is imprisoned, the rest of the group can carry on, and a new individual may even be recruited to replace the incarcerated individual. Criminologists call this recruitment the *replacement effect*, and they argue that it diminishes the incapacitation effect of imprisonment for crimes committed by groups. Some even say that the incapacitation effect of imprisonment for drug crimes is zero for this reason (Piquero & Blumstein, 2007).

Finally, it is well known that most street criminals "age out" of crime: Their criminal offending is highest in their late teens and early 20s, and it begins to decline as they age into their 30s and beyond. Many inmates thus stay in prison long after they would have been likely to stop their offending. For these offenders, says Christy Visher (2000), "part of the time incarcerated would not prevent any crimes" (p. 611). Daniel Nagin (1998) calls them "poor candidates for incarceration from an incapacitation perspective" (p. 364).

### Research on Incarceration and Crime Rates

We have examined several indications that the deterrent and incapacitation effects of the incarceration boom on crime rates are likely to have been fairly small. Several kinds of evidence tend to confirm this pessimistic expectation.

First, if incarceration has a strong crime-reduction impact, then as incarceration rose steadily after the early 1970s, crime should have declined steadily. As we saw in our brief discussion of crime and incarceration trends since the early 1970s, however, incarceration and crime have often not tracked in this manner.

Second, although crime rates finally fell nationally during the 1990s as incarceration rose, the states during the 1990s with the largest increases in incarceration rates generally had *smaller* declines in their crime rates than did states with smaller incarceration increases (Gainsborough & Mauer, 2000). New York State experienced a particularly sharp crime decline during the 1990s even though the rate of its imprisonment growth was the second lowest in the nation during this period, as its number of inmates grew by 9% while that in the whole nation grew by 63% (Jacobson, 2005).

Third, Canada's crime rate fell during the 1990s even though its incarceration rate did *not* rise (Webster & Doob, 2007). Thus, even though rising incarceration did accompany the U.S. crime drop in that decade, the Canadian experience leads some scholars to speculate that the U.S. crime rate might have dropped even if its incarceration rate had not risen (Zimring, 2006). At a minimum, the Canadian experience casts doubt on the importance of rising incarceration for the U.S. crime drop during the 1990s.

Fourth, several sophisticated studies of U.S. crime and incarceration rates conclude that incarceration has only a small impact, if that, on the crime rate (Stemen, 2007). Some find that a 10% rise in the incarceration rate produces (because of deterrence and/or incapacitation, with incapacitation probably the more relevant process) a decrease in the crime rate between less than 1% and 4% (Levitt, 1996; Marvell & Moody, 1994; Spelman, 2006; Western, 2006). Other research finds that an increase in incarceration would not affect the crime rate at all in most states (DeFina & Arvanites, 2002) and that increasing incarceration has not lowered the crime rate in Florida counties (Kovandzic & Vieraitis, 2006). One study even found that increases in incarceration would *raise* the crime rate in states with the highest incarceration rates (Liedka, Piehl, & Useem, 2006).

The crime reduction effect that some of these studies have found is small, and it comes at a great monetary cost. For example, using the 1% reduction figure favored by Western (2006), the United States would have to imprison about 160,000 more offenders (10% of the current 1.6 million in prison) to achieve a 1% reduction in the crime rate. Because each new prisoner costs about $29,000 a year to house (Warren, 2009), these new prisoners would cost about $4.6 billion (160,000 × $29,000) annually. Because prisons are already overcrowded, ideally 160 new prisons, housing 1000 inmates each, would have to be built. Assuming about $100 million per prison, the new prisons to house these new prisoners would cost $16 billion. Combining these cost estimates, a 1% reduction in crime achieved by increasing the prison population by 10% would cost many billions of dollars. A reasonable conclusion is that this tiny crime reduction would not be achieved in a cost-effective manner.

Recall that crime dropped sharply during the 1990s as imprisonment soared. Scholars have attempted to determine how much of the crime drop during that decade was due to rising imprisonment. Two studies concluded that the incarceration boom accounted for about one-fourth of the crime drop during the 1990s (Rosenfeld, 2006; Spelman, 2006), while another study concluded that it accounted for about one-tenth of the crime drop (Western, 2006). The authors of the first two studies concluded that the crime reduction they found was not cost-effective, with one saying that incarceration is "an incredibly inefficient means of reducing crime" (Spelman, 2006, p. 124). The other author noted that the huge cost of corrections diverts money from education, employment, and other areas in which spending might better reduce crime. Estimating that the 100 homicides averted annually during the 1990s by rising imprisonment cost more than $1 billion per year in prison costs, the author wondered, "Would we be better off if the $1 billion were spent on preschool programs, parent training, vocational training, drug treatment, and other promising prevention programs?" (Rosenfeld, 2006, p. 151). Thus, we can see that incarceration probably has, at most, a very small effect on the crime rate and is achieved in a most cost-ineffective manner.

## ■ Collateral Consequences of the Incarceration Boom

We have just seen that the massive increase in incarceration since the 1970s has probably produced little in the way of crime reduction. Not only has the incarceration boom failed to produce the positive consequence for which it was initiated, it has also had several negative collateral consequences that have attracted increasing attention from scholars and policy analysts (Clear, 2007; Garland, 2001; Mauer & Chesney-Lind, 2003; Petersilia, 2003; Travis, 2005). This section examines several of these consequences.

### The Huge Cost of Incarceration

A first collateral consequence involves the huge monetary cost of corrections. As we have seen, corrections now costs the United States tens of billions of dollars every year. Imprisonment accounts for almost 90% of the corrections cost that states pay, and incarceration costs occupy an increasingly higher proportion of state budgets and have made it difficult for state governments to pay for their many other responsibilities (Warren, 2009). As noted earlier, the many billions of dollars spent on corrections are diverted from pursuits that may have a much higher crime-reduction impact, including early childhood intervention programs involving home visits to new mothers whose children are at risk (because of the mother's young age or low income level, for example) for delinquency and other problems as they grow older (Welsh & Farrington, 2007). Studies show that every dollar spent on such programs produces a much larger crime-reduction effect than a dollar spent on corrections (Greenwood, 2006). One of the most unfortunate collateral consequences of the incarceration boom, then, has been its extraordinarily high economic cost, the financial burdens this cost has imposed on states and local governments, and the diversion of precious dollars from more effective (and more cost-effective) crime-prevention programs and policies.

### Effects on Families and Communities

Mass incarceration has several related, negative consequences for prisoners' families and communities (Clear, 2007). It removes inmates from their roles as breadwinners, making their families poorer, and from their roles as spouses and parents (because most inmates are males, the roles most affected are those of husband and father). This latter effect became more important as the incarceration boom increasingly placed less serious offenders (and thus better husbands and fathers) behind bars, and it has especially affected the "children of the prison generation" (Hagan & Dinovitzer, 1999, p. 153). A growing number of children, probably more than 1.5 million, have a parent in prison at any one time, and probably more than 7 million children have had a parent in prison at some point during their childhood. These children are more likely to commit delinquency and to experience "school-related performance problems, depression and anxiety, low self-esteem, and aggressiveness" (Clear, 2007, p. 97). Although the exact reasons that parental imprisonment has these effects remain to be determined, a recent review concluded that the effects "appear to be relatively strong, with multiple adverse outcomes" (Murray & Farrington, 2008, p. 186).

In a community-level problem, recall that about 700,000 prisoners are now being released back to their communities every year after serving their sentence (Eckholm, 2008). This flood of ex-inmates brings with it several problems of reentry (Travis, 2005). Ex-inmates arrive home with few prospects for employment or friendships with law-abiding citizens, and many arrive home with personal problems such as a history of being sexually and/or physically abused as children and a history of drug and alcohol addiction beginning in adolescence. Unfortunately, many inmates come out of prison with more personal problems than when they went in. While in prison they may have lived amid squalid conditions and been physically and/or sexually abused by other inmates or by guards; more generally, they were exposed to the criminogenic (crime-causing) influence of other inmates.

For all these reasons, it is no surprise that, as stated earlier, about two-thirds of ex-prisoners can be expected to be arrested within 3 years and more than half can be expected to be reincarcerated. Thus, the return of ex-prisoners to their communities may often raise the communities' crime rates. Exacerbating this problem, some research shows that incarceration makes it more likely, not less likely, that inmates will commit new offenses after being released from prison (Vieraitis, Kovandzic, & Marvell, 2007). Todd Clear (2007) explains why:

> Being in prison is a brutalizing experience, and people who are subjected to these experiences find it harder to adjust to free society again. . . . [No prison reformers] would be surprised to hear that people exposed to prison do less well adjusting to society than do those who have been exposed to a different kind of sanction. The uneasy conclusion from this thinking is that prison, far from being a deterrent, disables those who experience it. (p. 27)

Because of all these problems, mass imprisonment has impaired the well-being of children and their families and of whole communities. As Clear (2007) notes, "mass incarceration makes disadvantaged neighborhoods worse."

### Prison Living Conditions

In 1958 sociologist Gresham M. Sykes (1958) wrote a classic book, *The Society of Captives*, in which he discusses the "deprivations or frustrations of prison life" (p. 64) that he calls the *pains of imprisonment*. These deprivations include the loss of liberty, the loss of goods and services, the loss of heterosexual relationships, the loss of autonomy, and fear for one's physical safety.

At the time Sykes wrote his book, the incarceration boom had not yet begun. Half a century later, we have hundreds of thousands more prisoners and many more prisons than Sykes could have envisioned. Even though new prisons were built, prison overcrowding and related problems, such as inadequate medical care, became much worse (Ross, 2008). Many inmates now live amid squalid conditions, and several prisons, including California's system, have been under court order to reduce overcrowding and to improve inmates' living conditions in general. Although many Americans would not shed tears for the inmates living amid these conditions, they might consider that squalid conditions contribute to the brutalizing experience that makes offenders more dangerous when they are released from prison than they were when they entered it. These conditions, and the incarceration boom that has worsened them, thus put public safety at risk.

### Racial Inequality

A final, and perhaps the most unfortunate, collateral consequence is the greatly disproportionate incarceration of African Americans and other people of color (Mauer, 2006; Provine, 2007; Tonry & Melewski, 2008; Western, 2006). As the nation cracked down on crime and especially on drugs, African Americans became increasingly likely to be imprisoned out of proportion to their involvement in crime and even though they are no more likely than whites to use illegal drugs (Fellner, 2009). As Marc Mauer and Ryan S. King (2007) explain, "African Americans comprise 14% of regular drug users, but are 37% of those arrested for drug offenses and 56% of persons in state prison for drug offenses" (p. 2). According to one scholar, the war on drugs thus amounts to a "search and destroy" mission against African American males (Miller, 1996).

A few statistics indicate the enormity of this problem. Currently about one-third of African American males in their 20s are under some form of correctional control (in prison or jail or on probation or parole). According to one estimate, about one-third of African American men born in 2001 will be sent to prison at least once in their lifetimes, compared to only 6% of white men (Bonczar, 2003) (see **Figure 7–4**). Almost 50% of African American men who did not graduate

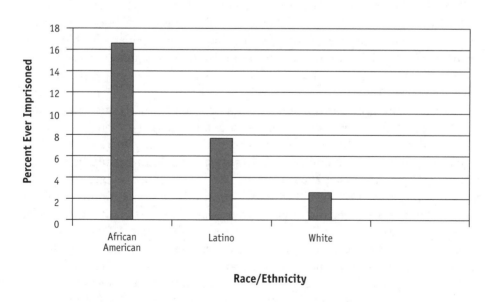

**Figure 7–4**  Race/Ethnicity and Percentage of Adult Males Ever Imprisoned as of 2001
**Source:** Adapted from T. P. Bonczar, "Prevalence of Imprisonment in the U.S. Population, 1974–2001," Bureau of Justice Statistics, August 2003.

high school are presently incarcerated, compared to less than 10% of young white men without a high school degree (Western, 2006).

Besides putting so many African Americans in prison or jail or under some other form correctional control, the incarceration boom has contributed to unequal treatment in other ways. First, because incarceration makes it more likely that an ex-inmate will be unemployed or, if employed, in a low-paying job, the incarceration boom has contributed to the economic inequality of African Americans and other people of color (Western, 2006). Second, many states take the right to vote away from convicted felons. Given their disproportionate rates of conviction and incarceration, about 8% of African Americans, including more than 12% of African American men, are currently or permanently disenfranchised (King, 2006; Manza & Uggen, 2006). With less voters, the political influence of African Americans in many states and communities is reduced. Finally, the many other problems that incarceration has created in communities have had an especially severe impact in African American communities because of the high levels of African American incarceration (Clear, 2007).

## ■ Conclusion

U.S. corrections policy since the 1970s has been both misguided and shortsighted. Originating in alarmist statements by public officials and inflammatory coverage by the news media, the incarceration boom has cost tens of billions of dollars while lowering crime by at most a very small degree, and certainly not in a cost-effective manner. It has also caused several serious collateral consequences, including the undermining of family well-being in communities with high rates of incarceration, higher crime rates in those communities, and an exacerbation of racial inequality because so many African Americans and other people of color have been incarcerated.

As the beginning of this chapter indicated, there is increasing recognition that many offenders, especially those who are nonviolent and first-time offenders, should receive probation or other forms of community corrections in lieu of incarceration and that similar offenders already in prison should be released earlier (Tonry & Melewski, 2008; Warren, 2009). There are also increasing calls for shortening sentences, for reducing the use of mandatory minimum sentencing, and for increased rehabilitation programming (Austin et al., 2007; Petersilia, 2008). These new strategies are in fact not new at all, as they would return the United States to the pre-1970s period when incarceration was not the preferred alternative for so many offenders. These strategies would save billions of dollars and, according to the best evidence, yield lower (or at least equal) recidivism rates than incarceration (Austin et al., 2007; Clear, 2007; Warren, 2009). They would also help reduce racial inequality and strengthen the social dynamics of many communities.

The United States, then, can have a safer and more equal society at a much lower cost than we have today. For all the reasons discussed in this chapter, the incarceration boom has done much more harm than good, and it is long past time to move in a new direction. Ironically, this is a direction that the United States followed before the 1970s, and it is one that Canada and western European nations still generally follow (Doob & Webster, 2006; Farrington, Langan, & Tonry, 2004). In considering a new direction for its corrections policy, the United States has much to learn from its own past and from its Western allies.

## ■ Discussion Questions

1. What factors have contributed to the dramatic increase in the U.S. incarceration rate since the 1970s?
2. Describe the role that the media and public officials have played in recent sentencing and imprisonment policies.
3. What factors should be considered when assessing the success or failure of the use of incarceration in the United States?
4. Discuss the "collateral consequences" caused by increases in the use of incarceration.
5. Is it inevitable that incarceration rates will continue to grow? What policies might help to reverse this trend?

## ■ References

Auerhahn, K. (2006). Conceptual and methodological issues in the prediction of dangerous behavior. *Criminology & Public Policy, 4,* 771–778.

Austin, J., Clear, T., Duster, T., Greenberg, D. F., Irwin, J., McCoy, C., et al. (2007). *Unlocking America: Why and how to reduce America's prison population.* Washington, DC: The JFA Institute.

Austin, J., & Irwin, J. (2001). *It's about time: America's imprisonment binge.* Belmont, CA: Wadsworth Publishing Company.

Barkan, S. E. (2009). *Criminology: A sociological understanding.* Upper Saddle River, NJ: Prentice Hall.

Beck, A. J. (2000). *Prisoners in 1999.* Washington, DC: Bureau of Justice Statistics, U.S. Department of Justice.

Beckett, K. & Sasson, T. (2004). *The politics of injustice: Crime and punishment in America.* Thousand Oaks, CA: Sage Publications.

Blumstein, A., & Cohen, J. (1973). A theory of the stability of punishment. *Journal of Criminal Law and Criminology, 64,* 198–207.

Bohm, R. M., & Walker, J. T. (2006). *Demystifying crime and criminal justice.* Los Angeles: Roxbury Publishing Company.

Bonczar, T. P. (2003). *Prevalence of imprisonment in the U.S. population, 1974–2001.* Washington, DC: Bureau of Justice Statistics, U.S. Department of Justice.

Chambliss, W. J. (1999). *Power, Politics, and Crime.* Boulder, CO: Westview Press.

Clear, T. R. (2007). *Imprisoning communities: How mass incarceration makes disadvantaged neighborhoods worse.* New York: Oxford University Press.

Cullen, F. T. (2007). Make rehabilitation corrections' guiding paradigm. *Criminology & Public Policy, 6,* 717–728.

Decker, S., Wright, R., & Logie, R. (1993). Perceptual deterrence among active residential burglars: A research note. *Criminology, 31,* 135–147.

DeFina, R. H., & Arvanites, T. M. (2002). The weak effect of imprisonment on crime: 1971–1998. *Social Science Quarterly, 83,* 635–653.

Doob, A. N., & Webster, C. M. (2003). Sentence severity and crime: Accepting the null hypothesis. *Crime and Justice: A Review of Research, 30,* 143–195.

Doob, A. N., & Webster, C. M. (2006). Countering punitiveness: Understanding stability in Canada's imprisonment rate. *Law & Society Review, 40,* 325–367.

Eckholm, E. (2008, April 8). U.S. shifting prison focus to re-entry into society. *The New York Times,* p. A23.

Farrington, D. P., Langan, P. A., & Tonry, M. (Eds.). (2004). *Cross-national studies in crime and justice.* Washington, DC: Bureau of Justice Statistics, U.S. Department of Justice.

Fellner, J. (2009). *Decades of disparity: Drug arrests and race in the United States.* New York: Human Rights Watch.

Gainsborough, J., & Mauer, M. (2000). *Diminishing returns: Crime and incarceration in the 1990s.* Washington, DC: The Sentencing Project.

Garland, D. (Ed.). (2001). *Mass imprisonment: Social causes and consequences.* Thousand Oaks, CA: Sage Publications.

Gottschalk, M. (2006). *The prison and the gallows: The politics of mass incarceration in America.* Cambridge: Cambridge University Press.

Greenwood, P. W. (2006). *Changing lives: Delinquency prevention as crime-control policy.* Chicago: University of Chicago Press.

Hagan, J., & Dinovitzer, R. (1999). Collateral consequences of imprisonment for children, communities, and prisoners. *Crime and Justice: A Review of Research, 26,* 121–162.

Jacobson, M. (2005). *Downsizing prisons: How to reduce crime and end mass incarceration.* New York: New York University Press.

King, R. S. (2006). *A decade of reform: Felony disenfranchisement policy in the United States.* Washington, DC: The Sentencing Project.

King, R. S. (2009). *The state of sentencing 2008: Developments in policy and practice.* Washington, DC: The Sentencing Project.

King's College London International Centre for Prison Studies. *World Prison Brief.* Available: http://www.kcl.ac.uk/depsta/law/research/icps/worldbrief/.

Kleck, G., Sever, B., Li, S., & Gertz, M. (2005). The missing link in general deterrence research. *Criminology, 43,* 623–659.

Kovandzic, T. V., & Vieraitis, L. M. (2006). The effect of county-level prison population growth on crime rates. *Criminology & Public Policy, 5,* 213–244.

Lappi-Seppälä, T. (2008). Trust, welfare, and political culture: Explaining differences in national penal policies. *Crime and Justice: A Review of Research, 37,* 313–387.

Levitt, S. (1996). The effect of prison population size on crime rates: Evidence from prison overcrowding litigation. *Quarterly Journal of Economics, 111,* 319–351.

Liedka, R. V., Piehl, A. M., & Useem, B. (2006). The crime-control effect of incarceration: Does scale matter? *Criminology & Public Policy, 5,* 245–276.

Maguire, K., & Pastore, A. L. (Eds). (2009). *Sourcebook of criminal justice statistics.* Available from the Bureau of Justice Statistics at: http://www.albany.edu/sourcebook

Manza, J., & Uggen, C. (2006). *Locked out: Felon disenfranchisement and American democracy.* New York: Oxford University Press.

Marvell, T. B., & Moody, C. E., Jr. (1994). Prison population growth and crime reduction. *Journal of Quantitative Criminology, 10,* 109–140.

Mauer, M. (2006). *Race to incarcerate.* New York: New Press.

Mauer, M., & Chesney-Lind, M. (Eds.). (2003). *Invisible punishment: The collateral consequences of mass imprisonment.* New York: The New Press.

Mauer, M., & King, R. S. (2007). *A 25-year quagmire: The war on drugs and its impact on American society.* Washington, DC: The Sentencing Project.

Michalowski, R. (2006). The myth that punishment reduces crime. In R. M. Bohm & J. T. Walker (Eds.), *Demystifyng Crime and Criminal Justice* (pp. 179–191). Los Angeles: Roxbury Publishing Company.

Miller, J. G. (1996). *Search and destroy: African American males in the criminal justice system.* New York: Cambridge University Press.

Moore, S. (2009a, February 10). Court orders California to cut prison population. *The New York Times,* p. A12.

Moore, S. (2009b, February 11). The prison overcrowding fox. *The New York Times,* p. A17.

Mumola, C. J., & Karberg, J. C. (2006). *Drug use and dependence, state and federal prisoners, 2004.* Washington, DC: Bureau of Justice Statistics, U.S. Department of Justice.

Murray, J., & Farrington, D. P. (2008). The effects of parental imprisonment on children. *Crime and Justice: A Review of Research, 37,* 133–206.

Nagin, D. (1998) Deterrence and Incapacitation. In M. Tonry (Ed.), *The Handbook of Crime and Punishment* (pp. 345–368). New York: Oxford University Press.

Petersilia, J. (2003). *When prisoners come home: Parole and prisoner reentry.* New York: Oxford University Press.

Petersilia, J. (2008). California's correctional paradox of excess and deprivation. *Crime and Justice: A Review of Research, 37,* 207–278.

Piquero, A. R., & Blumstein, A. (2007). Does incapacitation reduce crime? *Journal of Quantitative Criminology, 23,* 267–285.

Pratt, T. C. (2008). *Addicted to incarceration: Corrections policy and the politics of misinformation in the United States.* Thousand Oaks, CA: Sage Publications.

Provine, D. M. (2007). *Unequal under law: Race in the war on drugs.* Chicago: University of Chicago Press.

Raphael, S. (2009). Explaining the rise in U.S. incarceration rates. *Criminology & Public Policy, 8,* 87–95.

Raphael, S., & Stoll, M. A. (2009). Why are so many Americans in prison? In S. Raphael & M. A. Stoll (Eds.), *Do prisons make us safer? The benefits and costs of the prison boom* (pp. 27–72). New York: Russell Sage Foundation.

Robinson, P. H., & Darley, J. M. (2004). Does criminal law deter? A behavioral science investigation. *Oxford Journal of Legal Studies, 24,* 173–205.

Rosenfeld, R. (2006). Patterns in adult homicide: 1980–1995. In A. Blumstein & J. Wallman (Eds.), *The crime drop in America* (pp. 130–163). Cambridge: Cambridge University Press.

Ross, J. I. (2008). *Special problems in corrections.* Upper Saddle River, NJ: Prentice Hall.

Simon, J. (2007). *Governing through crime: How the war on crime transformed American democracy and created a culture of fear.* New York: Oxford University Press.

Spelman, W. (2006). The limited importance of prison expansion. In A. Blumstein & J. Wallman (Eds.), *The crime drop in America* (pp. 97–129). Cambridge: Cambridge University Press.

Spelman, W. (2009). Crime, cash, and limited options: Explaining the prison boom. *Criminology & Public Policy, 8,* 29–77.

Stemen, D. (2007). *Reconsidering incarceration: New directions for reducing crime.* New York: Vera Institute of Justice.

Sykes, G. M. (1958). *The society of captives: A study of a maximum security prison.* Princeton: Princeton University Press.

Tonry, M. (2007). Determinants of penal policies. *Crime and Justice: A Review of Research, 36,* 1–48.

Tonry, M. (2008a). Learning from the limitations of deterrence research. *Crime and Justice: A Review of Research, 37,* 279–311.

Tonry, M. (2008b). Preface. *Crime and Justice: A Review of Research, 37,* vii–viii.

Tonry, M., & Melewski, M. (2008). The malign effects of drug and crime control policies on black Americans. *Crime and Justice: A Review of Research, 37,* 1–44.

Travis, J. (2005). *But they all come back: Facing the challenges of prisoner reentry.* Washington, DC: The Urban Institute Press.

Tunnell, K. D. (1996). Let's do it: Deciding to commit a crime. In J. E. Conklin (Ed.). *New perspectives in criminology* (pp. 246–258). Boston: Allyn and Bacon.

Unnever, J. D., & Cullen, F. T. (2005). Executing the innocent and support for capital punishment: Implications for public policy. *Criminology & Public Policy, 4,* 3–37.

Vieraitis, L. M., Kovandzic, T. V., & Marvell, T. B. (2007). The criminogenic effects of imprisonment: Evidence from state panel data, 1974–2002. *Criminology & Public Policy, 6,* 589–622.

Visher, C. A. (2000). Career offenders and crime control. In J. F. Sheley (Ed.). *Criminology: A contemporary handbook* (pp. 601–619). Belmont, CA: Wadsworth.

Walker, S. (2006). *Sense and nonsense about crime and drugs: A policy guide.* Belmont, CA: Wadsworth Publishing Company.

Warren, J. (2009). *One in 31: The long reach of American corrections.* Washington, DC: Pew Center on the States.

Webster, C. M., & Doob, A. N. (2007). Punitive trends and stable imprisonment rates in Canada. *Crime and Justice: A Review of Research, 26,* 297–369.

Welsh, B. C., & Farrington, D. P. (Eds.). (2007). *Preventing crime: What works for children, offenders, victims and places.* New York: Springer.

West, H. C., & Sabol, W. J. (2008). *Prisoners in 2007.* Washington, DC: Bureau of Justice Statistics, U.S. Department of Justice.

Western, B. (2006). *Punishment and inequality in America.* New York: Russell Sage Foundation Publications.

Zimring, F. E. (2006). *The great American crime decline.* New York: Oxford University Press.

# Reentry as a Process Rather Than a Moment

# CHAPTER
# 8

Natasha A. Frost

## ■ Introduction

Prisoner reentry is hardly a new phenomenon. Prisoners have been returning to communities for as long as criminal offenders have been sent to prison. Natural life sentences, sentences of life without the possibility of parole, and death sentences—in other words, those sentences from which a person will never return—have always been and remain the exception. As in the past, almost all incarcerated persons (almost 95%) will eventually be released from prison (Travis, 2005). Nonetheless, with more than 725,000 formerly incarcerated persons returning to communities from state and federal prisons in 2007 (West & Sabol, 2008), the age-old problem of prisoner reentry has taken on a renewed sense of urgency. At no other point in U.S. history have so many individuals experienced prison and confronted the challenge of reentering communities following incarceration. Most are returning to those communities having spent time in facilities that have very little in the way of rehabilitative correctional programming. Few of those released have had any meaningful prerelease assistance, and most will encounter a series of obstacles to reintegration. Crucially, the communities to which inmates return are all too often among the most disadvantaged of communities.

About half of all prisoners released from prison will return to the very same neighborhood from which they were removed, and those returning to a different neighborhood typically return to a neighborhood that is very similar sociodemographically to the one from which they were removed (Visher & Farrell, 2005; Visher, Kachnowski, LaVigne, & Travis, 2004). The communities receiving these offenders are all too often characterized by significant social and economic disadvantage, as evidenced by higher than average poverty and unemployment rates (see, for example, Brooks, Solomon, Keegan, Kohl, &

Lahue, 2005; LaVigne & Kachnowski, 2003; Lavigne & Mamalian, 2003). In other words, many inmates returning from prison are returning to communities that are among the least able to accommodate them. These disadvantaged communities have diminished political and human capital and are able to offer little in the way of meaningful educational, work, or life opportunities (Clear, 2007).

Although renewed interest in prisoner reentry has been generated in part because of concerns about the sheer volume of prisoners returning to communities, the shifting characteristics of the inmate populations approaching release and the changing nature of release mechanisms have also stimulated concern. With changes in sentencing policy over the past several decades, prisoners approaching release are more likely to have served longer prison sentences, are less likely to have received any sort of educational or vocational programming while in prison, and are also increasingly less likely to receive adequate postrelease supervision.

In the meantime, it has become increasingly clear that a number of important policy shifts have changed the landscape of imprisonment and therefore of prisoner reentry. Over the past several decades, there has been steady and substantial annual growth in prison populations. In the three decades between 1977 and 2007, the prison population across the United States has grown more than fivefold, from less than 300,000 at year-end 1977 to more than 1.5 million at year-end 2007 (West & Sabol, 2008). The policy shifts responsible for that unprecedented growth are numerous and are discussed at length in Chapter 7. Just as policy shifts all but ensured the buildup of incarceration, policy shifts have also contributed to the growing challenge of prisoner reentry.

After setting the context and describing the challenges of prisoner reentry, this chapter focuses on three of those policy shifts most important to prisoner reentry and reintegration: (1) the shifting of prison release mechanisms, (2) the changing nature of postrelease supervision, and (3) the impact of legislatively imposed barriers to prisoner reintegration. The premise of the chapter is that despite a fairly substantial buildup of interest in the reentry process, a number of policy developments have meant that the transition from prison to community has become more, not less, difficult.

The news, however, is not all bad. Despite somewhat discouraging policy developments over the past couple of decades, there is reason for some optimism. The current prisoner reentry movement, first launched in the late 1990s, has gained substantial momentum over the past decade (see Travis, 2007; Visher, 2007). With hundreds of publications devoted to this relatively fledgling area of research, arguably no other aspect of the criminal justice process has received such sustained attention in recent years. Indeed, it seems fair to say that we have come full circle from a period of sustained interest in the *crisis* of mass imprisonment to a sustained interest in the *consequences* of that mass imprisonment in terms of reentry.

## ■ The Context for Reentry

After decades of relative stability (Blumstein, Cohen, & Nagin, 1977), in the mid-1970s, prison populations began a period of sustained growth that has continued unabated despite fairly substantial declines in crime since the mid-1990s (Zimring, 2007). At year-end 2007, close to 1.6 million people were serving time in U.S. prisons. When jail inmates were added, the total number incarcerated across the United States was approaching 2.4 million (West & Sabol, 2008).

As the prison population has grown, so has the number of prisoners returning to communities each year. Although prison admissions have generally outpaced prison releases, the number of prisoners released each year has increased dramatically over the past 30 years. As depicted in **Figure 8–1**, the annual number of releases from state and federal prisons has grown from just under 148,000 at year-end 1977 to more than 725,000 at year-end 2007. In other words, almost five times more persons were released from prisons in 2007 than in 1977.

Perhaps even more disturbing than these raw prison release figures are the prevalence statistics released in a Bureau of Justice Statistics (BJS) special report, "Prevalence of Imprisonment in the U.S. Population, 1974–2001" (Bonczar, 2003).[1] According to Bonczar, the increase in imprisonment over the past 30 years has meant that far more people have experienced imprisonment than ever before (prevalence) and that if rates remain unchanged, we can anticipate that far more will experience imprisonment in their lifetime (lifetime likelihood). Bonczar reports that in 2001 approximately 1 in every 37 adults had served time in prison,

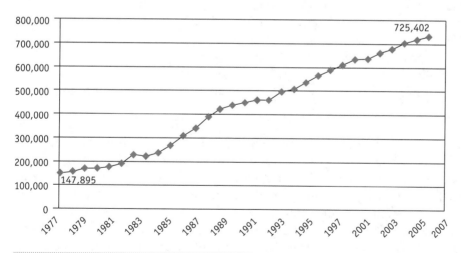

**Figure 8–1** State and Federal Prison Releases, 1977–2007
**Source:** Data from P. M. Harrison, "Total Sentenced Prisoners Released from State or Federal Jurisdiction, 1977–1998," Bureau of Justice Statistics, 2000; data from P. M. Harrison, "Prison Admissions and Releases by Gender, 1999 and 2000," Bureau of Justice Statistics, 2004; and data from H. C. West and W. J. Sabol, "Prisoners in 2007," Bureau of Justice Statistics, 2008.

and that if imprisonment rates remain unchanged, 6.6% of all U.S. citizens born in 2001 will have served time in a prison during their lifetime. Of course this "experience of prison" is not evenly distributed across the population. Males and minorities are more likely to have experienced prison than females or whites. Although men born in 2001 are far more likely to experience prison than are females (11.3% vs. 1.8%), the "imprisonment binge," as Austin and Irwin (2001) have called it, has certainly not left women behind. In fact, due to more rapidly growing imprisonment rates among women (Frost, Greene, & Pranis, 2006), the lifetime likelihood of imprisonment has increased more for women than it has for men. While men born in 2001, are just over three times more likely to serve time in prison than were those born in 1974, women similarly situated are six times more likely (Bonczar, 2003).

Some of the most disturbing figures in the BJS prevalence report concern the growing racial disparities in imprisonment and the likelihood that, without some substantial policy changes that will affect sentencing more generally, these disparities will likely not abate. Despite constituting only 13% of the U.S. population, Bonczar (2003) reported that "at year end 2001 nearly as many blacks (2,166,000) as whites (2,203,000) has ever served time in prison" (p. 5). Blacks and whites each constituted 39% of those who ever served time in a prison, while Hispanics accounted for the remaining 18%. Bonczar further reported that 16.6% of all black males had served time in prison, a number six times higher than the 2.6% of white men that had such an experience. If Bonczar's lifetime likelihood predictions bear out, the future does not look any brighter. Again, given that the 2001 rates of imprisonment remain unchanged, 1 in 3 black males born in 2001 can expect to go to prison in their lifetime (contrasted with 1 in 17 white males). While these statistics clearly suggest our incarceration policy has had profound consequences for black Americans, they also have important implications for prisoner reentry. If the projections bear out, just as more and more people will experience prison, more and more will have to negotiate the reentry process.

It is not simply the number of annual prison releases that causes concern. Since the mid-1970s, we have witnessed the decline of faith in the rehabilitative ideal exert its influence over the availability of programming within prisons. Those serving time in correctional facilities are less likely to have received prison-based correctional programming. Despite generally high levels of public support for rehabilitation as a correctional goal (see Cullen, Fisher, & Applegate, 2000; Nagin, Piquero, Scott, & Steinberg, 2006), politicians have all too often gauged faulty public opinion poll data gathered in the aftermath of exceptional cases and have been reluctant to support spending on rehabilitative programming in correctional contexts. By contrast, it has been relatively easy to pass legislation that further marginalizes this most stigmatized group.

# ■ The Challenge of Prisoner Reentry

Prisoner reentry and reintegration have always been challenging because ex-offenders are among the most stigmatized and least sympathetic of all marginalized groups. By virtue of their criminal convictions, ex-offenders are deemed less trustworthy, less honest, and among the least deserving. The stigma associated with having served time in prison remains long after release and can affect an ex-offender's life chances well into the future. Moreover, the stigma can affect an ex-offender across multiple domains of life. For the ex-offender, housing is more difficult to secure, jobs are more difficult to come by, and relationships are more difficult to mend. Although one might assume that the stigma of prison is attenuated somewhat as more and more people experience prison, research suggests that this is not necessarily the case. Formerly incarcerated persons are even stigmatized in neighborhoods where prison experience is relatively common (Clear, Rose, & Ryder, 2001). Equally troubling, not only is the probability of experiencing prison greater for African Americans, but because of the pairing of images of race with images of crime, the stigma associated with prison is amplified for African Americans as well (see Thompson, 2008).

Although we have long known that the stigma of a criminal conviction damages the life chances of formerly incarcerated persons and that the process of reentering society after a term of incarceration is challenging in and of itself, we have been more and more inclined in recent years to make this process more difficult for the ex-offender. Although this chapter focuses on a series of explicit policy initiatives that hinder prisoner reentry, changes in the mechanisms by which incarcerated persons are released have had important consequences for the reentry process. A brief discussion of the changing nature of release mechanisms and postrelease supervision will set the stage for a more thorough understanding of the context in which formerly incarcerated persons experience the reentry process.

## Shifting Release Mechanisms

Until the 1970s, it was fairly easy to describe sentencing and prison release practices across the United States—belief in the capacity for offender change and the promise of rehabilitation drove corrections, and all states relied on indeterminate sentencing systems (see Frost et al., 2006). Since that time, the focus has shifted away from rehabilitation and toward deterrence, incapacitation, and retribution. Indeterminate sentencing has, in almost half of the states, been supplanted by determinate sentencing. In many instances, the advent of the determinate sentence signaled the concomitant formal abolition of discretionary parole. Indeed, the shift from indeterminate to determinate sentencing has meant that discretionary parole release—a staple of indeterminate sentencing—has in many states been abandoned. The abolition of discretionary release has led to widespread changes in the way prison inmates earn and experience prison release.

Parole release is by definition an early release from prison and as such is a "conditional" release. One of the benefits of this conditional release system is that upon release, offenders are required to successfully complete a term of supervision typically lasting through the remainder of their court-imposed sentences. The move away from indeterminate sentencing and the abolition of discretionary parole in some states have meant that increasing numbers of prisoners are being released with finite (and often limited) terms of additional supervision. Some are released "unconditionally"—in other words, with no further supervision at all. Recent release statistics demonstrate that as many as one in five inmates leave prison with no further supervision at all (Travis, 2005, p. 44). Ironically, it is typically the inmates who have demonstrated the least progress in prison and who therefore expire their sentences who are released without condition and without further supervision. To further understanding of how shifting release mechanisms have affected the reentry process, it helps to briefly discuss each of the primary release mechanisms.

### Discretionary Release

Beginning in 1976, states began to switch from indeterminate sentencing structures, which stipulated a fairly broad range of sentence to allow for discretionary parole release upon evidence of rehabilitation, to determinate structures, with far narrower sentence ranges and release determined by calculating the sentence minus any good time earned. The advent of the determinate sentence came with the decline of faith in rehabilitation, which coalesced around the publication of Robert Martinson's now infamous 1974 article, "What Works? Questions and Answers about Prison Reform." In that article, Martinson asked what worked, reviewed an abundance of empirical work, and ultimately argued that with regard to rehabilitation, it seemed that "nothing worked" (i.e., there was not much evidence in support of the rehabilitative effects of programs). Prior to this article, rehabilitation, and the indeterminate sentencing scheme that accompanies the correctional goal of rehabilitation, had been attacked by those from the left and from the right in their political leanings (see Cullen, 1995; Feeley, 2003). Those on the right, who had begun to depict criminal offenders as intractable, irredeemable, and undeserving of our sympathies, felt that indeterminate sentences often led to lenience in punishment and lobbied for the pursuit of more retributively oriented punishment (see Wilson, 1975). Those on the left, dismayed with the amount of discretion afforded to judges and correctional authorities, argued that when the sentence was indeterminate and release was dependent upon the whim of a parole board, potential abuses would likely follow.

Beginning with California and Maine in 1976, states began to abandon indeterminate sentencing and replace it with determinate sentencing. By 2002, 19 states had adopted determinate sentencing as their overarching sentencing model, with 2 additional states adopting determinate sentences for a large number of offenses (see Stemen, Rengifo, & Wilson, 2006). Although most states have effectively abandoned rehabilitation as their correctional goal, the majority of states retain indeterminate sentencing and continue to allow for discretionary parole release. Even those states that have retained indeterminate sentencing, however, have moved toward determinate (mandatory) sentences for some offenses. Mandatory sentencing statutes stipulate that offenders sentenced for an offense for which there is a mandatory term be sentenced to that term without exception. According to Reitz (2001), the mandatory sentence approach is best understood as the adoption of determinate sentences offense by offense. Between the mid-1970s and mid-1980s, this offense-by-offense determinate approach was pursued by all states for certain offenses regardless of whether the state formally switched to a determinate sentencing scheme (Tonry, 1996).

### Mandatory Release

States that have formally switched to a determinate sentencing structure typically have adopted mandatory parole release. With mandatory release, a bookkeeper (or more commonly a computer) determines the release date through entering the minimum and maximum sentences and deducting any good time earned over the course of the incarceration. Although there is no parole board in states that have transitioned to mandatory release, mandatory release is a form of parole release in that in most cases the inmate is released early and further correctional supervision is typically attached to that early release. The distinction is between a parole board determining readiness for release (in discretionary release) and a formula determining the date of release (in mandatory release).

A key distinction must also be made between parole release and parole supervision. Whether through discretionary or mandatory release, parole is, by definition, an early release from prison. Under either system, that parole release may or may not be accompanied by parole supervision. While discretionary release was the dominant form of parole release until the mid-1970s, by the end of the 1990s, mandatory release had gained prominence, with more than three-quarters of all inmates released experiencing this type. Although it is tempting to assume that the discretionary system with its rehabilitative orientation would result in more reliance on parole supervision, evidence suggests the opposite. Parole supervision was imposed less in the era when discretionary release dominated and is more commonly imposed today when most inmates are released under mandatory release systems (Travis, 2005). Moreover, parole supervision varies greatly by state. Some states, such as California, subject almost all releases to some type of parole supervision; others, such as Rhode Island, impose parole supervision on very few releasees (Piehl & LoBuglio, 2005).

### Expiration Release

Expiration release occurs when inmates have served their entire judicially imposed prison sentence and can no longer be held under custody of correctional authorities. In discretionary release jurisdictions, these inmates are typically those who were never granted an early release by the parole board. Although it is often assumed that most inmates will eventually secure an early release under a discretionary system, historical data indicate that even during the period when indeterminate sentencing and discretionary release were the dominant structures, the number released unconditionally at the end of their sentence (i.e., not paroled) ranged from 30% to 50% (Travis, 2005, p. 42). In mandatory release jurisdictions, expiration release inmates are typically those who have not earned or were not eligible for the good time credits that are a staple of most correctional systems or who have lost those credits earned due to frequent disciplinary infractions during their incarceration. Some inmates expire their sentences as a direct result of mandatory minimum sentencing.

In contrast to mandatory sentencing, mandatory *minimum* sentencing provisions stipulate the minimum time to be served for the commission of a particular offense. Under mandatory minimum sentencing, a judge, while not required to impose the maximum sentence for a given offense, is not permitted to sentence to less than the required minimum. Additionally, some state provisions specifically restrict eligibility for parole to ensure that parole boards do not release the offender until the minimum term has been served (Reitz, 2001). Since 1970, all 50 states, regardless of overall sentencing structure, have adopted one or more mandatory minimum sentencing provisions. According to the "1996 Survey of State Sentencing Structures" (Bureau of Justice Assistance, 1998), mandatory minimum provisions typically apply to five classes of offenders or offenses: (1) repeat or habitual offenses, (2) drunk driving offenses, (3) drug offenses, (4) weapon offenses, and (5) sex offenses.

Although mandatory minimum sentences determine the minimum term of incarceration, some judges—presumably opposed to the lack of trust in the judiciary that such sentencing policies imply and concerned that the minimum sentences are too harsh—set the maximum term of incarceration just 1 day above the minimum (Massachusetts Sentencing Commission, 2004). Regardless of prison conduct, inmates sentenced under these conditions will be released without further supervision following their term of incarceration and may be ineligible for participation in prerelease programming by virtue of their mandatory minimum sentence (Brooks et al., 2005).

### Other Releases

There are instances in which inmates are released from custody under other conditions. Inmates might be released early to relieve overcrowding, courts might order immediate release, inmates might be released to the custody of other authorities or agencies (to serve another sentence), or inmates might be released to halfway houses or other transitional programs (including reentry programs). Persons in the latter two categories are often not technically released until they are released by these intermediary agencies. Finally, there are those persons who are not officially released at all, but rather who die in custody or escape from custody.

### The Changing Nature of Postrelease Supervision

Not only have the mechanisms for release shifted substantially since the 1970s, so has the nature of postrelease supervision. Although the release decision and the supervision decision used to be fairly tightly coupled, as states began to abolish discretionary parole (i.e., boards determining the timing of release), the decision to release and the decision to subject a formerly incarcerated person to parole supervision became more and more attenuated (see Travis, 2005; Petersilia, 2003). Today, ex-offenders are less likely to experience a parole board decision to release than they were 30 years ago, but they are more likely to experience postrelease parole supervision (see Travis, 2005, chap. 3). With increased supervision comes increased opportunity for reentry failure. Ex-offenders who would have never been monitored upon release in the past are now subject to increased surveillance and are expected to comply with all of the conditions of parole supervision. Parole officers who historically provided both services *and* surveillance now increasingly focus on the surveillance aspect of the job (see Petersilia, 2003, chap. 4). Increasingly, these ex-offenders are being returned to prison for violations of the conditions of their parole (i.e., technical violations like failing to report) rather than for actual reoffending (i.e., new crimes). This distinction becomes important in the study of prisoner reentry because although a person who returns to prison has failed in the reentry process, he or she has not necessarily failed through reoffending. Despite this important distinction, in most research, the return to prison will likely be counted as recidivism.

As a result of changing release mechanisms and the changing nature of postrelease supervision, the landscape of prisoner reentry has undergone a fairly substantial shift over the past several decades. Some of these changes have made navigating reentry to the community more difficult in and of themselves, but these difficulties have been compounded by legislation that has erected more and more barriers to full reintegration. In the mid- to late-1990s, when crime was a particularly salient issue and punitive criminal justice policy was the norm, the U.S. Congress enacted a series of legislative initiatives that would ultimately hinder rather than help ex-offenders as they negotiate the reentry process.

## ■ Legislative Barriers to Successful Reintegration

The challenges of prisoner reentry are many, and scholars concerned about the reentry process have particularly emphasized the importance of the periods immediately preceding and following release, with many arguing that release planning should begin on the day of admission (Travis, 2005). Most agree that the most pressing needs of prisoners returning to the community are in the areas of education, employment, housing, health care, and substance abuse. Perhaps not surprisingly, the chances of success in reentry are expected to diminish precipitously if a returning prisoner's needs in any of these areas are not addressed or cannot be met. For example, a returning offender who is unable to secure viable employment may resort quite quickly to criminal activity to "make ends meet." Similarly, the inability to secure housing may result in ex-prisoner homelessness and increased risk for re-offense. Although the risk of re-offense is heightened in the period immediately following release when the challenges are greatest, it is important that reentry initiatives focus not only on the point of release but also on the time leading up to release (reestablishing ties to the family and community) and on the months and years that follow release. Reentry is a process, not a moment.

Before turning to a discussion of policy developments that run counter to the renewed interest in addressing the problem of prisoner reentry, it seems imperative that the differences between reentry and reintegration are identified. The vast majority of prisoners sentenced to a period of incarceration will *reenter*—in that they are eventually released from custody and returned to communities. Although most will reenter, far fewer are able to successfully *reintegrate*. Success in terms of reintegration would require that the formerly incarcerated person become a fully functioning member of the community, eventually afforded most (if not all) of the rights and responsibilities of those who have never been incarcerated. A reintegrated ex-offender is that person who is able to remain crime-free, to secure and maintain employment and housing, to (re)establish positive ties to the community, and to successfully reconnect with family and friends after an often extended period of separation (Travis, 2005).

### Policies That Inhibit Reentry

Although it is widely recognized that offender reintegration is crucial to the successful reentry experience, there are multiple challenges faced by those trying to make this transition (and by organizations trying to assist offenders in navigating the process). Laws in most states restrict ex-offender access to benefits in the areas of housing, education, employment, and welfare assistance—the areas crucial to successful prisoner reentry (Lynch & Sabol, 2001). These restrictions, which have been referred to as "collateral consequences" and "barriers to reintegration," typically attach to a felony conviction and negatively impact the ability of returning prisoners to successfully make the transition to communities. Laws such as those restricting the extension of housing benefits to convicted drug or violent offenders may make sense from a public (housing) safety perspective but may prove counterproductive from a prisoner reentry perspective.

Most of these counterproductive policies were enacted in the mid- to late-1990s under the Clinton administration (shortly after crime rates had reached an all-time high and just before they began a steady and substantial decline (see Zimring, 2007). These policies have little to no effect on criminal defendants who are financially stable—all target offenders in the lowest socioeconomic strata (i.e., those in need of public housing, welfare, and food assistance). Due to structural inequalities in American society, these policies that target the most disadvantaged felony offenders will inevitably have a disparate impact on minorities. In explaining racially disparate increases in incarceration, Michael Tonry (1995) has argued that, whether intentional or not, "the rising levels of Black incarceration did not just happen. . . . They were the foreseeable effects of deliberate policies" (p. 4). It could similarly be argued that the collateral consequences of felony convictions would also have foreseeable disparate effects. Clinton-era policies in the areas of education, housing, public assistance, and parental rights are discussed in the sections that follow.

### Policies Limiting Access to Education

Although the link between employment and crime is complicated and not particularly well understood, few would deny that there is an important link between the two. Indeed securing and maintaining work is seen as particularly critical to successful reentry. Years of research have indicated that the stigma of a criminal conviction alone has a sizeable impact on a person's ability to secure employment (see Pager, 2007). The criminal conviction, however, is not the only reason that ex-offenders have a hard time securing employment. Those sentenced to prison, and thus those ultimately released from prison, have notable educational deficits (Petersilia, 2005) that impact their ability to secure and maintain employment postrelease. Recognizing that deficits in educational attainment limit opportunities for gainful employment and that lack of such employment is correlated to criminal behavior, educational programs were among the most common and most popular forms of prison programming. These prison-based education programs were financed, in part, through educational opportunity programs that have all but disappeared in recent years.

On November 8, 1965, Congress passed the Higher Education Act (HEA) of 1965, a piece of federal legislation that was designed to open the door to higher education for those who otherwise could not afford it. The HEA of 1965 faces regular congressional reauthorization and was most recently reauthorized in 2008 by the Higher Education Opportunity Act (HEOA). The original HEA of 1965 and subsequent amendments have had a profound effect on access to education for prisoners, both while in prison and upon their return to the community.

When the HEA of 1965 was passed, it established the education grant opportunity program that is commonly known as the Pell Grant program (Page, 2004). The Pell Grant program represents the single biggest investment in higher education that the federal government has ever made (Page, 2004). Between the 1970s and mid-1980s, Pell Grants served as a major funding stream for prison education, as prisoners tend to come from precisely the population that the Pell Grant program was intended to reach (i.e., those with demonstrable financial need). In the years since, the HEA of 1965 has been modified in ways that first limited and then completely eliminated prisoner access to Pell Grant funding for higher education (Page, 2004). With the passage of the Anti-Drug Abuse Act of 1988, drug offenders became the first class of offenders denied access to Pell Grant funding. Shortly thereafter, during the 1992 HEA reauthorization, those serving life sentences and those sentenced to death were denied access to Pell Grants. In 1994, prisoner access to Pell Grants was completely eliminated with the passage of the Violent Crime Control and Law Enforcement Act (Page, 2004).

During its 1998 reauthorization, the HEA of 1965 was amended to include a provision that suspended eligibility for federal grants, loans, and work assistance programs for those convicted of drug-related offenses. The Free Application for Federal Student Aid (FAFSA) includes a question that specifically asks about drug convictions—both an affirmative response and no response can trigger ineligibility (www.fafsa.ed.gov). Between the year 2000, when this question was added to the FAFSA, and 2005, at least 200,000 students (or 1 in 400 applicants) were denied federal student aid on account of drug convictions (Leinwand, 2006). Under the 1998 HEA Amendment (P.L. 105-244), the period of ineligibility following a drug conviction depends in part upon the type of conviction offense (possession or sale) and the offense history (with more substantial periods of ineligibility for second and subsequent offenses). A second offense for sale of controlled substances or third offense for possession of controlled substances results in indefinite suspension of eligibility. The suspension of eligibility, whether time-limited or indefinite, can only be lifted upon proof of rehabilitation or if the conviction is ultimately overturned or set aside. Although federal legislation often gives states the option to opt out of various provisions, states are not permitted to opt out of (or modify) the ban on federal student assistance (Legal Action Center, 2004). States, however, have denied *state aid* on the basis of *federal criteria* set out in the 1998 HEA Amendment (Coalition for Higher Education Act Reform, 2006).[2]

### Policies Restricting Access to Public Housing

Most would agree that securing housing upon release from incarceration is one of the most fundamental needs of prisoners facing reentry.[3] Despite the importance of securing suitable housing to the reintegration process, and with the full knowledge that the vast majority of offenders leaving prisons are drawn from the lowest socioeconomic strata, the federal government has made access to public housing all the more difficult for ex-offenders through a series of increasingly restrictive housing policy initiatives (Rubenstein & Mukamal, 2002; Thompson, 2008; Travis, 2002). The move to exclude ex-offenders from public housing began with the Anti-Drug Abuse Act (ADAA) of 1988 (mentioned earlier in the context of Pell Grant eligibility), which was passed during the height of the War on Drugs. The ADAA of 1988 was the second of two omnibus ADAAs to impose harsh penalties on drug offenders. The first, the ADAA of 1986 (P.L. 99-570, 100 Stat. 3207), had established the bulk of the mandatory minimum sentences for drug offenses that have become so inextricably associated with the War on Drugs, including those mandatory minimums providing for a 100:1 distinction in penalties for possession of crack versus powder cocaine (see Vagins & McCurdy, 2006).

In the 1988 ADAA, the federal government required that local public housing authorities include clauses in their public housing leases prohibiting tenants from engaging in drug-related criminal activity on or near public housing premises. Shortly thereafter, the law was amended to read that "any criminal activity that threatens the health, safety, or right to peaceful enjoyment of the premises by other tenants, or any drug related criminal activity" could result in eviction (42 U.S.C. 1437d(1)(5), 1990). Although the clause broadly references criminal activity, in its enforcement it has been particularly focused on evictions for drug-related criminal activity, which had been characterized as the scourge of public housing. After a series of revisions, the criminal activity provision applies not only to the tenant signing the lease, but also to "any member of the tenant's household, or a guest or other person under the tenant's control" and it applies to behavior not just in and around the public housing complex, but also "on or off such premises" (42 U.S.C. 1437d(1)(6)). As an enforceable provision of the tenancy lease, engaging in criminal activity can result in the termination of the lease at the discretion of the housing authorities. As a result, entire families could potentially see their public housing leases terminated based on the actions of one household member, a visiting relative, or even just a guest. These public housing laws were passed amid concern about growing crime and violence associated with the drug trade in and around public housing complexes.

The move to exclude convicted offenders from public housing continued through the 1990s as Congress strengthened the public housing policy when it passed the 1996 Housing Opportunity Program Extensions (HOPE) Act (P.L. 104-120, 110 Stat. 834, 1996) and the Quality Housing and Work Responsibility Act (Title V of P.L. 105-276, 1998). During his 1996 State of the Union Address, President Clinton championed strengthening the "one strike and you're out" eviction policy that was ultimately passed by Congress, saying:

> I challenge local housing authorities and tenant associations: criminal gang members and drug dealers are destroying the lives of decent tenants. From now on, the rule for residents who commit crime and peddle drugs should be one strike and you're out. (Woolley & Peters, n.d.)

Although the one-strike rule had technically been in the original legislation, the 1996 HOPE Act tied bonus federal funding to its enforcement (Harris, 1996). Crucially, no criminal *conviction* is required to trigger the eviction.

Stricter enforcement of the one strike and you're out policy led to a series of tragically sad one-strike eviction cases. The cases, ultimately appealed all the way to the U.S. Supreme Court, highlighted what some believed were overly broad criteria for whose behaviors could trigger an eviction under the new policy. The public housing eviction cases were described in detail on appeal to the U.S. Court of Appeals for the 9th Circuit in the case *Rucker v. Davis* (237 F.3d 1113, 2001).

The cases before the court were those of Pearlie Rucker, Barbara Hill, Willie Lee, and Herman Walker, four public housing residents from Oakland, California, who had been evicted by the Oakland Housing Authority on the basis of the criminal activity provision. Rucker, Hill, Lee, and Walker were all elderly tenants (aged 63 to 75 years) who had been residing in their public housing units for years. In the first three cases (those of Rucker, Lee, and Hill), the tenants had been evicted based on the drug use of their children or grandchildren. Rucker's mentally disabled daughter had been convicted of cocaine possession—not on the public housing premises, but rather three blocks away. Hill and Lee's grandsons were caught smoking marijuana in the parking lot of the public housing complex. Walker, a 75-year-old disabled man with a live-in caretaker, was evicted after that caretaker and some guests were found in possession of cocaine. In each of these cases, the criminal offender had resided in public housing, and in each of these cases the primary tenants denied any knowledge of drug use.

The question before the 9th Circuit Court of Appeals was whether the U.S. Congress, when it passed the ADAA of 1988 and subsequent housing policy bills, intended for the Department of Housing and Urban Development (HUD) to write guidelines that would result in the eviction of "innocents" (i.e., those who were not aware or in control of the conduct of the criminally offending party). Circuit Judge Michael Daly Hawkins, who presided over *Rucker v. Davis* (237 F.3d 1113, 2001), assessed the plight of those living in public housing that led to the passage of these provisions as follows:

> Many of our nation's poor live in public housing projects that, by many accounts, are little more than illegal drug markets and war zones. Innocent tenants live barricaded behind doors, in fear for their safety and the safety of their children. What these tenants may not realize is that, under existing policies of the Department of Housing and Urban Development ("HUD"), they should add another fear to their list: becoming homeless if a household member or guest engages in criminal drug activity on or off the tenant's property, even if the tenant did not know of or have any reason to know of such activity or took all reasonable steps to prevent the activity from occurring ("innocent tenants").

The Court of Appeals for the 9th Circuit ultimately concluded that Congress did not intend the eviction of innocent tenants. The decision of the 9th Circuit Court was overturned by the U.S. Supreme Court when it rendered its decision in *Department of Housing and Urban Development v. Rucker* (535 U.S. 125, 2002). In overturning the 9th Circuit Court, Chief Justice Rehnquist, writing for the majority, held that the statute "unambiguously requires lease terms that vest local public housing authorities with the discretion to evict tenants for the drug-related activity of household members and guests whether or not the tenant knew, or should have known, about the activity" (2002). Given the U.S. Supreme Court's 2002 ruling, public housing evictions for criminal activity are constitutionally permissible and can be initiated at the discretion of housing authorities.

The 1996 and 1998 housing policy acts also introduced stricter mandates that strengthened the ability of Public Housing Authorities (PHAs) to deny public housing to those with criminal histories (Rubenstein & Mukamal, 2002). Local PHAs can require criminal history checks for new applicants and can deny access to public housing to those who have engaged in criminal activity in the past. As Rubenstein and Mukamal (2002) have argued, these developments in public housing policy potentially have profound impacts on those attempting to reintegrate following prison release because "without access to decent, stable, and affordable housing, the likelihood of an ex-offender being able to obtain and retain employment and remain drug- and crime-free is significantly diminished" (p. 48).

### Policies Limiting Eligibility for Public Assistance

Closely related to the move to exclude ex-offenders (and particularly ex-drug offenders) from public housing is the move to permanently restrict their eligibility for public assistance. As with the policies strengthening restrictions in the areas of higher education and public housing, the move to deny welfare benefits to drug offenders was a policy initiative championed during the mid-1990s at the height of the Clinton administration. The "welfare ban" was included as part of the massive welfare reform initiative known as the Personal Responsibility and Work Opportunity Reconciliation Act (PRWORA) of 1996 (P.L. 104-193, 110 Stat. 2105). After almost two decades of efforts to reform welfare, Bill Clinton promised to "end welfare as we know it" during his 1992 presidential campaign and followed through on that promise with the signing of PRWORA in August of 1996 (Carcasson, 2006). With the passage of PRWORA, the welfare system in the United States underwent its largest overhaul since its establishment in the 1930s. Under PRWORA, the traditional welfare entitlement program known as Aid to Families with Dependent Children (AFDC) was replaced by the block-grant program Temporary Assistance to Needy Families (TANF). The PRWORA signaled the end of entitlement (TANF requires work and is often referred to as a "welfare to work" program) and limited TANF assistance to 5 years.

The public assistance eligibility ban imposed on drug offenders as part of PRWORA was introduced as an amendment to the initial welfare reform bill and was debated in Congress for just 2 minutes (Rubenstein & Mukamal, 2002). The provision allows states to impose a lifetime ban on public assistance for those convicted of felony drug offenses. Although states were allowed to opt out of the ban, according to the Legal Action Center (2004), only 12 states completely opted out of the ban. Moreover, while 21 states enforce the ban with some modifications, 17 states enforce the ban in full—permanently denying public assistance to those convicted of drug offenses. It is important to note that this particular ban applies only to those convicted of drug offenses—a conviction for an offense other than a drug offense has no bearing on access to public assistance or food stamps. Like the ban on public housing, this "collateral consequence" primarily affects ex–drug offenders of limited means; however, unlike the ban on public housing, which is (at least in theory) time limited, the ban on public assistance is permanent. Those released from prison for drug offenses in states that fully enforce the ban will never recover their eligibility for this type of assistance, no matter what steps they take to overcome their criminal histories and no matter how desperately they need it.

### Policies Affecting Child Custody

Another policy initiative associated with the Clinton administration also has a potentially damaging effect on the reintegration process. The Adoption and Safe Families Act of 1997 (P.L. 105-89) required criminal records checks for those seeking to be foster or adoptive parents and, more importantly, authorized the termination of parental rights for children who had been in state custody for 15 of the most recent 22 months. Although the ASFA does not specifically reference children in state custody because of parental imprisonment, the policy clearly might affect parents (and particularly women) who face periods of incarceration of more than 15 months. In 2000, it was estimated that 1.5 million children under the age of 18 years had a parent in prison and that more than 10% of those children were living in foster homes or agencies (Mumola, 2000). Although not all incarcerated parents had physical custody of their children in the period leading up to their incarceration, women in prison were more likely than men (64% versus 44%) to report having lived with their children immediately prior to their incarceration (Mumola, 2000). For the more than 150,000 children living in foster homes while a parent serves time in prison, permanent termination of parental rights is a real possibility (see Thompson, 2008, for a review of other policies affecting parental rights). The importance of maintaining (or if necessary reestablishing) family ties upon release from prison has been underscored by research that consistently shows that both maintaining those familial bonds and reestablishing parental roles after release increase the likelihood of successful reentry (Petersilia, 2005).

The policy initiatives ushered in during the mid- to late-1990s were not kind to those convicted of felony offenses (and particularly of felony drug offenses). Crucially, the consequences of these policies are not evenly felt. Most of these policies affect only those of limited socioeconomic means and—because of the focus on drug offenses and offenders—will have potentially profound effects on the poor, minority communities that have largely been the battleground for the War on Drugs. Although the War on Drugs led to increases in admissions to prison for drug offenses across the board, the increases in admissions for African Americans has been staggering (Travis, 2002, p. 32). Women, too, have been disproportionately affected by drug laws, with women of color bearing the brunt of the impact (see Thompson, 2008, p. 49). The stigma of a felony conviction has always meant that offenders exiting prison would come up against challenges when trying to reintegrate, but the erection of further barriers through the series of policy initiatives affecting education, housing, welfare, and child custody has only made that transition from prison to community more difficult.

## The Future of Reentry Policy

Although the policies just discussed are barely a decade old, most preceded the current surge in interest in prisoner reentry. Despite a longstanding interest in prison release and postrelease outcomes, the current prisoner reentry movement was first launched in earnest in the late 1990s when Jeremy Travis served as Director of the National Institute of Justice under Attorney General Janet Reno. The reentry movement has gained substantial momentum over the past decade. In his important book, *But They All Come Back*, Jeremy Travis (2005) opens with a sobering passage highlighting the importance of the renewed focus on reentry:

> Reentry is the process of leaving prison and returning to society. Reentry is not a form of supervision, like parole. Reentry is not a goal, like rehabilitation or reintegration. Reentry is not an option. Reentry reflects the iron law of imprisonment: they all come back. (p. xxi)

In practice, prisoner reentry is a collaborative endeavor. The collaborative nature of prisoner reentry is perhaps most evident in the funding of reentry initiatives. Although criminal justice initiatives receiving federal funding have traditionally been funded through the U.S. Department of Justice, recognizing the crucial role that education, employment, health, and housing will likely play in successful prisoner reentry, funding for reentry initiatives typically comes from multiple federal agencies and private foundations. Funding for the Serious and Violent Offender Reentry Initiative (SVORI), for example, came from five different federal agencies (the U.S. Departments of Justice, Labor, Education, Housing and Urban Development, and Health and Human Services). The federally funded SVORI program provided grantees in all 50 states with funds to develop collaborative relationships between correctional and community agencies (see Lattimore et al., 2004; Visher, 2007).

Similarly, in 2001, the Council of State Governments established the Reentry Policy Council (RPC) to develop bipartisan policy recommendations and facilitate information sharing around reentry initiatives. The RPC aims to assist and guide state and federal lawmakers, policy makers, and practitioners as they work to make the reentry process successful for formerly incarcerated persons. In 2005, the RPC published its landmark report, "Charting the Safe and Successful Return of Prisoners to the Community," providing practical guidance for developing innovative and comprehensive reentry initiatives. Although developing strategies for reducing recidivism is clearly at the forefront of the RPC's agenda, the recommendations simultaneously focus on helping formerly incarcerated persons "become productive, healthy members of families and communities." In other words, the RPC is interested in both the process of reentry and the end goal of meaningful reintegration. Just as with SVORI, multiple federal agencies and several private foundations support the work of the RPC. Prisoner reentry also factors prominently on the agendas at the Urban Institute, the Vera Institute of Justice, and other research organizations devoted to issues of crime and justice.

Just as there has been an abundance of scholarly work and a flood of non-profit activity in the area of prisoner reentry, there has been some recent legislative activity in support of the growing reentry movement. Most notably, Congress passed, and former President George W. Bush signed into law, the Second Chance Act (Public Law 110-199) on April 9, 2008 (Reentry Policy Council, n.d.). The Second Chance Act authorizes federal grant funding to government, community, and faith-based organizations providing various reentry-related services and programs (O'Hear, 2007). The explicit goal of the legislation is to reduce recidivism, but the funding supports organizations providing assistance across a number of domains, including employment and housing assistance, mental health and substance abuse treatment and counseling, and family and victim support services (see Pinard, 2006, 2007). Although Congress failed to pass an appropriations bill that would have funded the Second Chance Act in 2008, the Second Chance Act received the long-awaited funding with President Obama's signing of the Omnibus Appropriations Act 2009 on March 11, 2009 (White House, 2009). In February of 2009, the first Second Chance Act grant solicitations were issued by the U.S. Department of Justice (Bureau of Justice Assistance, 2009). State and local agencies confronting the issue of prisoner reentry and developing evidence-based practices to address it are eligible to apply for grant funding to develop demonstration projects under the Second Chance Act. The Act grant solicitation makes clear that in order to qualify for federal funding, jurisdictions must recognize prisoner reentry as a process and the demonstration projects must include both prerelease and postrelease interventions. Reduction of recidivism by at least 50% must be the goal of demonstration projects funded through the Second Chance Act (see Bureau of Justice Assistance, 2009).

## ■ Conclusion

Prisoner reentry is widely recognized as one of the most pressing policy challenges that penologists and policy makers of this generation face. Unprecedented growth in incarceration and a shift in the underlying rationale of corrections have meant an influx of returning prisoners who have received little in the way of rehabilitative programming. Without sufficient attention to the process by which these inmates return, most agree that both the returning prisoners and the communities receiving them will likely suffer.

In the current era of mass incarceration, prisoner reentry is not simply a correctional concern or a criminal justice system concern. Nor should concern about prisoner reentry center solely around offender recidivism (Lynch, 2006). Successful prisoner reentry initiatives need to recognize both the challenges faced by ex-offenders in trying to reintegrate and the challenges that ex-offenders pose to public safety as they try to navigate the reentry process. To that end, there are at least two crucial components to reentry initiatives: (1) a focus on the offender and maximizing the offender's potential for successful reintegration upon return from prison, and (2) a focus on communities and developing community capacity to successfully accept returning prisoners. Far more work has focused on the former than on the latter, although several recent books (Clear, 2007; Thompson, 2008) have brought communities to the forefront of the discussion.

This chapter concludes with an observation that, while obvious, is all too often taken for granted in works focusing on prisoner reentry: The crisis of prisoner reentry would not be nearly as substantial were it not for the magnitude of the crisis of mass incarceration. For more than 35 years now, politicians have introduced and backed policy initiatives that have led to incarceration rates of unprecedented scale. Thirty-five years ago it would have been utterly inconceivable that the population of incarcerated persons would exceed 2.5 million and that communities across the United States would be dealing with the return of more than 700,000 prisoners in a single year. Indeed, in the late 1960s and early 1970s, some of the most prominent punishment scholars were predicting an end to incarceration (see Tonry, 2004). As inconceivable as it might have once been, it would be a mistake to infer that we arrived at this point by accident. Our incarceration rate is the result of a series of policy choices that all but ensured nonstop growth in prison populations (see Frost & Clear, 2009). In the end, the best way to begin solving the problem of prisoner reentry might very well be to stop sending so many offenders to prison in the first place.[4]

## ■ Endnotes

1. All of the prevalence and lifetime likelihood estimates in the paragraphs that follow come from the Bonczar (2003) report.
2. In 2005, the Deficit Reduction Act limited the denial of federal student assistance to those convicted of drug offenses while receiving federal aid (Coalition for Higher Education Act Reform, 2006).
3. According to recent research, it is estimate that at least 10% of those leaving prison end up homeless (Roman & Travis, 2004).
4. For a series of policy recommendations to drastically reduce the prison population offered by prominent punishment scholars, see Austin et al. (2007).

## ■ Discussion Questions

1. Discuss the relationship between increases in the use of incarceration and the rise of interest in prisoner reentry.

2. What factors make prisoner reentry and reintegration policies particularly challenging for policy makers?

3. How do sentencing structures influence reentry and reintegration policies?

4. How do existing criminal justice and social policies create barriers for effective prisoner reentry and rehabilitation?

5. What steps can policy makers take to increase the likelihood that prisoner reentry and reintegration will be successful?

## ■ References

Austin, J., Clear, T., Duster, T., Greenberg, D. F., Irwin, J., McCoy, C., et al. (2007). *Unlocking America: Why and how to reduce America's prison population.* Washington, DC: JFA Associates, Inc.

Austin, J., & Irwin, J. (2001). *It's about time: America's imprisonment binge* (3rd ed.). Belmont, CA: Wadsworth.

Blumstein, A., Cohen, J., & Nagin, D. (1977). The dynamics of a homeostatic punishment process. *Journal of Criminal Law and Criminology, 67,* 317–334.

Bonczar, T. P. (2003). *Prevalence of imprisonment in the U.S. population, 1974–2001.* Bureau of Justice Statistics Special Report (NCJ 197976). Washington, DC: Bureau of Justice Statistics, U.S. Department of Justice.

Brooks, L. E., Solomon, A. L., Keegan, S., Kohl, R., & Lahue, L. (2005). *Prisoner reentry in Massachusetts.* Washington, DC: The Urban Institute.

Bureau of Justice Assistance. (1998). *1996 national survey of state sentencing structures* (NCJ 169270). Washington, DC: Government Printing Office.

Bureau of Justice Assistance. (2009). *Second Chance Act: Prisoner reentry initiative, FY 2009 competitive grant announcement.* Washington DC: U.S. Department of Justice.

Carcasson, M. (2006). Ending welfare as we know it: President Clinton and the rhetorical transformation of the anti-welfare culture. *Rhetoric & Public Affairs, 9*(4), 655–692

Clear, T. R. (2007). *Imprisoning communities: How mass incarceration makes disadvantaged neighborhoods worse.* New York: Oxford University Press.

Clear, T. R., Rose, D., & Ryder, J. (2001). Incarceration and the community: The problem of removing and returning offenders. *Crime and Delinquency, 47,* 335–351.

Coalition for Higher Education Act Reform. (2006). *Falling through the cracks: Loss of state-based financial aid eligibility for students affected by the federal Higher Education Act drug provision.* Washington, DC: Coalition for Higher Education Act Reform.

Cullen, F. T. (1995). Assessing the penal harm movement. *Journal of Research in Crime and Delinquency, 32,* 338–358.

Cullen, F. T., Fisher, B. S., & Applegate, B. K. (2000). Public opinion about punishment and corrections. In M. Tonry (Ed.), *Crime and justice: A review of research* (vol. 27, pp. 1–79). Chicago: University of Chicago Press.

Feeley, M. (2003). Review essay: Crime, social order and the rise of neo-conservative politics. *Theoretical Criminology, 7,* 111–130.

Frost, N. A., Greene, J., & Pranis, K. (2006). *Hard hit: The growth in the imprisonment of women, 1977–2004.* New York: Women's Prison Association.

Frost, N. A., & Clear, T. C. (2009). Understanding mass incarceration as a grand social experiment. *Studies in Law, Politics, & Society, 47,* 159–191.

Harris, J. F. (1996, March 29). Clinton links housing aid to eviction of crime suspects. *Washington Post,* p. A14.

Harrison, P. M. (2000). *Total sentenced prisoners released from state or federal jurisdiction, 1977–1998.* Washington, DC: Bureau of Justice Statistics.

Harrison, P. M. (2004). *Prison admissions and releases by gender, 1999 and 2000.* Washington, DC: Bureau of Justice Statistics.

Lattimore, P. K., Brumbaugh, S., Visher, C., Lindquist, C., Winterfield, L., Salas, M., et al. (2004). National portrait of SVORI: Serious and violent offender reentry initiative. Retrieved February 16, 2010, from the Urban Institute Web site: http://www.urban.org/publications/1000692.html

LaVigne, N. G., & Kachnowski, V. (2003). *A portrait of prisoner reentry in Maryland.* Washington DC: The Urban Institute.

LaVigne, N. G., & Mamalian, C. A. (2003). *A portrait of prisoner reentry in Illinois.* Washington, D.C.: The Urban Institute.

Legal Action Center. (2004). *After prison: Roadblocks to reentry.* New York: Author.

Leinwand, D. (2006, April 17). Drug convictions costing students their financial aid. *USA Today Online.* Retrieved February 16, 2010, from http://www.usatoday.com/news/nation/2006-04-16-drugs-students_x.htm

Lynch, J. P. (2006). Prisoner reentry: Beyond program evaluation. *Criminology & Public Policy, 5*(2), 401–412.

Lynch, J. P., & Sabol, W. J. (2001). *Prisoner reentry in perspective.* Washington, DC: The Urban Institute.

Martinson, R. (1974). What works? Questions and answers about prison reform. *Public Interest, 35,* 22–54.

Massachusetts Sentencing Commission. (2004). *Survey of sentencing practices, FY 2003.* Boston: Author.

Mumola, C. J. (2000). *Incarcerated parents and their children.* Washington, DC: Bureau of Justice Statistics.

Nagin, D., Piquero, A. R., Scott, E. S., & Steinberg, L. (2006). Public preferences for rehabilitation versus incarceration of juvenile offenders: Evidence from a contingent evaluation study. *Criminology & Public Policy, 5*(4), 627–652.

O'Hear, M. M. (2007) The Second Chance Act and the future of the reentry move-
      ment. *Federal Sentencing Reporter, 20*(2), 75–83.

Page, J. (2004). Eliminating the enemy: The import of denying prisoners access to
      higher education in Clinton's America. *Punishment & Society, 6*(4), 357–378.

Pager, D. (2007). *Marked: Race, crime, and finding work in an era of mass incarcera-
      tion.* Chicago: University of Chicago Press.

Petersilia, J. (2003). *When prisoners come home: Parole and prisoner reentry.* New
      York: Oxford University Press.

Petersilia, J. (2005). From cell to society. In J. Travis & C. Visher (Eds.), *Prisoner
      reentry and crime in America* (pp. 15–49). New York: Cambridge University
      Press.

Piehl, A. M., & LoBuglio, S. F. (2005). Does supervision matter? In J. Travis & C.
      Visher (Eds.), *Prisoner reentry and crime in America* (pp. 105–138). New York:
      Cambridge University Press.

Pinard, M. (2006). An integrated perspective on the collateral consequences of
      criminal convictions and reentry issues faced by formerly incarcerated indi-
      viduals. *Boston University Law Review, 86*, 623–690.

Pinard, M. (2007). A reentry centered vision of criminal justice. *Federal Sentencing
      Reporter, 20*(2), 103–109.

Reentry Policy Council. (2005). *Report of the Reentry Policy Council: Charting the
      safe and successful return of prisoners to the community.* New York: Council of
      State Governments.

Reentry Policy Council. (n.d.). The Second Chance Act. Retrieved February 16, 2010,
      from http://www.reentrypolicy.org/government_affairs/second_chance_act

Reitz, K. R. (2001). The disassembly and reassembly of U.S. sentencing practices.
      In M. Tonry & R. S. Frase (Eds.), *Sentencing and sanctions in Western countries*
      (pp. 222–258). New York: Oxford University Press.

Roman, C. G., & Travis, J. (2004). *Taking stock: Housing, homelessness, and prisoner
      reentry.* Washington, DC: The Urban Institute.

Rubenstein, G., & Mukamal, D. (2002). Welfare and housing—denial of benefits to
      drug offenders. In M. Mauer & M. Chesney-Lind (Eds.), *Invisible punishment:
      The collateral consequences of mass imprisonment* (pp. 37–49). New York: The
      New Press.

Stemen, D., Rengifo, A., & Wilson, J. (2006). *Of fragmentation and ferment: The
      impact of state sentencing policies on incarceration rates, 1975–2002.* Washington,
      DC: Final Report to the U.S. Department of Justice.

Thompson, A. C. (2008). *Releasing prisoners, redeeming communities: Reentry, race,
      and politics.* New York: New York University Press.

Tonry, M. (1995). *Malign neglect: Race, crime and punishment in America.* New
      York: Oxford University Press.

Tonry, M. (1996). *Sentencing matters.* New York: Oxford University Press.

Tonry, M. (2004). Has the prison a future? In M. H. Tonry (Ed.), *The future of imprisonment* (pp. 3–24). New York: Oxford University Press.

Travis, J. (2002). Invisible punishment: An instrument of social exclusion. In M. Mauer & M. Chesney-Lind (Eds.), *Invisible punishment: The collateral consequences of mass imprisonment* (pp. 15–36). New York: The New Press.

Travis, J. (2005). *But they all come back: Facing the challenges of prisoner reentry.* Washington, DC: The Urban Institute Press.

Travis, J. (2007). Reflections on the reentry movement. *Federal Sentencing Reporter, 20*(2), 84–87.

Vagins, D. J., & McCurdy, J. (2006). *Cracks in the system: Twenty years of unjust federal crack cocaine law.* New York: American Civil Liberties Union.

Visher, C. (2007). Returning home: Emerging finding and policy lessons about prisoner reentry. *Federal Sentencing Reporter, 20,* 93.

Visher, C., & Farrell, J. (2005). *Chicago communities and prisoner reentry.* Washington, DC: The Urban Institute.

Visher, C., Kachnowski, V., LaVigne, N. G., & Travis, J. (2004). *Baltimore prisoners' experiences returning home.* Washington, DC: The Urban Institute.

West, H. C., & Sabol, W. J. (2008). *Prisoners in 2007.* Washington, DC: Bureau of Justice Statistics.

Wilson, J. Q. (1975). *Thinking about crime.* New York: Basic Books.

White House. (2009, March 11). Statement from the President on the signing of H.R. 1105. Washington, DC: White House, Office of the Press Secretary.

Woolley, J. T., & Peters, G. (n.d.). William J. Clinton: Address before a joint session of Congress on the state of the union, January 23, 1996. *The American Presidency Project.* Retrieved February 16, 2010, from the University of California (hosted), at: http://www.presidency.ucsb.edu/ws/?pid=53091

Zimring, F. E. (2007). *The great American crime decline.* New York: Oxford University Press.

## ■ Court Cases Cited

*Department of Housing and Urban Development v. Rucker*, 535 U.S. 125 (2002)
*Rucker v. Davis*, 237 F.3d 1113 (2001)

# Prison Privatization Turns 25

# CHAPTER
# 9

Richard Culp

## ■ Introduction

In a 1989 study of privatization policy, John Donahue made the case that without ongoing competition among service providers and diligent contract monitoring by public managers, prison privatization would not generate any real cost savings or service quality improvements to the public. Absent healthy competition for contracts and well-monitored performance, any cost savings realized through privatization would go toward private profits rather than public savings, and quality of service would, at best, remain the same (Donahue, 1989). Donahue was writing in the waning days of the Ronald Reagan–Margaret Thatcher era, a time when market-based solutions to social problems and reducing the size of government were all the rage. The prison privatization idea was just taking off, and little outcome research had been completed, though a few preliminary studies suggested that governments could possibly save money by privatizing correctional facilities. Donahue was skeptical that the lofty claims of privatization advocates—of lower cost and better quality—were truly possible and cautioned that preliminary results "may not be representative of what a fully developed private corrections market would look like" (p. 158). Noting that competition was "far easier to praise than to arrange," Donahue posed a question that could only be answered in the future: "Will the incarceration industry, once it matures, be competitive?" (p. 165).

A quarter century has now passed since the birth of the modern prison privatization industry, giving us the perspective of hindsight to help us in answering Donahue's question. Do we have a competitive private prison industry that helps to lower costs and improve quality? Is the industry more innovative than the public sector? And has it brought service improvements to the prisons it runs?

Ultimately, has prison privatization been a good deal for the public? To help in answering these questions, we will look at the structure of the private prison industry that has evolved and the level of competition that exists in the marketplace for incarceration services. Placing the developing private prison business within an industrial life cycle framework, we will see that the realities of the mature prison privatization market do not match the promise of innovation and quality improvements voiced by privatization advocates 25 years ago. The private prison industry is dominated by a few large suppliers (an *oligopoly* of producers) and only a handful of buyers (an *oligopsony* of consumers), with the result being that any real cost advantage of privatization is marginal at best, private prison programs have become virtually indistinguishable from public prisons, and the promise of innovation remains unfulfilled.

We will walk through each of these points in turn, beginning with a brief review of privatization philosophy and a discussion of outcome research. Next we will look at the marketplace for incarceration services, examining how private incarceration services are becoming increasingly concentrated in fewer companies as time goes on. While the decrease in the number of viable incarceration companies limits competition (as well as cost and innovative advantage), it has also served to eliminate a number of incompetent providers. At the same time, few government agencies purchase incarceration services, further reducing competition, while the industry as a whole has not proved to be innovative at all.

## ■ Why Privatization?

The private prison "market" that emerged in the 1980s was qualitatively different than the privatization model that existed in an earlier era. The old model of privatization was based on convict leasing, or selling the labor of inmates on the open market. It was commonplace in early American prisons to contract out the labor of convicts to private entrepreneurs (Durham, 1989). While the practice extended well into the 20th century, it became increasingly disreputable as incidents involving abusive treatment of inmates and insider contracting arrangements were brought to light (Durham, 1989; Schneider, 1999), raising concerns about the appropriateness of private sector involvement in corrections. The notion of "market failure," that the government must step in to provide essential services when the private sector fails to adequately provide them, gained influence in public administration in the United States during the 1920s through the writings of Arthur C. Pigou. In Pigou's (1920) view, correctional services are a public good (as opposed to a private good) that can be produced in a free market. By definition, public goods are products or services that are characterized by nonrival and nonexcludable consumption. Correctional services are nonrival in that consumption of incarceration services by one inmate does not preclude another offender's consumption (nonrival consumption), while everyone in society (arguably) benefits from enhanced public safety whether they pay for it or not (nonexcludable consumption). This being the case, there is no true market for these goods to be traded and the government has the responsibility to allocate the goods administratively. And because everyone pays for public services (like corrections) through taxes, Pigou argued that tax revenue should be distributed to public employees and expended on facilities owned by the public.

The labor union movement also played a role in the demise of privatized prison labor. Organized labor lobbied against the practice, arguing that it unfairly depressed wages and cut into the private sector job market. The fate of private prisons and contract labor was sealed in 1929, when Congress passed the Hawes-Cooper Act and related legislation that banned prison-made goods from interstate commerce. By 1940, virtually every state had passed legislation banning the import of prison-made goods from other states as well (Clear & Cole, 1997).

A variation on the old prison privatization idea resurfaced in the United States 40 years later as part of a broader movement to downsize government and improve the efficiency of its operations. The idea that government should "steer and not row"—that it should oversee the delivery of services through contracts with third parties rather than deliver the services itself—was popularized in the new public management (NPM) (Kettl, 2000) and reinventing government (Osborne & Gaebler, 1992) philosophies. NPM sought to shrink government and impose market-style discipline on its operations (Kettl, 2000). The term originated in the 1980s to describe governmental reforms in New Zealand, but was broadened to include many Thatcher-era government reforms in the United Kingdom and reforms in the United States under the rubric of the "reinventing government movement" (Osborne & Gaebler, 1992). Both projects called for separating government's role as a purchaser of service (its policy function) from its role as a direct services provider (its service delivery function). The idea was that privatization of service delivery would introduce greater competition, thus containing costs while fostering innovation and higher quality in services delivered. Privatization advocates drew intellectual support from the public choice school of economics, arguing that government monopoly of service delivery leads to lower quality and higher cost (Girard, Mohr, Deller, & Halstead, 2009).

Advocates of the public choice perspective (Downs & Steiner, 1990; Niskanen, 1971) argued that the old view of the public sector advanced by Pigou was based on an erroneous assumption that public managers are motivated by public interest. They argued that all managers, whether public or private, are essentially motivated by self-interest. Where private managers seek to maximize self-interest through the generation of profits, public managers likewise pursue self-interest by seeking to maximize their authority, the number of staff under their control, and the size of their budgets. In the public choice view, inefficiency is the net effect of this self-interest; infusing competition into the delivery of public goods and services is the antidote.

Riding first the wave of public choice economics during the Reagan years and then the NPM and reinventing government movements under the Clinton administration during the 1990s, the number of inmates housed in private prisons in the United States grew from 3000 in 1987, to 15,000 in 1990, 50,000 in 1995, 90,000 in 2000, and 120,000 in 2007 (West & Sabol, 2008; Thomas, 1997). By 2000, there were 158 private correctional facilities operating in the United States (Austin & Coventry, 2001), and, collectively, private facilities held about 6% of all inmates in the country (West & Sabol, 2008). In 2000, however, the growth of private prisons stagnated, with the percentage of inmates held in private custody remaining relatively constant at between 6.5% and 7.5% (West, Sabol, & Cooper, 2009). As of 2008, the federal government and 27 states had pursued the policy of contracting with private prison companies for placement of a portion of their inmates (West & Sabol, 2009).

## ■ Research Synopsis

Twenty-five years of research on prison privatization has failed to determine which sector does a better job in managing inmates. Most early criminal justice studies of private and public prisons were designed to compare costs of operation (Legislative Budget Committee, 1996; Logan & McGriff, 1989; Sellers, 1989; Tennessee Select Oversight Committee, 1995; Texas Sunset Advisory Commission, 1991). Much of this research was sponsored by state-level agencies interested simply in finding out if states could save money by privatizing some of their prisons. While the studies generally found private prisons to be cheaper by approximately 8–15%, the findings are suspect because many jurisdictions stipulate that bids from the private sector must be lower than public sector costs in order for privatization to move forward. A Government Accountability Office (GAO) report noted the difficulty in making valid comparisons between public and private sector costs of running a prison and criticized the methodological validity of the body of existing privatization cost-comparison research (Government Accountability Office, 1996). McDonald, Fournier, Russell-Einhorn, and Crawford (1998) updated the GAO analysis of existing comparative studies and concluded that the jury is still out on whether private facilities are more cost-effective than public ones. But factors other than what sector manages a facility probably exert greater influence on cost-effectiveness. For example, Pratt & Maahs (1999) conducted a meta-analysis of 33 cost studies of private and public prisons and found that facility size, age, and security level were more predictive of facility costs than private versus public ownership.

Because not all government jurisdictions adopt a policy of prison privatization, researchers have tried to find characteristics that distinguish governments that do privatize from those that do not. The likelihood that a state will privatize prisons is positively correlated with the state's incarceration rate

(Schneider, 1999) and with a conservative electorate (Nicholson-Crotty, 2004; Price & Riccucci, 2005). However, despite labor union opposition to privatization in general, the likelihood of prison privatization in a state does not seem to be correlated with union strength (Nicholson-Crotty, 2004; Price & Riccucci, 2005). Overall, it is likely that a convergence of historical factors, including the popularity of free market political ideology, federal court limits on prison overcrowding, and state government budgeting woes have helped to move the policy forward (Culp, 2005).

Studies of the quality of services provided by private prisons have also not produced consistent results. The body of studies suggests that, systemwide, private prisons provide a quality of inmate care on par with public facilities (Austin & Coventry, 2001; Lukemeyer & McCorkle, 2006). However, Perrone & Pratt (2003) found considerable variation in the methodological quality of studies of conditions of confinement in public versus private facilities and questioned the wisdom of relying on the current body of work. As with the GAO (1996) meta-analysis of cost-effectiveness, no clear answer on which sector has the advantage in terms of program quality has emerged. Because most of the service quality research relies on case studies of one or two facilities, it is not possible to generalize to the sector in general (Perrone & Pratt, 2003). Nonetheless, it is reasonably safe to say that where low performance has been observed in private prisons, studies suggest that it is more a product of poorly written contracts or inadequate contract monitoring than inherent differences between the public and private sectors (Austin & Coventry, 2001; Collins, 2001; Culp, 2005).

Within the field of public administration, some researchers have studied privatization policy in general and government contracting in particular, but only a few have focused on private prisons specifically (and only as part of a broader public sector study—see, for example, Cooper, 2003; Donahue, 1989; Savas, 2005). Using data from the 1998 American State Administrators Project, a large-scale survey of agency heads from 95 different types of agencies in all 50 states, Brudney, Fernandez, Ryu, and Wright (2004) found that more than 70% of state government agencies contract out for the delivery of some types of services traditionally provided by the government. However, the goals of improved quality and lower costs—the underpinning of arguments in favor of privatization—are not being met across the board. Overall, only half of the state agencies surveyed acknowledged service delivery improvements, and just one-third reported decreased service costs. At the local level, where as much as 17% of all public services are provided by private companies, governments report a "rather lukewarm response of savings from privatization" (Girard et al., 2009, p. 388). Research by Van Slylce (2003) suggests that, at least in the social services area, a lack of competition among contract providers may be responsible for the general failure of privatization to achieve more promising results.

Just as research has failed to definitively prove whether private prisons are really cheaper and whether they provide better services than the public sector, the ethical issues regarding prison privatization also remain unsettled. A loosely organized but vocal antiprison privatization movement, drawing from the faith-based community, labor unions, and student groups, has advocated against the policy and has limited the extent of privatization in many jurisdictions (Culp, 2005). Many of these critics argue that the very idea of making money from the punishment of offenders is inherently unethical and immoral (Culp, 2005). Some critics of prison privatization warn that the imperative for private companies to turn a profit will lead to diminished quality in confinement conditions for prisoners as prison managers cut corners to maximize stockholder dividends (Christie, 1994; Lilly & Deflem, 1996; Robbins, 1988; Shichor, 1995). Others contend that prison privatization is fundamentally inconsistent with liberal democratic theory (Aman, 2007; Resig & Pratt, 2000). Much of the prison privatization debate is ideologically charged, as supporters tend to place a high value on limited government and free market solutions to social problems, while opponents tend to place greater value on the role of government in solving social problems. These are ideological positions that are not readily changed by empirical studies of costs and conditions. For now and probably for the foreseeable future, the ideological issues posed by the prison privatization debate will remain intractable.

## ■ The Private Prison Market

Many of the services that have been privatized by government (e.g., custodial services, food preparation, medical care) are provided by the private sector independently of government's decision to privatize or not. There is a free market analogue for many kinds of services that governments typically provide. Incarceration services are fundamentally different in that one cannot purchase incarceration services as a private individual.

The power to incarcerate someone—to hold a person against his or her will—is a defining characteristic of the state. The very identity of the modern state is rooted in its monopoly over the legitimate use of physical force within a given territory (Weber, 1946). As a form of coercion, the power to incarcerate is reserved for the government; only the state has the legitimate power to restrict a citizen's liberty. Individuals are prohibited by law from incarcerating another person under "false imprisonment" statutes. The government can delegate this power on a limited basis—for example, granting "merchant's privilege" to shopkeepers, which allows them to temporarily detain suspected shoplifters. But long-term incarceration is a different matter. Thus, when considering prison privatization, the question arises, how can the government allow a private company to incarcerate an individual for the full term of a lengthy sentence? Prison privatization opponents have pressed this issue in challenging the practice in federal court, arguing that the state has no authority to delegate its punishment function to private contractors (McDonald et al., 1998). But the courts have held that the government can contract out for incarceration services unless there is a prevailing law that specifically prohibits the practice (Bowman, Hakim, & Sidenstat, 1994). The state delegates only the management of inmate daily affairs to private prison officials; it retains custody over them and keeps control over who is admitted and released by the private prison. At the same time, prisoners possess, at minimum, all rights extended to them in a public facility. In effect, contracting out of custody services is more properly defined as "partial privatization" (Benson, 1998); the government never abdicates its ultimate responsibility for the inmates, only specific functions related to their daily care and management.

Contracting out of noncustody prison services such as medical care, food service, maintenance, education, and mental health services has been practiced for a long time and with little controversy. As noted earlier, the contracting out of the custody function had essentially disappeared by the 1940s. However, the deinstitutionalization movement in juvenile corrections during the 1970s prompted a number of experiments in privatized services *and* custody. In 1975, the RCA Corporation won a contract to provide a 20-bed, high-security facility for juvenile delinquents in Weaversville, Pennsylvania (Sellers, 1989). In 1983, the U.S. Bureau of Prisons contracted with Eclectic Communications, Inc., to operate a prison for youthful offenders near San Francisco (Clear & Cole, 1997). And with the creation of the Job Corps in 1964, the Department of Labor began contracting with the private sector to operate residential centers with educational and job training services for at-risk youth, ages 16 through 24 years. Some of the early entrants into the private prison industry, including Management and Training Corporation (MTC) and Wackenhut Services, Inc., the forerunner of Wackenhut Corrections Corporation, operated Job Corps centers before venturing into the adult prison business.

A major step toward contracted custody of adults occurred when the Immigration and Naturalization Service (INS, the forerunner to today's Immigration and Customs Enforcement [ICE]) decided in 1983 to partially outsource the detention of illegal aliens in its custody (Knowlton, 1985). In the summer of 1993, the INS issued a request for proposals to build and operate a 350-bed detention facility for illegal aliens in Houston, Texas. The newly formed Corrections Corporation of America (CCA) submitted the winning bid and was awarded the contract in October of that year. CCA purchased a site in Houston, hired a contractor, and had the new facility ready by April 1984 (Knowlton, 1985).

Likewise, the community corrections movement in the adult system involved contracting out of the custody function in many low-security adult facilities.

Minnesota passed the Community Corrections Act in 1971, and 25 states followed suit with similar legislation over the next 12 years. Community corrections legislation transferred funding from state-level departments of correction to local governments who in turn used the funds for halfway house programs and other services for lower-level offenders (Shilton, 1995). Many jurisdictions turned to private contractors to operate these facilities. In 1986, MTC secured a contract to operate a community correctional facility in Eagle Mountain, California. Also in 1986, the state of Kentucky contracted out the development and operation of a 200-bed minimum-security facility in Marion County. Another newly formed company, the U.S. Corrections Corporation, was awarded the contract. The company purchased an old seminary and converted it into a correctional facility (Bowman, Hakim, & Sidenstat, 1994). Similarly, the U.S. Bureau of Prisons began contracting out the operation of low-security halfway house programs to the private sector during this time. Another early private prison company, Correctional Services Corporation, began business in 1989 with two contracts from the Federal Bureau of Prisons (BOP) to operate halfway houses in Manhattan and Brooklyn in New York City (Sullivan & Purdy, 1995). The same year, Correctional Services Corporation won a contract to manage an INS-owned detention center in Seattle, Washington.

These early entrants to the private prison business found a political climate supportive of privatization in several states. Along with Kentucky and Tennessee, the states of Texas, Florida, Arizona, New Mexico, and Louisiana all began experimenting with correctional privatization in the late 1980s. By 1994, there were approximately 20 companies actively seeking contracts to either build and manage company-owned prisons or manage existing facilities owned by federal, state, and local jurisdictions (Ramirez, 1994).

## ■ Concentration in the Private Prison Industry

The traditional industrial life cycle model suggests that new markets move through a process marked by four stages: fragmentation, shakeout, maturity, and decline (Klepper & Graddy, 1990). The fragmentation period occurs when the industry is new, many entrepreneurs enter the market, and they operate at low volume and tend to serve narrow geographic areas. Eventually, a shakeout period occurs as some companies become more efficient and the less efficient ones fail to stay in business. Growth and competition are strongest during this period, as are the profits of the largest companies. The industry reaches maturity when growth slows and surviving companies try to solidify their positions in the industry. At some later time, the industry moves into decline, as the demand for the product or service drops and companies seek new ways to recover profitability.

**TABLE 9-1 Annual Growth Rate of Private Prison Capacity**

| Year | Annual Growth Rate | Year | Annual Growth Rate |
|------|--------------------|------|--------------------|
| 1992 | 33.7% | 2000 | –2.4% |
| 1993 | 57.4% | 2001 | 0.6% |
| 1994 | 51.0% | 2002 | 2.1% |
| 1995 | 29.4% | 2003 | 1.9% |
| 1996 | 34.0% | 2004 | 3.1% |
| 1997 | 25.5% | 2005 | 9.4% |
| 1998 | 24.0% | 2006 | 5.3% |
| 1999 | 9.5% | 2007 | 4.0% |

Source: Data from C. W. Thomas, "Private Adult Correctional Facility Census," December 1997. Previously available: http://web.crim.ufl.edu/pcp/; and data from H. C. West and W. J. Sabol, "Prison Inmates at Midyear 2008—Statistical Tables," Bureau of Justice Statistics, 2009.

The decade of the 1990s witnessed the rapid growth of prison privatization, consolidation of the industry, and the cooling off of growth rates as the decade ended. The industry seems to have moved through the traditional life cycle stages of fragmentation to maturity during this period (growth rates for the private prison industry are depicted in **Table 9–1**). The industry experienced tremendous growth from 1992 to 1998, averaging an increase in capacity of 36% each year. But between 1999 and 2007, growth declined to an average rate of under 4% per year. Market maturity is commonly defined as reaching a state of equilibrium marked by the absence of significant growth or innovation (Samuelson & Nordhaus, 1989); it would appear that the industry reached maturity as the new millennium began.

The shakeout stage of the industry is reflected in the difference between the companies doing business in 1996 and those doing business in 2006. In 1996, there were 14 companies with private prison contracts in the United States. **Table 9–2** depicts these companies, their rated capacity, and their market share of U.S. private prison business. A common measure of the dominance of companies in a given market is the market concentration ratio (CR), defined as "the percentage of total industry sales (or capacity, or employment, or value added, or physical output) contributed by the largest few firms, ranked in order of market shares" (Scherer & Ross, 1990, p. 71). The CR is typically reported in terms of the market share of the leading four firms. Generally, a market is considered to have a high CR when the four leading firms control over two-thirds of market share, a moderate concentration when the ratio falls between one-third and two-thirds, and a low concentration when the top four firms control less than a third of total market share (Samuelson & Nordhaus, 1989).

**TABLE 9–2** Private Prison Industry Market Concentration in 1996

| | Capacity | 1996 Market Share | CR |
|---|---|---|---|
| Corrections Corporation of America | 40,365 | 52.3% | 52.3% |
| Wackenhut Corrections Corporation | 19,479 | 25.2% | 77.5% |
| U.S. Corrections Corporation | 4,038 | 5.2% | 82.7% |
| Management and Training Corporation | 2,978 | 3.9% | 86.1% |
| Cornell Corrections, Inc. | 2,611 | 3.4% | |
| The Bobby Ross Group | 2,164 | 2.8% | |
| Correctional Services Corporation | 2,150 | 2.8% | |
| Capital Correctional Resources | 1,908 | 2.5% | |
| Maranatha Production Company, LLC | 500 | 0.7% | |
| The GRW Corporation | 302 | 0.4% | |
| Dove Development Corporation | 295 | 0.4% | |
| Fenton Security, Inc. | 228 | 0.3% | |
| Avalon Community Services, Inc. | 144 | 0.2% | |
| Correctional Systems, Inc. | 82 | 0.1% | |
| | | 100.0% | |

Source: Data from C. W. Thomas, "Private Adult Correctional Facility Census, Tenth Edition" 1997. Previously available: http://web.crim.ufl.edu/pcp/

If we total the market share of the top four companies, we obtain a four-firm market concentration ratio ($CR_4$) of 86% in 1986, a very high level of concentration. According to economic theory, when production is highly concentrated in very few companies, the market is an oligopoly and is inherently less competitive and innovative than a market with more broad-based representation (Samuelson & Nordhaus, 1989). As a market form, an oligopoly is characterized by interdependence, avoidance of competition, and a rigid attachment to the status quo among the leading firms (Brock, 2006). By the end of 2006, only six companies remained in the private prison business, and the market share of the top four firms increased to 99% (**Table 9–3**). In addition to becoming more oligopolistic, this suggests that the market has moved into the mature stage of its life cycle.

Another way of looking at the growth in concentration is by using the Herfindahl-Hirschman Index (HHI) (Rhoades, 1993). A problem with the $CR_4$ measure is that it does not take into account the relative size of all the firms involved in a market. The HHI resolves this problem. It is calculated by including the market shares of all firms in the market, squaring the market share of each competing firm, and then summing the resulting numbers. The HHI approaches zero when a market consists of a large number of firms of relatively equal size. The HHI increases

**TABLE 9-3** Private Prison Industry Market Concentration in 2006

| | Capacity | 2006 Market Share | CR |
|---|---|---|---|
| Corrections Corporation of America | 58,978 | 59.0% | 59.0% |
| GEO Group (Wackenhut) | 26,385 | 26.4% | 85.4% |
| Cornell Corrections, Inc. | 8,002 | 8.0% | 93.4% |
| Management and Training Corporation | 5,596 | 5.6% | 99.0% |
| Community Education Centers | 500 | 0.5% | |
| The GRW Corporation | 493 | 0.5% | |
| | | 100.0% | |

Source: Data from American Correctional Association. *The 2006 Directory of Juvenile and Adult Correctional Departments, Institutions, Agencies, and Paroling Authorities.* American Correctional Association, 2007./

both as the number of firms in the market decreases and as the disparity in size between those firms increases. Markets in which the HHI exceeds 1800 points are considered concentrated. Horizontal mergers that increase the HHI by more than 100 points in concentrated markets presumptively raise antitrust concerns under the Horizontal Merger Guidelines issued by the U.S. Department of Justice and the Federal Trade Commission (Federal Trade Commission, 1997). A comparison of the HHI of the private prison market in 1996 and 2006 is presented in **Table 9-4**.

In the past 10 years, market concentration as measured by the HHI has increased by 24%, from 3343 to 4274. The rise in CR means the market has become more tightly concentrated in the past decade and that the industry as a whole has become less competitive. Four of the companies that disappeared between 1996 and 2006 (U.S. Corrections Corporation, Correctional Services Corporation, Fenton Security, Inc., and Correctional Systems, Inc.) were acquired by larger companies. Collectively, these four firms held 8.4% of the private prison market in 1996, a share that added to the increase in industry concentration in 2006. Four other firms went out of business (The Bobby Ross Group, Capital Correctional Resources, Dove Development Corporation, and Maranatha Production Company, LLC). These failed firms controlled 6.3% of the private prison market in 1996. A brief review of the merged and failed companies is illustrative of the private prison industry's movement through the shakeout life cycle stage during the late 1990s.

### The Big Get Bigger

The largest company—Corrections Corporation of America (CCA)—increased its market share by 6% during the period, largely by acquiring the third-place company, U.S. Corrections Corporation, in 1998. U.S. Corrections, based in Kentucky, had much in common with Tennessee-based CCA. Both companies were

**TABLE 9–4 Estimated Change in Herfindahl–Hirschman Index (HHI), 1996–2006**

| | Capacity | 1996 Market Share | MS² |
|---|---|---|---|
| Corrections Corporation of America | 40,365 | 52.3% | 2731.1 |
| Wackenhut Corrections Corporation | 19,479 | 25.2% | 636.0 |
| U.S. Corrections Corporation | 4,038 | 5.2% | 27.4 |
| Management and Training Corporation | 2,978 | 3.9% | 14.9 |
| Cornell Corrections, Inc. | 2,611 | 3.4% | 11.4 |
| The Bobby Ross Group | 2,164 | 2.8% | 7.8 |
| Correctional Services Corporation | 2,150 | 2.8% | 7.7 |
| Capital Correctional Resources | 1,908 | 2.5% | 6.1 |
| Maranatha Production Company, LLC | 500 | 0.7% | 0.4 |
| The GRW Corporation | 302 | 0.4% | 0.2 |
| Dove Development Corporation | 295 | 0.4% | 0.1 |
| Fenton Security, Inc. | 228 | 0.3% | 0.1 |
| Avalon Community Services, Inc. | 144 | 0.2% | 0.0 |
| Correctional Systems, Inc. | 82 | 0.1% | 0.0 |
| | | 100.0% | |
| | | **1996 HHI** | **3443.3** |

| | Capacity | 2006 Market Share | MS² |
|---|---|---|---|
| Corrections Corporation of America | 58,978 | 59.0% | 3481.0 |
| GEO Group (Wackenhut) | 26,385 | 26.4% | 697.0 |
| Correctional Services Corporation | 8,002 | 8.0% | 64.0 |
| Management and Training Corporation | 5,596 | 5.6% | 31.4 |
| Community Education Centers | 500 | 0.5% | 0.3 |
| The GRW Corporation | 493 | 0.5% | 0.3 |
| | | 100.0% | |
| | | **2006 HHI** | **4273.8** |

Source: Data from C. W. Thomas, "Private Adult Correctional Facility Census, Tenth Edition" 1997. Previously available: http://web.crim.ufl.edu/pcp/; and data from American Correctional Association. *The 2006 Directory of Juvenile and Adult Correctional Departments, Institutions, Agencies, and Paroling Authorities.* American Correctional Association, 2007.

founded by entrepreneurs who were well connected with the political establishment in their state. One of CCA's founders, Thomas W. Beasley, was a former Chair of the state Republican Party and managed the successful 1978 gubernatorial campaign of Lamar Alexander (Knowlton, 1985). Honey Alexander, the governor's wife, was an early investor in CCA (Hallinan, 2001). U.S. Corrections Corporation, based in neighboring Louisville, did all of its business in Kentucky prior to 1995. Executives of the company contributed over $77,000 to political campaigns in the state between 1987 and 1993, including $23,000 to Governor Wallace G. Wilkinson and his wife Martha. Clifford Todd, the CEO of U.S. Corrections, was implicated in a payoff scandal involving the company's contract to run the county jail in Louisville. He was arrested by the FBI in 1994, pleaded guilty to mail fraud, and was sentenced to 6 months in jail and fined $250,000. He sold his stake in the company for $15 million in 1994 (Hallinan, 2001). At the time of the buyout by CCA, U.S. Corrections owned a total of four facilities in Kentucky, Ohio, and North Carolina, with a capacity of 5275 beds, and had contracts to manage publicly owned facilities in Kentucky, Florida, and North Carolina, with a capacity of 5743 beds. The acquisition increased the number of facility beds owned by CCA by 43% (PRNewswire, 1998).

The second-largest company in 1996, Wackenhut Corrections Corporation (WCC), was founded in 1954 by George Wackenhut, a former FBI agent. Over the ensuing years, WCC grew to become one of the largest private security companies in the world. As noted earlier, the company became involved in residential services for at-risk juveniles and young adults under the Job Corps program. WCC was incorporated in 1999 as a wholly owned subsidiary of the Wackenhut Corporation and went public in 1994 (with Wackenhut Corporation as the majority shareholder). In 2002, the Danish company Group 4 Falck merged with the Wackenhut Corporation and, indirectly, became the owner of 57% of the WCC stock. In 2003, WCC bought out Group 4 Falck's interest in the company and changed its name to the GEO Group, Inc. (GEO Group, 2009). The GEO Group increased its market share by acquiring the industry's seventh-largest company, Correctional Services Corporation (CSC), in 2005.

Founded in 1989, CSC operated under several different names, including Esmor Correctional Services, Inc. The company was founded by the owners of a welfare hotel in New York City and began its prison business operating halfway houses under contract with the BOP. CSC also managed the INS Detention Center in Elizabeth, New Jersey, where detainees were involved in a high-profile disturbance and subsequent lawsuit in 1995, protesting conditions in the facility (Sullivan & Purdy, 1995). CSC has a checkered history of ethical and legal challenges: In February 2003, the New York State Lobbying Commission fined CSC $300,000 for failing to report free transportation, meals, and gifts to a dozen state legislators in Brooklyn and the Bronx in an effort to keep contracts for the placement of recently released prisons. The fine was the largest the state had ever imposed on a company for violating the state's lobbying laws (McKinley, 2003). In May 2008, the Securities and Exchange Commission (SEC) filed charges against three doctors in Fort Lauderdale, Florida, alleging insider trading in GEO Group's 2005 acquisition of CSC. According to the complaint, the three illegally bought shares in CSC immediately before the company announced in July 2005 that it was being bought out by GEO. One of the brothers worked as a consultant to GEO, and his son worked in GEO's finance department, where he allegedly learned of the pending deal. The three doctors were charged with illegally purchasing $390,000 in CSC stock before the acquisition (Rugaber, 2008).

Cornell Corrections, Inc., moved from fifth place in 1996 to third place in 2006 and increased its market share from 3.4% to 8%, due in part to acquiring Correctional Systems, Inc. in 2005. Cornell was incorporated in 1996 and built its business primarily with youth services and community-based rehabilitation programs for adults. At the time of its acquisition, Correctional Systems managed eight jails, six community corrections facilities, and five alternative sentencing programs in California, New Mexico, Texas, and Kansas, with a combined total of 986 corrections beds (Business Wire, 2005). In 2004, the hedge fund firm Pirate Capital, LLC acquired a 13% interest in Cornell, became the company's largest shareholder, and replaced several board members and the company's CEO with financiers more focused on short-term growth (Fishman, 2004; Hansard, 2005). Pirate Capital unloaded its shares of Cornell in 2006 in the midst of an SEC investigation over its stock sales practices, several bad investment decisions, and company downsizing (Anderson, 2006).

A fourth company on the 1996 list, Fenton Security, Inc., was acquired by CiviGenics in 1996 and operated as a subsidiary, but under its original name. The company was founded by Charles Fenton, a former federal prison warden. It managed four correctional facilities when it was acquired in 1996. CiviGenics, in turn, was acquired by Community Educations Centers, Inc. in 2007.

### The Losers

Among the failed firms, The Bobby Ross Group exemplified some of the early entries into the privatization business in Texas—local entrepreneurs who capitalized on the prison building boom of the 1990s, using local connections and business experience but lacking correctional expertise. Founded by a former local sheriff in Texas in 1993, the company landed several management contracts to run facilities built with public funds. At its peak, it operated seven prisons in Texas and a juvenile facility in Georgia, with contracts for the placement of out-of-state inmates from Colorado, Missouri, Montana, Oklahoma, Virginia, and Hawaii (Kakesako, 1997). Former FBI Director William Sessions served as a company consultant, helping to secure contracts (Pierce, 1995). About the only legacy of the company is a record of being the defendant in numerous federal lawsuits filed by inmates. It encountered serious problems in several of its facilities in the late 1990s, including the escape of two sex offenders from a group of 500 Colorado inmates placed in the company's Karnes County facility, and a major disturbance between Hawaii and Montana inmates in the company's Dickens County Correctional Facility that left one Montana inmate dead and resulted in the state pulling its inmates from the facility. Also, 11 inmates escaped from the company's prison in Newton, Texas, in 1998. In the Newton incident, the escaping inmates released another 300 inmates and set fire to a prison building. The company's contracts to operate the Karnes County and Dickens County facilities expired in early 1998 and were taken over by Wackenhut, now GEO. Later in the year, Corrections Corporation of America took over the company's contract to run a medium-security prison in Webb County (Laredo). And after an influence-buying scandal in Georgia, the company's contract juvenile prison was turned over to the adult parole department and eventually closed in 2002 (Kakesako, 1997; Texas Prison Bid'ness, 2007).

The second failed company, Capital Correctional Resources, Inc. (CCRI) is enshrined as one of the most incompetent private prison upstarts. Like The Bobby Ross Group, CCRI formed alliances with county sheriff's in Texas to use county jail development capital to expand local facilities, provide overall management of operations, and contract with other states willing to place their inmates out of state to relieve in-state overcrowding. The Mississippi-based company secured county jail management contracts with Brazoria and Limestone Counties in Texas, as well as contracts with the states of Missouri and Oklahoma to house their inmates in Texas. The interstate prisoner business also included public facilities. For example, Navarro County Jail in Corsicana also contracted for the placement of Missouri inmates. In 1997, Texas facilities were accommodating some 5500 prisoners from 11 other states, with several public facilities engaging in these income-generating contracts (Bell, 1997). In an incident foreshadowing the Abu Ghraib prison ignominy, a video came to light in 1997 that graphically recorded scenes from a September 1996 incident at CCRI's Brazoria County facility. The video depicted "guards kicking seemingly compliant prisoners in the head and groin, swearing at them, beating them with riot sticks and electric prods, forcing them to crawl on their bellies, some with their pants down around their ankles, and a German shepherd biting the legs of at least one inmate" (Gillespie, 1997). Oddly enough, the video was recorded by CCRI staff for use as a training film. Not only were CCRI personnel involved in the abuse, but County Sheriff's officials as well. In response, Missouri immediately cancelled its contracts and removed its prisoners from the facility. The incident prompted considerable litigation against the company, including a federal district court case that, in ruling against the company, offered the opinion that "experience, training, and temperament may become expendable virtues when their associated costs threaten the bottom line" (*Kesler v. King*, 1998). The court used the terms "quack" to describe CCRI's approach and "commodities" to describe how it treated inmates under its care (Blakely & Bumphus, 2005). The incident forced the company out of business.

Dove Development Corporation, like The Bobby Ross Group and CCRI, went into business in Texas and experienced similar problems dealing with mass transfers of inmates from other states. Dove was awarded a contract to manage the

Frio County detention center in 1992 and secured another contract in Crystal City 2 years later. In 1995, the company was one of three that submitted a bid to operate the women's annex of the Bexar County Jail in San Antonio (the other two were CCA and Wackenhut), but county officials rejected its bid due to its relative lack of experience. Dove contracted with the state of Utah for the placement of 100 inmates in 1995, but within a year the contract was in the process of termination. The Frio County facility experienced numerous escapes by the Utah inmates, including four inmates serving time for murder. A violent disturbance in September 1996 required the intervention of local law enforcement. The state of Utah pulled its inmates, and in 1997, Frio County officials signed a contract with CSC of Florida to take over operation of the 286-bed facility.

The fourth failed company was a small operation with only one facility, the Maranatha Production Company, LLC. The company built and secured a contract with the California Department of Corrections for the 500-bed Victor Valley Modified Community Correctional Facility in Adelanto, California (Cate, 2004). The company struggled with financial problems from the start, beginning with a lawsuit over allegedly unpaid wages to construction workers (Private Corrections Institute, 2008) and ending with the loss of its contract with the State of California after a dispute over inmate telephone call revenues (Cate, 2004). In 1999, the company made a proposal to Hawaii officials to build a 2300-bed prison on Kauai and in 2001 proposed to build and operate a prison in Wyoming. Neither plan came to fruition. The company went out of business in 2005 and sold the Adelanto facility to San Bernardino County (Private Corrections Institute, 2008).

## ■ Competition and Innovation?

The total market share represented by the four failing companies accounted for 6.3% of the total, which, along with the 8.4% market share of four merged companies, resulted in each of the remaining six companies increasing their individual share of the market. One indicator of the effect of this is an apparent decline in the number of firms responding to requests for proposals from contracting governments. In 1995, for example, a total of seven companies submitted bids in response to a 1995 Arkansas Department of Corrections request for proposals for two new privately operated prisons. CCA, Wackenhut, U.S. Corrections Corp, The Bobby Ross Group, GRW Corporation, MTC, and CCRI all submitted bids, with Wackenhut winning out (Pierce, 1995). By contrast, a 2006 request for proposals to manage a private prison in Pennsylvania yielded only two responses.[1] With market maturation and increased concentration, it is increasingly difficult for new companies to get into the business and for marginal performers to stay afloat. Given the record of incompetence, public safety consequences, and human rights abuses of the failed companies, the public is better off with these companies no longer in business. At the same time, however, oligopoly theory and industry life cycle theory suggest that the remaining companies will be less competitive over-all, seeking interdependent relationships with their competitors to help secure their position now that industry growth rates have cooled. As this comes to pass, we should expect to see that any cost savings from privatization will be marginal at best.

There is preliminary evidence that these savings may already be declining as existing private prison contracts are renewed. While a brand new private prison staff may be cheaper to operate than an existing public prison, personnel costs will move toward parity over time and the cost advantage will begin to disappear. As personnel costs level out over time, the cost advantage decreases. For example, the average cost of operating three BOP prisons, based on two independent stud-ies, increased from $38 per day in 1999 to $41.77 in 2002, an increase of 10%. A comparable private facility, with BOP inmates, increased from $34.12 to $38.50 over the same time span, an increase of 13%. A cost-per-day savings of $3.88 in 1999 declined to $3.27 by 2002.[2] This decline in cost advantage is likely the norm with most private prison contracts over time.

The record of serious incidents involving the failing companies and the cloud of SEC investigations and financial sector speculation involving some of the sur-vivors serve as a reminder that the ethical objections to prison privatization are not going away, despite 25 years of correctional privatization experience and a body of research that has yet to suggest that either sector has an upper hand in running a good prison program.

The demand side of prison privatization, like the oligopoly of private prison companies, is also highly concentrated. The potential consumers in the private prison market are government agencies at the federal, state, and local level that operate jail and prison programs. At the federal level, the BOP, the U.S. Marshals Service, and the ICE all manage a variety of custody facilities. The four branches of the military also operate facilities for military personnel sentenced under the Uniform Code of Military Justice. Of course, the 50 states and the District of Columbia each have departments of correction. According to the BJS, there are some 1821 state and federal correctional facilities, not counting facilities operated by the U.S. Marshals Service and the ICE. At the local level, counties and cities operate about 2875 jail facilities in the United States (Minton & Sabol, 2008). In sum, the potential consumer base of the private prison industry numbers some-where in the area of 4700 facilities.

After 25 years of correctional privatization, there are fewer than 200 private correctional facilities in the United States, accounting for only about 4% of all facilities. The federal government is the most actively involved in privatization, with 16.3% of federal inmates serving time in private facilities. State governments are next, with 6.6% of state inmates in private prisons (West & Sabol, 2009). However, whereas 60% of all correctional facilities in the United States are at the local level, only about two dozen city and county jurisdictions (1.7% of the total) contract with private companies to operate their prisons and jails (International City/County Manager Association, 2007).

In practice, there are very few buyers of privatized incarceration services, and the federal government is the largest single customer. Between 2000 and 2008, the number of state inmates placed in private prisons increased by about 25%, from 75,018 to 93,537. In the federal system, however, the number increased from 15,525 to 32,712, or about 110% (West & Sabol, 2009). During the same period, the number of states placing some portion of their prisoners in private facilities actually *declined* from 30 to 27.

There are a total of only 54 customers who are buying incarceration services from the private prison industry—the 3 federal agencies, 27 state-level departments of correction, and 24 local jurisdictions. Within this small customer base, the federal government plus seven states (Texas, Florida, Arizona, Oklahoma, Colorado, Tennessee, and Mississippi) collectively account for greater than 70% of all private prison business. In effect, the market of buyers constitutes an *oligopsony*, a market in which only a few customers buy a certain good and therefore possess the power to affect pricing (Meyer, 2009). The three largest publicly traded private prison companies all recognize this dependency on a limited number of governmental customers as a threat to their profitability and include a warning to their stockholders to this effect in their annual reports. Cornell notes that 34.2% of their total revenue for 2008 came from contracts with the BOP; GEO Group states that while they have a total of 45 governmental clients (customers), 4 of these customers accounted for over 50% of their consolidated revenue (the BOP, ICE, U.S. Marshals Service, the State of Florida). Among these, the three federal agencies combined are responsible for 27.7% of GEO Group's total revenue and 39% of the revenues generated by CCA.[3]

The oligopsony of governmental consumers serves not only to limit competition but to discourage innovation as well. Indeed, prisons are not generally thought of as innovative places, and research suggests that a control-oriented management style produces greater order, amenity, and service in a prison than more progressive management styles, such as those derived from responsibility or consensus models (DiIulio, 1991). According to DiIulio, successful prison leaders rarely innovate, as innovation tends to erode correctional officer loyalty (DiIulio, 1991). Genders (2003) observed that hiring new staff without prison experience—ostensibly in the interest of breaking free of the old way of doing things—can contribute to prison disturbances.

In practice, governmental purchasers of incarceration services have sufficient power over the sellers to require that private prison companies duplicate policies and procedures practiced in public prisons. Many jurisdictions have specifically done this, to the effect that the standard operating procedures of most private prison programs closely mirror those of public prisons in the same state

(Culp, 2005). Private prison companies encourage the adoption of public prison practice, rather than the development of innovative practice, by actively recruiting management-level staff from within the public sector. As a case in point, a review of the background of CCA's management staff at the facility level suggests a widespread practice of mining the public corrections system for managers. CCA offers the names and background experience of the wardens of 63 of the company's correctional facilities around the United States. Nearly two-thirds of all CCA wardens formerly worked in state departments of correction (36 wardens) or the federal BOP (5 wardens). The most warden-rich jurisdiction was the Texas Department of Criminal Justice, from whence 28% of all CCA wardens were recruited. These experienced staffers bring a degree of order and control to the private prisons, which was lacking in many of the failed companies, but they are not likely to be hired for their spirit of experimentation and innovation.

Arguably, private prisons are not looking to be innovative, unless it is a way of cutting costs. The most common way for these companies to make money from government contracts is by reducing personnel costs (Cooper, 2003), and because labor represents approximately 80% of the operating cost of a prison (Genders, 2003), much of the cost savings in private prisons is a result of paying private correctional officers less than comparable public correctional officers receive (Camp & Gaes, 2001). As noted, however, this advantage begins to erode in a market where private companies are dependent upon contract renewals (with more experienced staff) rather than new facilities (with new, entry-level staff). Even as labor rates vary among the states, public sector correctional officer starting salaries average $28,000 across all states with a (one standard deviation) range between $23,000 and $34,200 (Corrections Compendium, 2007). By comparison, the Bureau of Labor Statistics reports a mean annual salary of $42,270 for all occupations in the United States (in May 2008). Although comparable figures are not available for private sector correctional officers, public sector prison staff salaries are very low already, suggesting that it is not that easy for the private sector to undercut the government in personnel costs.

One of the more innovative ideas to help cut correctional officer wages even more was advanced in 2005. Post-September 11, concern about illegal immigration temporarily raised the business prospects of the private prison companies as ICE increased its reliance on contracts with the private sector. In Arizona and California, concern over the cost of incarcerating illegal aliens convicted of crimes in the United States prompted legislators to float the novel idea of creating a "Foreign Private Prison Commission" that could contract with a private company to build and staff a prison in Mexico for placement of Mexican nationals convicted and sentenced for crimes committed in the United States.[4] The idea raised the interesting prospect of building prisons in other countries, where labor is cheaper, to incarcerate inmates convicted in the United States. The legislation failed to pass in either state, due in major part to the daunting task of first securing a treaty between the United States and Mexico that would legitimize the international relations necessary for such a plan to move forward. But given the history of prison privatization in the United States, which began with and continues to receive major funding from federal immigration agencies and is most popular in southern and western states, such an offshore prison may become a reality some day.

## ■ Conclusion

Despite the promises of less cost and better services made by private prison advocates in the 1980s, a quarter century of privatization experience has not brought those promises to fulfillment. The private prison industry that has emerged over the years exists in a market with limited government consumers and even fewer private company producers. This oligopsony of buyers stifles innovation, while the oligopoly of sellers limits competition. Balkanized in southern and western states, dependent upon federal government agencies, and with little competition and no reason to innovate, the mature private prison industry of today is assuredly a disappointment to new public management advocates who thought that creating a free market for incarceration services would help foster a more efficient prison system. It just has not happened that way. At the same time, the worries of privatization critics—of wholesale violations of prisoner rights and dangerous lapses in prison security and safety—have not come to fruition either. The companies that survived the shakeout of the 1990s rely on staff who gained prison management experience in the public sector and who operate facilities that essentially mimic the policies and procedures of public prisons. Even though we have highlighted some cases where incompetent private companies have endangered prison inmates, staff, and the public, we could just as easily offer as many anecdotes of incompetent practice in public sector prisons; when it comes to running good prisons, the public sector certainly has no monopoly on virtue.

A new twist on the old argument that greater competition can improve the quality of prison services has recently emerged with calls for increasing the role played by the nonprofit sector in the operation of secure custody facilities for adults (Low, 2003; Moran, 1997). The idea is that nonprofit agencies can "maximize the advantages of public and private prisons, while minimizing their disadvantages" (Low, 2003, p. 4). According to this view, a major disadvantage of the private prison business is that profits are paid out to stockholders rather than reinvested in service improvements and program innovations. Private companies make reinvestments only to the extent that they will enhance profitability. As mission-driven and "non-owned" organizations (Mintzberg, 1996), donors and volunteers would direct surplus revenue into furthering the reform-oriented goals of nonprofit organizations. In juvenile corrections, where a majority of private juvenile correctional facilities are operated by nonprofit organizations rather than for-profit companies (Culp, 1998), research suggests that nonprofits are providing a high level of service quality. For example, Bayer and Pozen (2005) followed the case records of more than 5000 released juvenile offenders who were placed in juvenile correctional facilities in Florida. More than half of the juveniles were placed in facilities operated by nonprofit agencies, some 1000 were placed in for-profit facilities, and another 1500 were placed in programs operated by state and county governments. They found that youths placed in the public and nonprofit facilities were less likely to be charged with a criminal offense within a year of release than were juveniles released from programs operated by for-profit companies. Although the for-profit facilities initially operated at a lower cost to the government (in the year that the juveniles were placed with them), cost–benefit analysis found that nonprofit agencies had the most favorable cost outcome in the long run due to reduced recidivism rates for juveniles released from nonprofit programs. Perhaps such promising outcomes might be replicated in the adult system by encouraging the nonprofit sector to become more involved in the delivery of adult custody services. The government could help facilitate the entry of nonprofits into the private prison market by making available loan guarantees or tax-exempt bonds to help raise the capital required for building secure correctional facilities.

Notwithstanding the promising advantages of greater nonprofit agency involvement in adult prison operation, there is a possibility that a nonprofit oligopoly would also emerge in the adult system and that it would counterbalance the competitive and innovative forces we would hope to encourage. In studying the admittedly limited scope of private juvenile providers in Massachusetts, Armstrong (2002) found that the small, public-spirited, and innovative nonprofits that responded to the deinstitutionalization movement of the 1970s subsequently evolved into an oligopoly of five or six multimillion-dollar nonprofit corporations managed by risk-averse "entrepreneurial bureaucrats." Unfortunately, greater involvement by the nonprofit sector is unlikely to be a panacea for our incarceration problem.

Our problem is overreliance on incarceration, and this will not be solved through policies that simply redistribute the number of inmates serving time in public, for-profit, or nonprofit prisons. The United States has the highest incarceration rate in the world (Walmsley, 2009), due in large part to our penchant for the custodial sentence as the first choice among sanctioning alternatives. Considering that incarceration is the most expensive of sentencing options, the smarter policy would be to invest in research and development of noncustodial alternatives.

## ■ Endnotes

1. This finding is based on the author's consultant experience with a county government in Pennsylvania.
2. Findings are based on secondary analysis of cost data presented in Gaes, G. (2008). Cost, performance studies look at prison privatization. *NIJ Journal, 259*, 32–36.
3. Sources of revenue for the three largest companies are based on company statements provided in each company's Form 10-K, filed with the SEC for the fiscal year ended December 31, 2009.
4. AZ House Bill 2709, Forty-seventh Legislature, 2005; and CA Assembly Bill 1686, introduced February 22, 2005.

## ■ Discussion Questions

1. What factors contributed to the emergence of a market for private prisons in the United States?
2. Discuss the relationship between politics, ideology, and private prisons.
3. What factors should policy makers take into account when considering the viability of private prisons?
4. Is the growth of private immigration detention facilities a positive or negative development? How should this determination be made?
5. Can the private prison industry be described as competitive and innovative? Should the industry be restructured? How can policy makers optimize the use of private prisons in the criminal justice system?

## ■ References

Aman, A. C., Jr. (2007). An administrative law perspective on government social service contracts: Outsourcing prison health care in New York City. *Indiana Journal of Global Legal Studies, 14*(2), 301–328.

Anderson, J. (2006, September 29). Hedge fund shrinks staff and faces S.E.C. inquiry. *The New York Times*, p. 4.

Armstrong, S. (2002). Punishing not-for-profit. *Punishment & Society, 4*(3), 345–368.

Austin, J., & Coventry, G. (2001). *Emerging issues on privatized prisons.* Washington, DC: Bureau of Justice Assistance.

Bayer, P. & Pozen, D. E. (2005). The effectiveness of juvenile correctional facilities: Public versus private management. *Journal of Law and Economics, 48*(2), 549–596.

Bell, K. (1997, August 24). Missouri says Texas attack was planned. *St. Louis Post-Dispatch* (Missouri), p. A1.

Benson, B. (1998). Partial privatization: The level and scope of contracting out in criminal justice. In B. Benson (Ed.), *To serve and protect: Privatization and community in criminal justice*. New York: New York University Press.

Blakely, C.R. & Bumphus, V.S. (2005). An analysis of civil suits filed against private and public prisons: A comparison of Title 42: Section 1983 litigation. *Criminal Justice Policy Review, 16*(1), 74-87.

Bowman, G. W., Hakim, S., & Sidenstat, P. (Eds.). (1994). *Privatizing correctional institutions*. New Brunswick, NJ: Transaction Publishers.

Brock, J. W. (2006). Antitrust policy and the oligopoly problem. *The Antitrust Bulletin, 51*(2), 227–280.

Brudney, J. L., Fernandez, S., Ryu, J. E., & Wright, D. S. (2004). Exploring and explaining contracting out: Patterns among the American states. *Journal of Public Administration Research and Theory, 15*(3), 393–419.

Business Wire. (2005, February 01). Correctional Systems Inc. to be acquired by Cornell Cos. In S. D. Camp & G. G. Gaes, *Growth and quality of U.S. private prisons: Evidence from a national survey*. Washington, DC: Federal Bureau of Prisons.

Camp, S.D. & Gaes, G.G. (2001). *Growth and quality of U.S. private prisons: Evidence from a national survey*. Washington, DC: Federal Bureau of Prisons, Office of Research and Evaluation.

Cate, M. L. (2004). *Review of inmate telephone revenues at the Victor Valley Modified Correctional Facility*. Sacramento, CA: California Department of Correction, Office of the Inspector General.

Clear, T. R., & Cole, G. F. (1997). *American corrections*. Belmont, CA: Wadsworth Publishing Company.

Collins, W. (2001). *Contracting for correctional services provided by private firms*. Washington, DC: Corrections Program Office, Office of Justice Programs, United States Department of Justice.

Cooper, P. J. (2003). *Governing by contract: Challenges and opportunities for public managers*. Washington, DC: CQ Press.

Corrections Compendium. (2007). Correctional officers: Hiring requirements and wages. *Corrections Compendium, 32*(3), 12–27.

Christie, N. (1994). *Crime control as industry*. London: Routledge.

Culp, R. F. (1998). Privatization of juvenile correctional facilities in the US: A comparison of conditions of confinement in private and government operated programs. *Security Journal, 11*, 289–301.

Culp, R. F. (2005). The rise and stall of prison privatization: An integration of policy analysis perspectives. *Criminal Justice Policy Review, 16*(4), 412–442.

DiIulio, J. J., Jr. (1991). No escape: The future of American corrections. New York: Basic Books.

Donahue, J. (1989). *The privatization decision: Public ends, private means*. New York: Basic Books.

Downs, A., & Steiner, R. (1990). *An economic theory of democracy*. Reading, MA: Addison-Wesley Publishing Company.

Durham, A.M. (1989). Origins of interest in the privatization of punishment: The nineteenth and twentieth century American experience. *Criminology, 27*(1), 107–139.

Federal Trade Commission. (1997). Horizontal merger guidelines (Issued: April 2, 1992; Revised: April 8, 1997). Washington, DC: U.S. Department of Justice, Federal Trade Commission.

Fishman, S. (2004, November 15). Get richest quickest. *New York Magazine*.

Genders, E. (2003). Privatisation and innovation—rhetoric and reality: The development of a therapeutic community prison. *The Howard Journal of Criminal Justice, 42*(2), 137–157.

GEO Group. (2009). Corporate milestones. Retrieved September 15, 2009, from company Web site: http://www.thegeogroupinc.com/index.asp

Gillespie, N. (1997). Prisoners' dilemma: Private jails are easier to reform. Retrieved June 3, 2010, from http://reason.com/archives/1997/11/01/prisoners-dilemma

Girard, P., Mohr, R. D., Deller, S. C., & Halstead, J. M. (2009). Public-private partnerships and cooperative agreements in municipal service delivery. *International Journal of Public Administration, 32*(5), 370–392.

Government Accountability Office. (1996). *Private and public prisons: Studies comparing operational costs and/or quality of services*. Washington, DC: U.S. Government Printing Office.

Hallinan, J. T. (2001). *Going up the river: Travels in a prison nation*. New York: Random House.

Hansard, S. (2005, October 31). Hedge fund activism concerns advocates of corporate governance: Groups question focus on short-term gain. *Investment News*.

International City/County Management Association. (2007). *Profile of local government service delivery choices, 2007*. Washington, DC: International City/County Management Association.

Kakesako, G. K. (1997, November 24). Lone star lock-up. *Honolulu Star Bulletin*, p. 1.

Kettl, D. F. (2000). *The global public management revolution: A report on the transformation of governance*. Washington, DC: The Brookings Institution.

Klepper, S., & Graddy, E. (1990). The evolution of new industries and the determinants of market structure. *Rand Journal of Economics, 21*, 27–44.

Knowlton, W. (1985) Corrections Corporation of America. Kennedy School of Government Case Program. Cambridge, MA: The President and Fellows of Harvard College.

Legislative Budget Committee for the State of Washington. (1996). *Department of corrections privatization feasibility study*. Olympia, WA: Legislative Budget Committee.

Lilly, J. R.& Deflem, M. (1996). Profit and penalty: An analysis of the corrections-commercial complex. *Crime and Delinquency, 42*(1), 3-20.

Logan, C. H., & McGriff, B. W. (1989). *Comparing costs of public and private prisons: A case study.* Washington, DC: National Institute of Justice, U.S. Department of Justice.

Low, D. L. (2003). Nonprofit private prisons: The next generation of prison management. *New England Journal on Criminal & Civil Confinement, 29,* 1–66.

Lukemeyer, A., & McCorkle, R. C. (2006). Privatization of prisons. Impact on prison conditions. *American Review of Public Administration, 36*( 2), 189–206

McDonald, D., Fournier, E., Russell-Einhourn, M., & Crawford, S. (1998). *Private prisons in the United States: An assessment of current practice.* Cambridge, MA: Abt Associates.

McKinley, J. C., Jr. (2003, February 27). Company gets record fine for its giving to lawmakers. *The New York Times,* p. 1.

Meyer, P. (2009). Online glossary of economics research. Retrieved February 18, 2010, from http://econterms.com/

Minton, T.D. & Sabol, W.J. (2008). *Jail inmates at midyear 2007.* Washington, DC: U.S. Department of Justice, Bureau of Justice Statistics.

Mintzberg, H. (1996). Managing government, governing management. *Harvard Business Review, 74*(3), 75–83.

Moran, R. (1997, August 23). A third option: Nonprofit prisons. *The New York Times,* p. 23.

Nicholson-Crotty, S. (2004). The politics and administration of privatization: Contracting out for corrections management in the United States. *Policy Studies Journal, 32*(1), 41–57.

Niskanen, W. A., Jr. (1971). *Bureaucracy and representative government.* Chicago: Aldine-Atherton.

Osborne, D., & Gaebler, T. (1992). Reinventing government: *How the entrepreneurial spirit is transforming the public sector.* Reading, MA: Addison-Wesley.

Perrone, D., & Pratt, T. C. (2003). Comparing the quality of confinement and cost effectiveness of public versus private prisons: What we know, why we do not know more, and where to go from here. *The Prison Journal, 83*(3), 301–322.

Pierce, R. (1995, August 3). Seven companies apply to run state prisons. *Arkansas Democrat-Gazette,* p. 1B.

Pigou, A. C. (1920). *The economics of welfare.* London: Macmillan.

Pratt, T. C., & Maahs, J. (1999). Are private prisons more cost-effective than public prisons? A meta-analysis of evaluation research studies. *Crime & Delinquency, 45*(3), 358–371.

Price, B. E., & Riccucci, N. M. (2005). Exploring the determinants of decisions to privatize state prisons. *The American Review of Public Administration, 35*(3), 223–235.

Private Corrections Institute. (2008). The truth about correctional privatization. Retrieved February 18, 2010, from Private Corrections Institute Web site: http://www.privateci.org/index.html

PRNewswire. (1998, April 20). Corrections Corporation of America and CCA Prison Realty Trust to merge in $4 billion transaction; Companies Acquire U.S. Corrections Corporation. Retrieved September 15, 2009, from: http://www2.prnewswire.com/cgi-bin/stories.pl?ACCT=104&STORY=/www/story/04-20-1998/0000634740&EDATE=

Ramirez, A. (1994, August 14). Privatizing America's prisons, slowly. *The New York Times*, p. 31.

Reisig, M. D. & Pratt, T. C. (2000). The ethics of correctional privatization: A critical examination of the delegation of coercive authority. *The Prison Journal 80*(2), 210-222.

Rhoades, S.A.. (1993). The Herfindahl-Hirschmann Index. *Federal Reserve Bulletin, 79*(3), 188-189.

Robbins, I. P. (1988). *The legal dimensions of private incarceration.* Washington, DC: American Bar Association.

Rugaber, C. S. (2008, May 12). SEC charges 3 doctors with insider trading. *Associated Press, Business News*.

Samuelson, P. A., & Nordhaus, W. D. (1989). *Economics* (13th ed.). New York: McGraw-Hill.

Savas, E. S. (2005). *Privatization in the city: Successes, failures, lessons.* Washington, DC: CQ Press.

Scherer, F. M., & Ross, D. (1990). *Industrial market structure and economic performance* (3rd ed.). Boston: Houghton Mifflin Company.

Schneider, A. L. (1999). Public-private partnerships in the U.S. prison system. *American Behavioral Scientist, 43*(1), 192–208.

Sellers, M. P. (1989). Private and public prisons: A comparison of costs, programs, and facilities. *International Journal of Offender Therapy and Comparative Criminology, 33*(3), 241-256.

Shichor, D. (1995). *Punishment for profit.* Thousand Oaks, CA: Sage.

Shilton, M. K. (1995). Community corrections acts may be Rx systems need. *Corrections Today, 57*, 32-36.

Sullivan, J., & Purdy, M. (1995, July 23). A prison empire: How it grew—a special report; Parlaying the detentions business into profit. *The New York Times*, p. 11.

Tennessee Select Oversight Committee on Corrections. (1995). *Comparative evaluation of privately-managed CCA prison and state-managed prototypical prisons.* Nashville, TN: Tennessee Legislature.

Texas Prison Bid'ness. (2007). Texas prison bid'ness blog. Retrieved February 18, 2010, from http://www.texasprisonbidness.org/about-blog

Texas Sunset Advisory Commission. (1991). *Information report on contracts for correctional facilities and services: Recommendations to the Governor of Texas and members of the Seventy-Second Legislature, final report.* Austin, TX: Texas Sunset Advisory Commission.

Thomas, C. W. (1997). *Private adult correctional facility census.* Gainsville, FL: University of Florida, Center for Studies in Criminology and Law, Private Corrections Project.

Van Slylce, D. M. (2003). The mythology of privatization in contracting for social services. *Public Administration Review, 63*(3), 296–315.

Walmsley, R. (2009). *World prison population list* (8th ed.). London: International Centre for Prison Studies, King's College London.

Weber, M. (1946). *From Max Weber: Essays in sociology* (H. H. Gerlh & C. Wright Mills, Eds.). New York: Oxford University Press.

West, H. C. & Sabol, W. J. (2008). *Prisoners in 2007.* Washington, DC: U.S. Department of Justice, Bureau of Justice Statistics.

West, H. C., & Sabol, W. J. (2009). Prison inmates at midyear 2008—statistical tables. Washington, DC: U.S. Department of Justice, Bureau of Justice Statistics.

West, H. C., Sabol, W. J., &. Cooper, M. (2009). *Prisoners in 2008.* Washington, DC: U.S. Department of Justice, Bureau of Justice Statistics.

■ **Court Cases Cited** ▬▬▬▬▬▬▬▬▬▬▬▬▬▬▬▬▬▬

*Kesler v. King*, 29 F. Supp. 2d 356, 371 (S.D. Tex. 1998).

# The U.S. Juvenile Justice Policy Landscape

# CHAPTER 10

Janeen Buck Willison, Daniel P. Mears, and Jeffrey A. Butts

## ■ Introduction

State governments enacted sweeping changes in law and policy in recent decades, profoundly affecting the juvenile justice landscape in the United States. Many of the changes mirror those made to the adult justice system (Butts & Mears, 2001; Howell, 2003; Katzmann, 2002; Wool & Stemen, 2004). Reduced sentencing discretion, increased information sharing among juvenile and adult justice systems, greater public access to juvenile records, and the transfer of juveniles to adult courts for prosecution are just a few examples. The last two decades have also been marked by innovation aimed at preventing delinquency and promoting the fair treatment and rehabilitation of young offenders (Butts & Mears, 2001). Recent advances include the proliferation of problem-solving courts (e.g., juvenile drug, mental health, and truancy court programs), developmentally based diversion programs, competency assessments, and increased efforts to integrate treatment and evidence-based approaches into the fabric of juvenile justice (Bishop, 2006; Mihalic, Fagan, Irwin, Ballard, & Elliott, 2006; Snyder & Sickmund, 2006).

New juvenile justice laws and policies continue to emerge as state lawmakers respond to fluctuations in crime, public sentiment, and changing policy priorities. Between 2005 and 2007, states proposed over 1000 juvenile justice measures, with roughly 100 new laws enacted annually (Buck Willison, Mears, Shollenberger, Owens, & Butts, 2008). In 2007 alone, roughly 70% of the states (35) passed a combined total of 113 juvenile justice-related laws (National Conference of State Legislators, 2008). Like the reforms of the preceding decade, the substance of these measures varies considerably. Many seek to advance public safety and crime reduction through increased penalties and harsher sentences, particularly for specific classes of offenders. Others mandate the use of specific approaches, including case management, evidence-based practices, and restorative justice principles. Some laws clearly enhance the rehabilitative capacity of the juvenile justice system, while others diminish the system's influence and further erode the barrier between juvenile and criminal justice. Each of these new laws affects the administration of justice for young offenders, either contributing to or detracting from the goals of the juvenile justice system.

The diverse mix of policies and practices introduced in recent years raises important questions about the posture of juvenile justice today. Most scholars agree that decades of "get-tough" reforms diminished the influence of the juvenile court (Bernard, 2006; Butts & Mears, 2001; Fagan, 2008; Scott & Steinberg, 2008). Many contend that these changes rendered the criminal (adult) and juvenile justice systems largely indistinguishable (Butts & Mitchell, 2000; Feld, 1993). Recent research, however, calls these claims into question (Bishop, 2006) and suggests that rehabilitation remains a critical goal for juvenile justice professionals (Buck Willison et al., 2008).

An accurate description of the present orientation of the juvenile justice system is essential to guide future decisions about policy, practice, and the role of the juvenile court. This chapter examines the state of juvenile justice policy nationally. The goal is to explore whether juvenile justice today is uniformly punitive in its orientation or whether it reflects the founding tenets of the original juvenile court. The extent to which states differ in their approach to juvenile justice is also examined to determine whether juvenile justice is as uniform as is implied. We draw on analyses of recent legislation and practice to identify emerging policy trends at the national level, and we review findings from a national survey of juvenile justice practitioners.

The remainder of the chapter proceeds as follows: First, we briefly review the origin and mission of the juvenile court and the founding tenets of that system to set the present analysis in context and provide a framework to evaluate both the present orientation of juvenile justice and the significance of various changes to the system. Second, we consider the circumstances under which policy reforms of the last two decades occurred to identify key factors prompting these changes. Third, we review recent juvenile justice legislation to identify emerging policy trends. Fourth, we examine current practice, as defined by the prevalence of selected policies and practices across the states, to evaluate the prevailing approach (punitive or rehabilitative) to juvenile justice. Finally, we present evidence from practitioners about their views on current policy and practice and how they would improve both. These analyses draw heavily from and greatly expand on findings from the recently completed Assessing Policy Options (APO) project, conducted by the authors under the auspices of the National Institute of Justice, U.S. Department of Justice.

# ■ Juvenile Justice in the United States: Then and Now

A brief view of the origin, evolution, and original mission of juvenile justice in the United States is required to fully appreciate the changes, challenges, and opportunities facing the system today, and to properly assess its present orientation.

Each of the 50 states and the District of Columbia operates its own juvenile justice system with distinct laws, policies, and practices for handling legal matters involving children and youth (King, 2006). While the structure and organization of these systems vary from state to state, common components exist. Juvenile court (called family court in some jurisdictions) is the heart of the juvenile justice system. It is dedicated to resolving matters involving individuals not legally defined as adults, including delinquency, status, and dependency matters. Other components of the juvenile justice system include detention facilities to hold and house youth charged or adjudicated for law infractions (these facilities are similar to adult jails, but because they serve children, most operate full-day classrooms) and juvenile probation and parole services, including aftercare, to supervise young offenders in the community. Many police departments and prosecutor's offices also have special units that handle cases involving juvenile delinquents.

Juvenile justice in the United States has evolved over time. In its current form, it is, arguably, quite different from what its founders envisioned. Notably, in today's juvenile court, judges possess less discretion, procedures are more formal and, thus, less flexible, and the focus is primarily on the offense (crime) as opposed to the young offender (Feld, 1993; Feld, 2006; Urban, St. Cyr, & Decker, 2003). Furthermore, the juvenile court is at a critical crossroads. Opponents question its necessity, citing the lack of evidence of its effectiveness. Proponents point to the growing body of evidence from the field of developmental psychology, substantiating the significant ways in which adolescents differ from adults, as justification for a separate justice system.

The concept of juvenile justice in the United States emerged near the turn of the 20th century. Until that time, children as young as 7 years old were typically handled in adult court and often received adult punishments, including prison sentences and the death penalty (Bernard, 2006; Snyder & Sickmund, 2006; Urban et al., 2003). Formation of the first juvenile court in Chicago, Illinois, in 1899 reflected the philosophy of European reformers that children were developmentally different—less cognitively and morally developed—than adults. Reformers saw immaturity, not calculated criminal intent, behind the law-breaking behavior of young offenders (Scott & Steinberg, 2008, p. 16). Early reformers also believed children were more amenable to change than adults, in part because their identities were not yet fully formed (Steinberg & Scott, 2006). As such, the law-breaking behavior of children and youth was thought to be more appropriately addressed in a separate legal process, one that took into account their cognitive capacities, moral development, and potential for rehabilitation (Bernard, 2006).

The doctrine of *parens patriae* (commonly translated "state as parent") guided early juvenile court policies and operations. Focused on the child, not the offense as in the adult court, the juvenile court assumed the role of benevolent parent, weighing the best interests of the child in its decisions and facilitating the child's rehabilitation through treatment and other services (Butts & Mitchell, 2000). In keeping with *parens patriae,* the juvenile system sought to advance the welfare, rehabilitation, and fair treatment of the errant child. Meting out appropriate consequences was an equally important function. These goals stood in stark contrast to the adult criminal court, which focused on the severity of the offense, the culpability of the offender, retribution, and public safety.

The most prominent distinction between the two systems is jurisdiction. Originally, the age of the offender defined the system's scope of legal authority; the age of majority—i.e., the age at which an individual is recognized as having the rights and privileges assigned to adults (right to vote, enter into binding legal agreements, etc.)—was the determining factor for jurisdiction. Today, the severity of the offense frequently determines which system (juvenile or adult) a young offender enters. Additionally, the juvenile court addresses a range of behaviors and issues (status, dependency, delinquency) that would fall outside the purview of the legal system if the defendant were an adult. **Figure 10–1** presents key legal stages in the juvenile court process.

The juvenile court concept quickly spread across the United States in the early 1900s. By 1910, virtually every state in the Union (32 of the 46 established states) had a juvenile court (Bernard, 2006; Butts & Mitchell, 2000; Snyder & Sickmund, 2006). At that time, most juvenile courts had exclusive jurisdiction over youth ages 17 years and younger, and juvenile court judges possessed a great deal of discretion regarding how best to handle delinquency cases; due process protections were scarce. The juvenile court operated largely unfettered until the 1940s when critics questioned the court's efficacy and necessity (Butts & Mitchell, 2000). Over the next 60 years, the pendulum of juvenile justice reform would swing between punishment and rehabilitation in keeping with public sentiment (Bernard, 1992).

As discussed in subsequent sections, reform and innovations, like juvenile crime, reached new levels during the 1990s. Unlike juvenile arrests, which peaked in 1994 then steadily declined, new laws and policies continued to emerge as state lawmakers sought to improve the system's effectiveness.

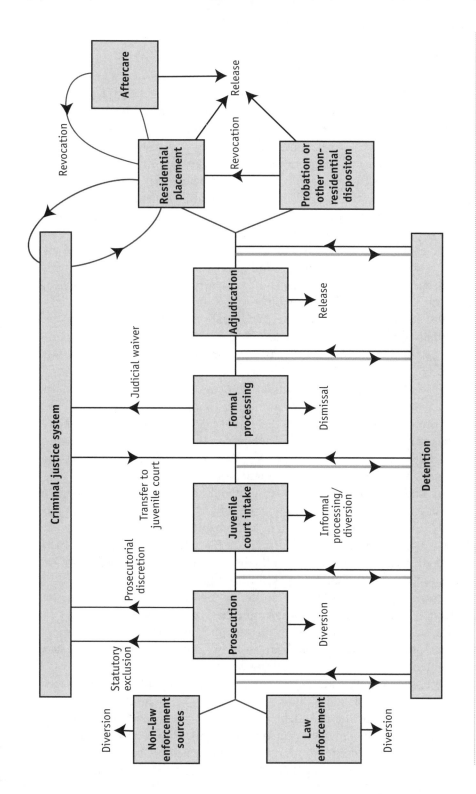

**Figure 10–1** Juvenile Court Case Flow Chart
**Source:** Reproduced from Office of Juvenile Justice and Delinquency Prevention. *Statistical Briefing Book [Online].* U.S. Department of Justice, 2010.

## ■ Reform and Innovation in the 1900s

In the late 1980s, juvenile crime, especially violent crime, began to increase dramatically (Bishop, 2004; Butts & Travis, 2002; McCord et al., 2001; Snyder & Sickmund, 1999). The increases, fueled in part by public concern about crime (Roberts, 2004), led states to introduce far-reaching and historically unprecedented changes to their juvenile justice systems (Fagan and Zimring, 2000; Feld, 1999; Harris, Welsh, & Butler, 2000; Howell, 2003). These changes included new and expanded laws for transferring youth to the adult system and for bridging gaps between the juvenile and adult justice systems; mandatory minimum sentence statutes, sentencing guidelines, and graduated sanctions models that attempt to create greater consistency in sentencing; laws to reduce the confidentiality of juvenile records and hearings; and efforts to target youth charged with violence, drugs, and weapon offenses (Butts & Harrell, 1998; Butts & Mitchell, 2000; Howell, 2003; National Criminal Justice Association, 1997; Snyder & Sickmund, 2006; Torbet et al., 1996; Torbet & Szymanski, 1998). Despite the scope and potentially significant costs of these changes, they remain largely unexamined (Bishop, 2004; Butts & Mears, 2001; Guarino-Ghezzi & Loughran, 2004; Krisberg, 2005; Mears, 2002; Mears & Butts, 2008).

The range of these recent policy initiatives defies simple characterization. In some cases, they clearly contradict or significantly modify the traditional philosophy and approach of the juvenile court (Feld, 1999). For example, many new policies emphasize more punitive and get-tough orientations that mirror those of the criminal (adult) justice system (Howell, 2003; Wool & Stemen, 2004). The expansion of transfer laws and the creation of mandatory minimum sentences for serious, violent, and gang-related crimes, drug offending, and gun-related offenses illustrate this shift (Butts et al., 2002; Kupchik, 2003). Others, however, attempt to strike a new balance between punishment and rehabilitation (Guarino-Ghezzi & Loughran, 2004; Katzmann, 2002; Mears, 2000). Blended sentencing laws, which allow terms of incarceration to begin in the juvenile justice system and continue in adult prison, are just one example (Fagan & Zimring, 2000).

Some recent policy efforts have addressed perceived problems with the administration of justice. Critics, for example, have pointed to inconsistency in sanctioning across judges and between jurisdictions, and have noted that intermediate sanctions are often ignored. In response, policy makers have created juvenile sentencing guidelines and graduated sanctioning models to increase consistency in sanctioning and to create more appropriate and effective dosages of punishment, treatment, and services (Howell, 2003; Krisberg, 2005).

Many states reduced the confidentiality of juvenile court records and proceedings (Sanborn, 1998; Snyder & Sickmund, 2006). Historically, the juvenile court process was private and court records were sealed or expunged when a youth reached the age of majority. Recent concerns about juvenile crime, however, have led to new laws that broaden access of juvenile court records to victims, the community, the media, and the criminal court system. Some of the laws limit the sealing and expunging of records while others encourage fingerprinting and photographing of youth and the creation of offender registries and statewide data repositories. Almost every state has enacted such laws, motivated by the idea that, as emphasized by the National Criminal Justice Association (1997), "juveniles should be held accountable when their criminal behavior has an impact on the community as a whole" (p. 37). Such laws allow, among other things, for greater sharing of information of a youth's legal and social history among prosecutors, corrections, probation, law enforcement, educational, and social service agencies, and others with a "need to know" (Torbet et al., 1996, p. 35).

Other states have enacted laws giving victims greater rights and a more prominent role in juvenile justice proceedings. For example, some states created a range of new provisions allowing victims of juvenile crime to be notified when a youth has a disposition hearing or is to be released from custody. Other provisions created opportunities for victims to participate in juvenile court hearings and to submit victim impact statements, and many established services for victims of crime (Torbet et al., 1996, p. 48).

Perhaps more than any period in juvenile justice, the last two decades have witnessed an almost ceaseless effort to develop new ways of preventing juvenile crime while holding young people accountable. Certainly, a recurring cycle, involving swings between two distinct orientations—get-tough punishment and rehabilitative treatment—exists (Bernard, 1992). Yet, in recent years, there is increasing evidence of multiple new ways of structuring and achieving juvenile justice (Butts & Mears, 2001).

Nationally, many state and local jurisdictions emphasize providing specialized services to particular populations of youth with unique needs, such as youths who have drug problems or those who suffer from mental illness or co-occurring disorders (Grisso, 2004). In some cases, the result has been the emergence of specialized modes of processing, such as drug or mental health courts. In others, the result has been an attempt to create greater integration of court and justice system efforts with those of social service agencies and treatment providers (Katzmann, 2002). However, few juvenile justice systems have systematically integrated specialized service delivery as a basic feature of everyday court processing, or assessed the impacts of related efforts, such as the widely touted wraparound initiatives (Howell, 2003, p. 236), on court operations. More relevant for debates about juvenile justice is the fact that such efforts, even though they emerged during an era widely characterized as punitive, do not really fit within a get-tough approach to juvenile justice.

Still other initiatives exist that illustrate the theme of change and also point to increasing challenges confronting the juvenile justice system. For example, parental accountability laws and case processing standards have become common

(Butts & Sanborn, 1999; National Criminal Justice Association, 1997). Accountability laws make parents civilly or criminally liable for the behavior of their children. Case processing laws and standards place limits on the amount of time the courts can allow for the adjudication and/or disposition of delinquency cases. Alongside such changes are new federally sponsored efforts that promote a range of other policies, such as expanding the use of risk and needs assessment and restorative justice programs (Andrews & Marble, 2003). Collectively, these and other emerging efforts aim to improve juvenile justice through a myriad of ways, some punitive, some rehabilitative, and some not neatly fitting into either category. These efforts also can create burdens on juvenile courts and practitioners that may undermine their effectiveness.

Some issues generate prolonged debate. Blended sentencing laws, for example, continue to raise the question of whether there should be two separate justice systems, one for juveniles and one for adults. Opponents argue that blending harms juveniles, while proponents argue that it brings greater efficiencies, protection of juvenile rights, and ultimately improved outcomes for young offenders (Fagan & Zimring, 2000). Other laws—including those that exclude entire age groups from the purview of the juvenile justice system (Torbet et al., 1996) and others that "criminalize" the juvenile justice system by modeling it increasingly after the adult justice system (Feld, 1999)—have led to similar debates. Although no state has begun to formally unify its juvenile and adult justice systems, it remains unclear how many court practitioners would oppose or embrace such a change (Butts & Harrell, 1998). The debate continues, as do attempts to modify various sentencing laws to better achieve such goals as deterrence, retribution, rehabilitation, and accountability.

## ■ Analysis of Juvenile Justice Legislation

As we have discussed, reform and innovation characterized juvenile justice over much of the last two decades. The content and purpose of these reforms varied greatly. Many instituted harsher penalties in an effort to reduce crime and enhance public safety. Other reforms sought to improve the effectiveness and accountability of the system through enhanced treatment and greater transparency. Innovation flourished. New approaches and practices (e.g., balanced and restorative justice, graduated sanctions, specialized courts) emerged to satisfy the system's often competing goals of rehabilitation, accountability, and public safety. Whether these developments signal a return to the original philosophy of the juvenile court (rehabilitation model) or an advance of the adult model (punitive model) is not entirely clear. Our central question—What is the prevailing approach for handling young offenders today?—remains unanswered.

To arrive at a clearer picture of today's juvenile justice landscape, we examine the scope and nature of state-level juvenile justice legislation proposed and enacted between calendar years 2005 and 2007, and the composition of the measure enacted.[1] Here, we take a closer look at the composition of recent legislation falling within four broad topics: jurisdiction, sentencing and penalties, confidentiality, and treatment and rehabilitation. The extent to which these measures contribute to or detract from primary juvenile justice goals—rehabilitation of the child, offender accountability, and public safety—provides clues about prevailing policy orientations among states.

## Juvenile Court Jurisdiction

Age specifications delineate the parameters for juvenile court original jurisdiction (legal authority), determining which children are the responsibility of the juvenile court and which are the responsibility of the adult criminal justice system, and define case processing policies for petitioning, transfer, and sentencing. It is difficult to overstate the significance of age in juvenile justice policy. Therefore, legislation that amends the upper age limit of the juvenile court—either lowering or increasing the age boundary—or the statutory definition of the age of majority affects the welfare of literally thousands of children (Sickmund, 2008).

Historically, juvenile courts had exclusive jurisdiction over children aged 17 years and younger involved in delinquency and status offenses and dependency matters (Butts & Mears, 2001; Snyder & Sickmund, 2006). As noted earlier, the proliferation of get-tough policies in the 1990s chipped away at the court's jurisdiction. Laws increasing the number of transfer-eligible offenses and reducing the upper age limit for juvenile court jurisdiction further eroded the court's jurisdiction (Bernard, 2006; Butts & Mitchell, 2000; Redding, 2008; Snyder & Sickmund, 2006). By 2004, 13 states[2] had set the upper age for juvenile court jurisdiction over delinquency offenses at 15 and 16 years of age (Snyder & Sickmund, 2006, p. 103).

Recent legislation suggests this trend may be reversing. Since 2005, more than half the states introduced measures to restore and expand the jurisdiction of the juvenile court. Rhode Island and Connecticut, for example, passed laws[3] that expanded juvenile court jurisdiction to youth aged 18 years of age and under. Connecticut's bill was significant not only because it expanded the original jurisdiction of the juvenile court in that state from youth 16 years and under to 18 years and under, but also because its passage reduced the number of states to just two—New York and North Carolina—in which 16 year olds qualify as adults and are automatically handled in the criminal justice system.[4]

States also expanded the jurisdiction of juvenile courts in more subtle ways. Idaho[5] lawmakers, for example, returned status offenders aged 18 years and younger to the jurisdiction of the juvenile court; previously, juvenile court jurisdiction in Idaho was limited to status offenders aged 14 years and under (Schmid, 2005, p. 52). Arkansas amended its juvenile code under House of Representatives

Bill 1475 to "ensure that a felony or misdemeanor committed by a juvenile before age 18 may be prosecuted as a delinquency offense in juvenile court."[6] Additionally, a number of states adopted measures to extend the period for which juvenile courts could retain jurisdiction over adjudicated youth. New Hampshire (H.R. 627) and Rhode Island (S. 1141), for example, passed bills delineating circumstances under which family courts in those states could retain jurisdiction for youth over the age of 18 years. Legislation seeking to limit the use of mandatory minimums, life without parole, and the death penalty also figured prominently on state legislative agendas; these measures are discussed in more detail in subsequent sections.

In summary, juvenile court jurisdiction figures prominently in recent state legislation, with more states passing measures to expand the court's jurisdiction than to reduce it—although several states introduced measures with the latter objective in mind. Measures that appropriately expand the reach of the juvenile court are generally considered to be progressive in nature because they support the goals of juvenile justice.

### Transfer, Sentencing, and Penalties

Perhaps the most prominent change in juvenile justice in recent decades has been the explosion of new laws for transferring juveniles to adult court. These laws allow juvenile courts to waive jurisdiction, or, as some accounts describe it, to allow children to be certified adults and prosecuted in criminal court. Whatever the terminology—transfer, waiver, or certification—the laws create diverse mechanisms for placing young offenders into the criminal justice system. Motivation for enacting ever-new ways of sending young people to adult court stems from several concerns, including real and perceived increases in violent crime and frustration with some of the limits of sanctions available in juvenile court (Zimring, 2005). The popularity of transfer as a focus of policy change is reflected in trends nationally; as Snyder and Sickmund (2006) noted, "since 1992, all states but Nebraska have changed their transfer statutes to make it easier for juveniles to be tried in criminal court" (p. 113).

Transfer laws remain popular with state lawmakers despite the recent movement to reclaim and expand juvenile court jurisdiction. Roughly 43 states considered transfer measures between 2005 and 2007, of which 17 states adopted these measures. Many of the transfer laws enacted during the study period focused on weapons offenses and violent crimes such as murder; some sought to expand the laws' reach by lowering the minimum age for transfer on already transfer-eligible crimes, while others, interestingly enough, restricted the use of transfer.

During the years examined, at least six states passed transfer-enhancing measures. In 2006, for example, Oklahoma legislators enacted Senate Bill 1760 providing for the automatic transfer to adult court of children as young as 13 years charged with first-degree murder, unless previously certified as a youthful offender; a related portion of the bill excludes youths between the ages of 15 and 17 years charged with first-degree murder from being considered as juveniles. California and Louisiana both expanded the number of transfer allowances (terms) under which a child could be transferred to criminal court; California policy makers also lowered the age (from 16 to 14 years of age) under which a youth could be found unfit for juvenile court (S. 520) if charged with selected violent crimes.

Some states restricted the number of transfer-eligible offenses during the review period. Delaware limited criminal court jurisdiction to youth with prior felony adjudications and abolished the automatic transfer of youth charged with first-degree robbery to criminal court (S. 200, 2006). In 2007, Virginia (H.R. 3007) stipulated that juveniles tried as adults must be convicted of the crime to be recognized or processed as an adult in future legal actions.

These legislative changes suggest increasing support among lawmakers, and arguably the public, for reserving transfer to adult court for serious violent crimes such as murder and leaving less serious delinquency offenses to the juvenile court. In short, it appears legislators are working to strike a more balanced approach with respect to transfer and sentencing (Snyder & Sickmund, 2006). Changes to other laws governing sentencing and penalties also support this observation.

## Confidentiality and Juvenile Records

Confidentiality has been the tacit hallmark of the juvenile justice system since its inception more than 100 years ago. Historically, juvenile court hearings were closed to the public, the names of juvenile offenders protected, and records expunged once the delinquent child reached the age of majority (National Criminal Justice Association, 1997). Designed to protect the vast majority of children it served from stigma and labeling that could inhibit meaningful rehabilitation, these measures reflected the traditional mission of the juvenile court to act in the best interest of the child. In keeping with the increased emphasis on public safety and individual accountability in recent decades, the confidentiality of juvenile court proceedings and records has eroded. States increasingly open juvenile proceedings to the public and press, mandate the collection of DNA and other personal data from young offenders, and retain the records of adjudicated youth offenders in administrative databases accessible to law enforcement, social services, schools, and the general public.

Juvenile records were the policy focus of legislation in 33 states between 2005 and 2007. Proposed and enacted measures addressed a variety of issues, including (1) access to juvenile court records by victims and the public; (2) the type of personal data that may be collected (e.g., DNA, fingerprints, photographs) from juveniles at various stages in the legal process, and how the data could be used; (3) conditions under which records would be sealed or expunged; and (4) information sharing between justice agencies and youth-serving systems. The most notable trend observed during this period was the expansion of DNA collection from juveniles. In 2006 alone, six states (Arkansas, Arizona, Colorado, Illinois, Kansas, and Ohio) passed measures governing the collection, storage, and use of DNA specimens collected from juveniles adjudicated for serious or violent offenses. Some states (Ohio, Kansas, Michigan) also passed legislation allowing juvenile arrest records to be added to and retained in statewide criminal records databases. Changes in confidentiality laws served multiple purposes, often highlighting the tension between the system's conflicting objectives: protecting public safety and promoting the welfare of the children it serves.

As noted earlier, collecting such information from juvenile offenders ran counter to the mission of the juvenile court and its stated goals. While many states passed laws expanding the collection and retention of personal data from juveniles, some states enacted measures restricting these practices. North Carolina (S. 1211) specified a limited set of circumstances under which juveniles could be photographed or fingerprinted prior to actual adjudication. Montana (S. 119) and several other states strengthened provisions for sealing and expunging juvenile records.

### Treatment and Rehabilitation

Increased awareness of and growing empirical evidence about the use of drugs and alcohol and mental health needs of young offenders have propelled treatment to the forefront of juvenile justice (Grisso, 2007). During the study period, states appropriated funding for substance abuse and mental health treatment and services (Arkansas, Colorado, Indiana, Massachusetts, Nebraska, Tennessee), specialized courts (Indiana, Tennessee), and pilot programs to increase capacity for community-based treatment and services (IL, S. 1145; MD, S. 882). These bills mandated a wide range of treatment-related practices, including screening, assessment, individualized treatment plans, treatment services, counseling, the use of evidence-based treatment and programming, and multidisciplinary service teams. Much of the legislation enacted in 2007, in particular, focused on the treatment needs of juvenile justice-involved youth.

Among the most innovative provisions were those mandating evidence-based practices, including screening and assessment, integrated treatment approaches, and multidisciplinary teams. In 2007, Colorado (H.R. 1057) legislation, for example, called for a system of care and services for juveniles with mental health needs. The year before, Colorado (S. 06-122) increased funding statewide for substance abuse treatment based on data indicating that between 60% and 80% of juvenile justice-involved youth required some level of treatment. Also, in 2006, Massachusetts (H.R. 5097) directed its Commissioner of Public Health to fund a "comprehensive and accessible continuum of substance abuse treatment and prevention programming" (National Juvenile Defender Center [NJDC], 2007, p. 106). West Virginia (S. 517, 2006) mandated multidisciplinary teams at juvenile diagnostic centers as part of initial assessment at facilities where juveniles adjudicated delinquent are in custody, and Hawaii (S. 3207, 2006) called for the development of "community-based programs to encourage positive youth outcomes" (NJDC, 2007, p. 41).

Other notable treatment measures passed during the period of review include the following:

- California (S. 81) established the Youthful Offender Block Grant in 2007 to build county capacity to provide "appropriate rehabilitative and supervision services to youthful offenders" (NCSL, 2008).
- Washington (H.R. 1483, 2006) established goals and objectives for early intervention and services to juvenile justice-involved youth under the Reinvesting in Youth program.
- Oklahoma (H.R. 2999) identified the Office of Juvenile Affairs as the coordinating and oversight agency for services to juvenile delinquents (alleged and adjudicated) and defined the range of "core community-based" services to be provided to juveniles as the continuum of screening, evaluation, assessment, treatment planning, and case management (NJDC, 2007, p. 110).
- Oregon (H.R. 2149, 2007) authorized courts to mandate assessment and treatment for alcohol use.
- Tennessee (H.R. 1871) provided juvenile courts with the authority to "develop and operate drug treatment programs" (National Conference of State Legislatures, 2008).

Mental health treatment measures were equally diverse, although one theme emerged: States are willing to fund mental health treatment and services. Under Senate Bill 1455 in 2006, Idaho allocated $700,000 for services to juvenile offenders diagnosed with mental illness, and Senate Bill 1389 called for a comprehensive mental health center to provide intensive support services and transitional housing for juveniles (and adults) with mental illness or addiction disorders (NJDC, 2007, p. 113) Also in 2006, Alaska appropriated $600,000 for juvenile mental health services. On a slightly different note, Colorado (S. 6005) mandated insurance plans to pay for court-ordered "medically necessary mental health services" if such services were deemed necessary as a result of the youth's contact with the criminal or juvenile justice systems (NJDC, 2007, p.113).

This brief analysis suggests that many states are moving toward more progressive reforms and away from the sweeping punitive get-tough responses that dominated juvenile justice policy in the previous decade (Bernard, 2006; Bishop, 2006; Snyder & Sickmund, 2006; Urban et al., 2003). Notable shifts include restoring the jurisdiction of the juvenile court, instituting more stringent confidentiality measures to protect young offenders, increasing focus on the treatment needs of juvenile justice-involved youth who use drugs and alcohol or who are mentally ill, and emphasizing research-based programming and services that, together, build a continuum of care for at-risk, as well as delinquent youth. Several states appear to be headed in the direction of adopting more strength-based, not just risk-oriented, approaches to remediate the unique circumstances of the individual youth served by the juvenile justice system. These shifts suggest a return to the founding tenets of the juvenile court: rehabilitation through individualized interventions focused on the unique circumstances of the youth (Bernard, 2006; Butts & Mitchell, 2001; Snyder & Sickmund, 2006).

Transfer continues to enjoy support among policy makers, as evidenced by the number of measures introduced. Although widely used in some jurisdictions (Olsen, 2009), others appear to be setting limits on its use. This observation, coupled with the trend toward expanding the jurisdiction of the juvenile court, suggests that policy makers and legislators are working to strike a proper balance in sentencing and handling delinquent youth, a challenge that underscores the tension of balancing the system's multiple goals. The diversity of measures proposed and enacted between 2005 and 2007 further underscores the dynamic nature of juvenile justice policy, a theme discussed in earlier sections of this chapter.

## ■ Is Juvenile Justice Punitive or Rehabilitative?

Analysis of legislation offers a national perspective on which direction policy is leaning at a specific time; it may be best interpreted as an indicator of the current policy climate and probable future direction. The preceding analysis of recent juvenile justice legislation, for example, suggests that states are moving toward more progressive reforms. While a valuable indicator, it alone cannot describe the entire landscape and orientation of juvenile justice. We must also consider policy activity prior to the period of the legislative analysis to obtain a more complete picture.

In this section, we examine the prevalence of selected practices and policies—some punitive, some progressive—enacted across the states to determine whether juvenile justice today is primarily punitive, as some assert, or progressive in nature.[7] Two composite indicators are used to rank the states on the relative punitiveness and progressiveness (i.e., rehabilitativeness) of their approach to juvenile justice. Indicators of "punitive" juvenile justice consist of 10 practices that emphasize accountability, penalties, and punishment. Indicators of "progressive" juvenile justice consist of 7 practices that facilitate rehabilitation through treatment and services, and that take into account a child's developmental capacities. States received a point for each policy enacted. For the punitive indicator, 10 points were possible, and scores ranged from 1 to 7. For the progressive indicator, 7 points were possible, and scores ranged from 1 to 5.

### Punitive Indicators

Punitive indicators cover three broad categories: age specifications related to the jurisdiction of the juvenile court; transfer mechanisms; and penalties similar in nature to those traditionally reserved for adults. Each is briefly described in the following list. In general, these practices and policies are considered to be punitive because they facilitate the processing and punishment of children as adults. **Figure 10–2** indicates how a relative punitiveness score is calculated.

- *Age specifications.* The first three indicators in the composite address age specifications—exclusion of all 16 year olds, exclusion of all 17 year olds, and no minimum age specified for which a child could be found culpable for a delinquent act (i.e., no child is considered too young to adjudicate). The lower the upper age limit for processing juveniles, the more punitive a state's approach. Again, states receive a point for each applicable policy.
- *Juvenile transfer.* Transfer-related indicators include legislative transfer, prosecutorial transfer, and blended sentencing (also called concurrent jurisdiction [Snyder & Sickmund, 2006]). These mechanisms are the most common by which transfer to adult criminal court occurs. Legislative transfer refers to laws that remove certain cases from juvenile court jurisdiction; here, again, such laws are limited by age and offense criteria. Prosecutorial transfer laws allow prosecutors to determine independently whether to file a case in juvenile or adult court; in such cases, this arrangement, referred to as "concurrent jurisdiction," imposes age and offense criteria on prosecutorial decision making. Blended sentencing allows the juvenile court to impose criminal (adult) sentences while retaining jurisdiction over the youth; actual imposition of the adult sentence typically hinges on the youth's compliance with his or her juvenile disposition (Snyder & Sickmund, 2006). As of 2004, 15 states had concurrent jurisdiction provisions, and 29 had statutory exclusion provisions (Snyder & Sickmund, 2006, p. 113). Every state employs at least one form of transfer. Transfer provisions are primarily punitive in nature.
- *Penalties.* Life without parole and DNA collection are the final indicators in the punitiveness composite. Until relatively recently, both practices were reserved for adults. With the U.S. Supreme Court's *Roper v. Simmons* decision (2005) declaring the juvenile death penalty unconstitutional, life without parole is now the most severe punishment possible for juvenile delinquents. DNA collection, like the trend toward photographing and fingerprinting juveniles, is a relatively recent innovation that mirrors the adult system; such practices were frowned upon by early juvenile justice advocates because they increase the risk of labeling and stigma that may inhibit the child's rehabilitation.

## Punitive Scores

Scores ranged from 1 to 7. No state scored a 0, indicating that every state had at least one punitive policy in place. Roughly 25% of the states (12) scored 6

| Indicators of "Punitive" Juvenile Justice | |
|---|---|
| State excludes all 17-year-olds from juvenile court jurisdiction | 1 point |
| State excludes all 16-year-olds from juvenile court jurisdiction | 1 point |
| State does not specify a minimum age for juvenile court jurisdiction | 1 point |
| State mandates Life Without Parole for crimes committed under age 18 | 1 point |
| State permits Life Without Parole for crimes committed under age 18 | 1 point |
| State uses "automatic" (i.e., legislative) exclusion for criminal courts | 1 point |
| State uses prosecutor discretion for criminal court transfer | 1 point |
| State juvenile courts permitted to "blend" adult-system sanctions | 1 point |
| State permits collection of juvenile DNA | 1 point |
| | **10 points possible** |

| Indicators of "Progressive" Juvenile Justice | |
|---|---|
| State juvenile court hearings closed to the public | 1 point |
| State juvenile court records are confidential | 1 point |
| State juvenile code has provisions for competency determination | 1 point |
| State juvenile code includes provisions for treatment (mental health, drug and alcohol, or sex offender) | 1 point |
| State juvenile code has provisions for specialized juvenile courts (teen, drug, mental health, reentry, etc) | 1 point |
| State juvenile Purpose Clause specifies adherence to BARJ principles | 1 point |
| State juvenile Purpose Clause specifies adherence to Child Welfare principles | 1 point |
| | **7 points possible** |

**Figure 10–2** Indicators of "Punitive" and "Progressive" Juvenile Justice
**Source:** Data from recent issues of Szymanski, L., *NCJJ Snapshot*. Pittsburgh, PA: National Center for Juvenile Justice; data from "State Data," National Juvenile Defender Center. Available: http://www.njdc.info/state_data.php; and data from Snyder, Howard N., and Sickmund, Melissa. 2006. *Juvenile Offenders and Victims: 2006 National Report*. Washington, DC: U.S. Department of Justice, Office of Justice Programs, Office of Juvenile Justice and Delinquency Prevention.

or higher on the punitive scale. Four states (Florida, Georgia, Illinois, Michigan) scored 7, securing the distinction as most punitive. As **Figure 10–3** illustrates, almost half the states (21) scored 3 or lower on the punitive composite measure. Kansas was the only state to score a 1; provisions authorizing DNA data collec-

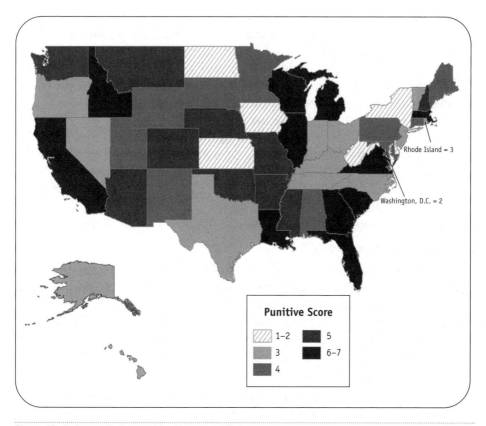

Rhode Island = 3

Washington, D.C. = 2

**Punitive Score**

1–2
3
4
5
6–7

**Figure 10–3** Composite Scores of Punitiveness by State

tion from juveniles was that state's only punitive policy among those included in the composite measure. Of the six states (Delaware, Washington, D.C., Maine, New York, North Dakota, West Virginia) that scored 2 on the punitive composite, only one (ME) had provisions for juvenile DNA collection. The most prevalent policy among these six largely nonpunitive states was the lack of a specified lower age limit for juvenile court jurisdiction; five of the six states did not have provisions for this at the time of our review.

Among the 32 states scoring in the moderately punitive range (3–5), most employed a mix of punitive practices. Twenty-two states received a point because they did not specify a lower age limit at which a child may be presumed too young to act with purposeful criminal intent and, thus, unfit for prosecution (Snyder & Sickmund, 2006). Twenty-one of the 32 states earned a point for laws that permitted DNA collection from juveniles, while 15 states also had mandatory life-without-parole penalties for juveniles. Indeed, juvenile life without parole is the most prevalent indicator across all 50 states: 41 states permit it as a sentencing option for crimes committed under the age of 18 years; 26 of those states have mandatory life-without-parole requirements for certain crimes, namely violent offenses. Least prevalent among these 32 states was the use of blended sentencing (7 states). Next to provisions excluding 16 year olds from the juvenile court, blended sentencing was the least prevalent practice across the 50 states.

The distribution of states' scores suggests that the orientation of juvenile justice is not predominantly punitive. Just a quarter of the states (12) employed a wide range of punitive adultlike criminal penalties. This finding, however, should not be interpreted as evidence that juvenile justice is primarily progressive. The two constructions are not inversely related. To determine the relative "progressiveness" of the states, and thus juvenile justice as a whole, we must examine the prevalence of discrete progressive practices.

### Progressive Indicators

Progressive indicators consist of seven practices grouped into four broad categories: confidentiality of juvenile court proceedings and records; competency standards; treatment, including specialized courts; and juvenile justice purpose clause. These practices and policies were selected for this measure because they are consistent with the original mission and goals of juvenile justice: acting in the best interest of the child, providing treatment, and facilitating rehabilitation. Each is briefly described here.

- *Confidentiality of juvenile court proceedings and records.* Some states' laws governing juvenile court proceedings are more open than others. Until the 1980s, juvenile court proceedings were typically closed (Snyder & Sickmund, 2006). Closing hearings to the general public was designed to protect delinquent youths from the stigma and labeling that could preclude meaningful rehabilitation. Some states also "seal" or restrict access to a juvenile's record, or in some instances expunge the record once the juvenile reaches adulthood. These practices are considered protective or progressive, not punitive, in orientation.

- *Competency proceedings and standards.* The issue of juvenile competence concerns a youth's ability to fully participate in legal proceedings owing to emotional and cognitive immaturity (MacArthur Foundation, 2008).[8] A growing body of research supports the notion that children and youth are developmentally different from adults, thus official responses to their lawbreaking behaviors should consider their cognitive capacities and maturity of judgment, as well as the severity of the offense. The number of states with laws establishing procedures to assess or address juvenile competence in juvenile court proceedings is growing (MacArthur Foundation, 2006). Such measures are generally viewed as progressive and consistent with the traditional mission of juvenile justice because they recognize that children differ from adults and facilitate the appropriate treatment of young offenders.

- *Specialized courts.* Specialized courts are generally viewed as a progressive measure that supports rehabilitation. States received a point if they had legislation establishing at least one of the following specialized courts: teen court, juvenile drug court, truancy court, or mental health court; the legislation represents a threshold of support that is different from a county starting a specialized court on its own initiative.

- *Juvenile justice purpose clause.* A state's purpose clause delineates its philosophical approach toward the handling of young offenders and, theoretically, guides policy decisions (Bishop, 2006; Feld, 1993; Snyder & Sickmund, 2006). Some emphasize rehabilitation while others focus on accountability and punishment (Snyder & Sickmund, 2006), or some combination of the two. The National Center on Juvenile Justice (NCJJ) groups the states' purpose clauses into five categories depending on their emphasis. Two categories are primarily progressive, supporting the rehabilitation goals of the juvenile justice system: balanced and restorative justice (BARJ), and child welfare. A state's purpose clause may emphasize several constructs; therefore, for this measure, a state could receive a point for both BARJ and child welfare if its purpose clause includes language that emphasizes both concepts.

## Progressive Scores

States' scores ranged from 1 to 5, indicating that every state had at least one progressive policy in place. Roughly 25% of the states (13) scored 4 or higher on the progressive composite indicator. Two states (Florida and Kentucky) scored 5, securing the distinction as most progressive relative to the other states. Eight states (Hawaii, Iowa, Louisiana, Maine, Missouri, Nevada, North Dakota, South Dakota) scored just one point on the composite; the most prevalent policy across these states was confidentiality of juvenile court records. As **Figure 10–4** illustrates, 60% percent of the states (31) scored 3 or higher on this composite measure of progressive juvenile justice.

The distribution of states' scores suggests a solid progressive orientation. Statutes protecting the confidentiality of juvenile court records (found in 46 states), legislation establishing or affirming specialized courts (found in 30 states), and provisions in support of treatment (found in 18 states) were the most prevalent progressive policies in place across the 50 states. In contrast, just 15 states have firm policies that close juvenile hearings to the public and media (Snyder & Sickmund, 2006), and 4 incorporate specific child welfare principles in their respective purpose clause.

These findings, however, should be viewed with some caution. Confidentiality in juvenile court proceedings is the most prevalent progressive policy across the states, and while most states' statutes declare juvenile records closed to the

Rhode Island = 3

Washington, D.C. = 4

**Progressive Score**

1

2

3

4

5

**Figure 10–4** Composite Scores of Progressiveness by State

public, many also provide exceptions that open the files of violent or predicate offenders. While only a handful of states (Arizona, Iowa, Kansas, Washington, West Virginia) do not afford juvenile records any confidentiality, it would be misleading to conclude the majority of states are upholding juvenile confidentiality standards as envisioned by the court's founders.

Additional cautions and limitations should be noted. First, this analysis focuses on state law as the sole indicator of a state's orientation toward policy and practice; it does so, in part, because formal legislation connotes priority. Many practices and policies, especially those associated with programming and treatment, are not formally legislated; because sentencing and punishment involve issues of due process and the rule of law, they will be legislated and therefore easier to systematically identify. Thus, our analysis runs the risk of failing to detect a state's emphasis on largely unlegislated progressive measures and potentially overstating a state's emphasis on punitive measures. Second, our measures merely indicate if the state has a law in place; it does not, except as noted, offer details about the provisions of the policy itself, nor does it reveal if the policy is widely used or enforced. Because the scope of implementation is unknown, the true impact of the policy is likewise unknown.

Nonetheless, these composite measures are highly illustrative and offer clues about the current orientation of juvenile justice and how individual states approach it. Briefly, the distribution of states' relative punitiveness scores suggests that the states' collective orientation toward juvenile justice is not primarily punitive. Just a quarter of the states (12) employ a wide range of punitive adult-like criminal penalties, as demonstrated by the selected punitive laws across the states. Half (21) scored within the low-to-moderate range. Practices in these states were largely limited to policies permitting life without parole for juveniles, transfer to adult court, and failure to specify a minimum age for juvenile court jurisdiction. These patterns are consistent with trends identified by our legislative analysis. Interestingly, a similar percentage of states achieved high scores on the composite measure of progressiveness. Thirteen states scored either a 4 or a 5, indicating they employ more than half the practices comprising the composite measure.

Some states scored high on both measures (California, Florida, Georgia, Illinois, Idaho, Wisconsin), which suggests these states have a well-developed continuum of practices that embrace both punitive and progressive measures. Others scored high on one measure and low on the other, suggesting one orientation prevails over the other. For example, Iowa and Louisiana both scored 6 on the punitive composite measure and 1 on the progressive composite; this suggests a proclivity toward legislating punitive measures. As mentioned previously, progressive measures including treatment and programming that supports rehabilitation may simply not be legislated in these states but may be prevalent in the community nonetheless. The same ambiguity about a states' punitive orientation is unlikely because most punitive indicators, as discussed earlier, must be established by a legislative process. The risk of undercounting punitive measures is low.

Regardless, states that scored high on the progressive composite (California, Colorado, District of Columbia, Florida, Georgia, Idaho, Illinois, Kentucky, Minnesota, New Jersey, Texas, Wisconsin) typically scored high on the punitive composite, suggesting these states not only have a well-developed continuum of practices but an equally well-developed legislative agenda for juvenile justice (i.e., policies and practices are formalized).

States with low scores on both composites tend to have a greater number of punitive measures in place. Exceptions include the District of Columbia and Kansas.

On balance, however, the distribution of states scores suggests two conclusions. First, most states employ a range of measures but trend slightly toward a progressive orientation. Second, analysis suggests that states employ different combinations of these practice and policies, presumably to address the unique local context in which juvenile justice is administered. In short, analysis suggests that the orientation and landscape of juvenile justice today varies depending on the state.

## ■ Practitioner Perspectives

Thus far we have examined recent legislation and the prevalence of various practices across the states to gain a better understanding of the present posture of juvenile justice. We have concluded that some practices and policies are more prevalent than others and that juvenile justice is tougher in some states and more progressive in others. In either case, it may be reasonable to presume the practices and policies employed have been selected because they are effective in achieving specific outcomes. Very little of the extant research, however, focuses on the effectiveness of juvenile justice policies and practices (Mears, 2000). Indeed, the rapid growth of new juvenile justice policies stands in marked contrast to the dearth of studies examining their implementation or impacts, or the opinions of juvenile justice practitioners toward these policies. Most studies focus on a few specific juvenile justice policies, most notably transfer (Butts & Mears, 2001), while opinion surveys typically focus on the public (e.g., Schwartz, Guo, & Kerbs, 1993; Sprott, 1998) or individuals under correctional supervision (Landsheer & Hart, 2000; Miller & Foster, 2002; Veneziano, Veneziano, & Gill, 2001). As Bishop (2006) noted, "there is a paucity of research on the contemporary juvenile court, and on the philosophies and practices of intake officials, judges, prosecutors, defense attorneys, and those who administer and work in the juvenile correctional system" (p. 58).

In absence of empirical evidence, practitioner perceptions offer an informed perspective about the effectiveness and necessity of recent changes. Charged with implementing diverse new measures, seasoned practitioners are uniquely positioned to remark on the effectiveness, necessity, and unintended consequences of these changes. With this in mind, researchers at the Urban Institute, Florida State University, and the Chapin Hall Center for Children at the University of Chicago conducted an online survey of juvenile justice professionals to measure their impressions of recent policy changes and the critical needs facing today's juvenile justice system. The survey also sought to garner recommendations for improving policy administration and effectiveness. Between March and August 2007, juvenile court judges, chief probation officers, court administrators, prosecutors, and juvenile defenders from the nation's 285 most-populated jurisdictions completed a 15-minute survey about the critical needs facing the juvenile justice system and the perceived effectiveness of 17 prominent policies and practices; the survey also solicited their recommendations for improving practice. In this last section, we consider practitioner perceptions of effectiveness and their recommendations. To what extent do practitioner opinions align with current practice? What practices do juvenile justice professionals rate as effective?

## Effective Policies and Practices

The survey asked practitioners to rate the ability of 17 prominent policies and practices to achieve six different outcomes: (1) reduce crime in the community, (2) reduce recidivism, (3) provide appropriate punishment, (4) facilitate fair treatment, (5) enhance the efficiency of the system, and (6) support the traditional mission of juvenile justice. **Figure 10–5** lists the policies and practices practitioners assessed.

---

### APO National Practitioner Survey

The APO National Practitioner Survey was an online self-administered questionnaire consisting of four sections. Respondents used a unique "username" and private password assigned by the Urban Institute to log-on to the survey. Most participants completed the survey in about 15 minutes.

**Section 1: Demographics**

**Section 2: Critical Needs (13 items; 5 domains)**
Staff development & training • Gender responsive services • Culturally relevant services • Developmentally appropriate services • Resources for non-English speakers • Policymaker support for rehabilitation • Public support for rehabilitation • Effective juvenile defense counsel • Effective prosecution • Alternatives to secure detention • Disproportionate minority contact • Information technology • System capacity to measure performance and evaluate programs

**Section 3: Policy Outcomes (17 items, 4 domains)**
Curfew laws • Parental accountability laws • Reduced confidentiality • Victim participation • Restorative justice • Time limits on proceedings • Specialized courts • Transfer • Graduated sanctions • Risk and needs assessments • Coordination with social services • Effective mental health treatment • Effective substance abuse treatment • Effective sex offender treatment • Targeting gang-involved youth • Community-based alternatives to secure detention • Reentry services and planning

**Six outcome measures of policy effectiveness included**
Less crime • Less recidivism • Fair treatment • Appropriate punishment • System efficiency • Traditional mission

**Section 4: Practitioner Recommendations for System Improvement**

---

**Figure 10–5** APO National Practitioner Survey
**Source:** Data from Urban Institute. (2007). APO Practitioner Survey was designed for the Assessing the Past, Present and Future of Juvenile Justice Policy (APO) Project under NIJ grant #2005-IJ-CX-0039.

The following five practices garnered the most support across practitioner groups, with the largest percentage of respondents *strongly agreeing* that these were effective in achieving all six outcomes: mental health treatment (54%), substance abuse treatment (51%), sex offender treatment (47%), reentry services and planning (46%), and coordination of juvenile justice with wraparound services (37.5%).

In contrast, respondents consistently identified a set of five punitive practices as largely ineffective by reporting that they *strongly disagree* the following practices were effective in producing any of the six desired outcomes: reduced confidentiality of juvenile records (23%), transfer of juveniles to adult court (19%); time limits on delinquency proceedings (13%); parental accountability laws (9%); targeting of gang-involved youth for special prosecution (8%). While these findings indicate that most practitioners view the majority of practices and policies as effective, these five measures garnered the least support. These trends generally held across practitioner groups, although some differences were observed between prosecutors and practitioners, reflecting their differing roles and responsibilities. Prosecutors, for example, were more likely to favorably rank the effectiveness of these five measures than were juvenile defenders (Mears, Shollenberger, Buck Willison, Owens, & Butts, 2008).

### Practitioner Recommendations

Practitioner recommendations addressed a number of issues that often fall outside the scope of policy evaluation but are critical nonetheless to the health and welfare of the system. Collaboration, increased information sharing, greater coordination among system stakeholders, and adequate funding to train and retain staff were frequent suggestions.

Recommendations specific to programming and services called for implementation of evidence-based practices that would yield a balanced continuum of approaches. Practitioners offered an array of suggestions focused on prevention and intervention, and the application of graduated sanctions and restorative justice principles. In keeping with this latter recommendation, practitioners also directed policy makers to expand their focus to include treatment and rehabilitation in addition to appropriate punishment.

These findings are consistent with those trends discussed in earlier sections of this chapter and lend additional support to the notion that juvenile justice policy is moving in the direction of rehabilitation. Practitioner perceptions of treatment and specialized services programming as effective suggests that policies and practices advancing rehabilitation will be embraced and implemented by practitioners, increasing the likelihood that such practices may positively affect young offenders.

# ■ Conclusion

Claims abound that juvenile justice today is primarily punitive. Decades of wide-scale reforms, many punitive in nature, are frequently cited as evidence. Little effort, however, has been devoted to substantiating the accuracy of this claim. This chapter has examined the state of juvenile justice policy nationally to investigate this claim, particularly whether juvenile justice today is uniformly punitive in its orientation or whether it reflects the founding tenets of the original juvenile court. Analysis of state legislation and current practice indicates that juvenile justice today clearly represents a mix of punitive and rehabilitative approaches and that states vary dramatically in the extent to which they lean toward greater punitiveness or rehabilitation.

On the whole, our review of proposed and enacted legislation (2005–2007) across the states suggests that lawmakers are embracing a broad expanse of progressive measures aimed at advancing rehabilitation and retaining the balance of young offenders in the juvenile court. That is not to suggest a scarcity of punitive measures. Transfer to criminal court remains popular. Many states sought to expand the number of potential transfer mechanisms during the period examined, as well as increase penalties for violent crime or habitual offending. Yet, on the whole, it appears recent policy is trending toward rehabilitation.

We find additional support for our observation when we examine current practice. Again, we find evidence that states employ a range of measures, both punitive and progressive, but that most trend slightly toward a progressive orientation. Clearly, juvenile justice is tougher in some states (Louisiana, Missouri, Nevada) and more progressive in others (Kentucky, Colorado). Second, this analysis suggests that states employ different combinations of these practices and policies, presumably to address the unique local context in which juvenile justice is administered. Third, looking at a mix of different practices and the extent to which states employ these practices allows us to examine the extent to which the states' approaches to policy differ. Some clearly have well-developed policy agendas, as evidenced by the extent to which both punitive and progressive practices are legislated. Policy agenda is arguably less formal, if not less developed, in states with fewer laws governing juvenile justice practices and policies. In short, analysis suggests that the orientation and landscape of juvenile justice today varies depending on the state.

Practitioner input garnered from a recently completed survey of juvenile justice professionals suggests wide-scale support for progressive practices that support the original mission and ethos of the juvenile justice system. Treatment and services were consistently identified by juvenile justice professionals as effective, while punitive measures such as transfer were viewed by practitioners as less likely to achieve critical outcomes. In turn, practitioners offered a host of recommendations for improving the juvenile justice system, many of which reinforced the need for a broad range of responses to facilitate offender rehabilitation and accountability.

These observations underscore the importance of providing more balanced assessments of the state of juvenile justice by examining a broad spectrum of policies and practices, and consulting a variety of data sources. Focusing on only one policy issue or consulting only one perspective provides a distorted view of juvenile justice. Our analysis consulted several sources and examined a range of policies and practices to determine the present posture of juvenile justice. We conclude that juvenile justice is neither rigidly uniform—it varies greatly from state to state and even within states—nor predominantly punitive in its orientation. Rather, most states employ a range of measures. While some punitive measures such as transfer remain popular, analysis of recent legislation and current practice suggests that states employ many progressive reforms. Findings from a recent survey of juvenile justice practitioners indicate strong support for progressive practices like treatment and evidence-based programming and services.

Harkening back to the beginning of this chapter, we recall that juvenile justice in the United States consists of 51 different juvenile justice systems with unique structures, policies, and practices tailored to address the distinct issues and needs of their respective jurisdictions. Our analysis reaffirms the accuracy of this observation. Likewise, the diverse mix of laws, policies, and practices employed by each state suggests that the innovation, flexibility, and emphasis on the individual that characterized the early juvenile justice system remains central today. Future assessments of juvenile justice policy must examine not only national trends but state-specific trends to determine the status of practice, policy, and progress, and juvenile justice practitioners should be consulted on a regular basis to comment on these issues. Together, these will ensure that an accurate portrait of juvenile justice emerges.

## ■ Endnotes

1. Multiple sources were consulted for this analysis, including the National Juvenile Defender Center (NJDC), the National Conference of State Legislatures (NCSL), and the National Center for Juvenile Justice (NCJJ).
2. These states include Connecticut, Georgia, Illinois, Louisiana, Massachusetts, Michigan, Missouri, New Hamphsire, New York, North Carolina, South Carolina, Texas, and Wisconsin.
3. Illinois Senate Bill 458 passed in 2005; the same Illinois bill also specified that all juvenile drug offenses must be charged in juvenile court but are subject to presumptive waiver.
4. See www.ncsl.org/programs/cj/07jjsummary.htm.
5. Idaho passed House of Representatives Bill 205 in 2005.

6. See www.ncsl.org/programs/cj/07jjsummary.htm.

7. The present analysis focuses on state law as the sole indicator of a state's orientation toward policy and practice; many practices and policies, especially those associated with programming and treatment, are not formally legislated. In turn, because sentencing and punishment involve issues of due process and rule of law, they will be legislated and therefore easier to systematically identify. Thus, this analysis risks failing to detect a state's emphasis on progressive measures.

8. See www.adjj.org/downloads/9805issue_brief_1.pdf.

## ■ Discussion Questions

1. In what ways has the original mission of juvenile justice in the United States changed over time?

2. Discuss the various policies that have been implemented in the area of juvenile justice since the 1990s. Which policy trends are positive, and which are negative? What criteria should be used to make this assessment?

3. Is current juvenile justice policy punitive or rehabilitative? What evidence can help to make this determination?

4. To what extent should policy makers consider the views of practitioners when developing juvenile justice policy?

5. What explains the variation in juvenile justice policy across the United States?

## ■ References

Andrews, C., & Marble, L. (2003.) *Changes to OJJDP's juvenile accountability.* Retrieved February 19, 2010, from http://www.ncjrs.gov/pdffiles1/ojjdp/200220 .pdf

Bernard, T. J. (1992). *The cycle of juvenile justice.* New York: Oxford University Press.

Bernard, T. J. (Ed.). (2006). *Serious delinquency: An anthology.* Los Angeles: Roxbury Publishing Company.

Bishop, D. M. (2004). Injustice and irrationality in contemporary youth policy. *Criminology and Public Policy 3*, 633–644.

Bishop, D. M. (2006). Public opinion and juvenile justice policy: Myths and misconceptions. *Criminology and Public Policy, 5*, 653–664.

Buck Willison, J., Mears, D. P., Shollenberger, T., Owens, C., & Butts, J. A. (2008). *Past, present, and future of juvenile justice: Assessing Policy Options (APO) final report.* Washington, DC: Urban Institute.

Butts, J. A., Coggeshall, M., Gouvis, C. Mears, D. P., Travis, J., Waul, M., et al. (2002). *Youth, guns, and the juvenile justice system.* Washington, DC: The Urban Institute.

Butts, J. A., & Harrell, A. V. (1998). *Delinquents or criminals: Policy options for young offenders.* Washington, DC: The Urban Institute.

Butts, J. A., & Mears, D. P. (2001). Reviving juvenile justice in a get-tough era. *Youth and Society 33,* 169–198.

Butts, J. A., & Mitchell, O. (2000). Brick by brick: Dismantling the border between juvenile and adult justice. In Charles M. Friel (Ed.), *Criminal Justice 2000: Boundary Changes in Criminal Justice Organizations* (vol. 2, pp. 167–213). Washington, DC: National Institute of Justice.

Butts, J. A., & Sanborn, J. A. (1999). Is juvenile justice just too slow? *Judicature 83,* 16–24.

Butts, J. A., & Travis, J. (2002). *The rise and fall of American youth violence: 1980 to 2000.* Washington, DC: The Urban Institute.

Fagan, J., & Zimring, F. E. (Eds.). (2000). *The changing borders of juvenile justice.* Chicago: University of Chicago Press.

Fagan, J. (2008). Juvenile crime and criminal justice: Resolving border disputes. *Juvenile Justice 18*(2), 81–118.

Feld, B. C. (1993). Criminalizing the juvenile court: A research agenda for the 1990s. In I. M. Schwartz (Ed.), *Juvenile justice and public policy: Toward a national agenda.* (pp. 59–88). New York: Lexington Books.

Feld, B. C. (1999). *Bad kids: Race and the transformation of the juvenile court.* New York: Oxford University Press.

Feld, B. C. (2006). The honest politician's guide to juvenile justice in the twenty-first century. In T. J. Bernard (Ed.), *Serious delinquency: An anthology* (pp. 225–234). Los Angeles: Roxbury Publishing Company.

Grisso, T. (2004). *Double jeopardy: Adolescent offenders with mental disorders.* Chicago: University of Chicago Press.

Grisso, T. (2007). Progress and perils in the juvenile justice and mental health movement. *The Journal of the American Academy of Psychiatry and the Law, 35,* 158–167.

Guarino-Ghezzi, S., & Loughran, E. J. (2004). *Balancing juvenile justice.* New Brunswick, NJ: Transaction Publishers.

Harris, P. W., Welsh, W. N., & Butler, F. (2000). A century of juvenile justice. In G. LaFree (Ed.), *Criminal justice 2000: The nature of crime: Continuity and change* (vol. 1, pp. 359–425). Washington, DC: National Institute of Justice.

Howell, J. C. (2003). *Preventing and reducing juvenile delinquency: A comprehensive framework.* Thousand Oaks, CA: Sage.

Katzmann, G. S. (Ed.). (2002). *Securing our children's future: New approaches to juvenile justice and youth violence.* Washington, DC: Brookings Institution Press.

King, M. (2006). *Guide to the state juvenile justice profiles.* Pittsburgh, PA: National Center for Juvenile Justice.

Krisberg, B. A. (2005). *Juvenile justice: Redeeming our children*. Thousand Oaks, CA: Sage.

Kupchik, A. (2003). Prosecuting adolescents in criminal court: Criminal or juvenile justice? *Social Problems 50*, 439–460.

Landsheer, J. A., & Hart, H. (2000). Punishments adolescents find justified: An examination of attitudes toward delinquency. *Adolescence 35*, 683–693.

MacArthur Foundation. (2006). *MacArthur Foundation research network on adolescent development and juvenile justice*. Retrieved February 19, 2010, from http://www.adjj.org/downloads/552network_overview.pdf

MacArthur Foundation. (2008). *Adolescent legal competence in court: Issue brief 1*. Retrieved February 19, 2010, from http://www.adjj.org/downloads/9805issue_brief_1.pdf

McCord, J., Widom, C. S., & Crowell, N. A., (Eds.). (2001). *Juvenile crime, juvenile justice*. Washington, DC: National Academy Press.

Mears, D. P. (2000). Assessing the effectiveness of juvenile justice reforms: A closer look at the criteria and the impacts on diverse stakeholders. *Law and Policy 22*, 175–202.

Mears, D. P. (2002). Sentencing guidelines and the transformation of juvenile justice in the twenty-first century. *Journal of Contemporary Criminal Justice 18*, 6–19.

Mears, D. P., & Butts, J. A. (2008.) Using performance monitoring to improve the accountability, operations, and effectiveness of juvenile justice. *Criminal Justice Policy Review, 19*, 264–284.

Mears, D. P, Shollenberger, T., Buck Willison, J., Owens, C. O., & Butts, J. A. (2008). Practitioner views of priorities, policies, and practices in juvenile justice. Retrieved February 19, 2010, from Sage Journals Online at: http://cad.sagepub.com/cgi/content/abstract/0011128708324664v2

Mihalic, S., Fagan, A., Irwin, K., Ballard, D., & Elliott, D. (2006). Blueprints for violence prevention. In T. J. Bernard (Ed.), *Serious delinquency: An anthology* (pp. 171–186). Los Angeles: Roxbury Publishing Company.

Miller, F., & Foster, E. (2002). Youth's perception of race, class, and language bias in the courts. *The Journal of Negro Education, 71*, 193–205.

National Conference of State Legislatures. (2008). *Juvenile justice state legislation in 2007*. Retrieved February 19, 2010, from http://www.ncsl.org/programs/cj/07jjsummary.htm

National Criminal Justice Association. (1997). *Juvenile justice reform initiatives in the states: 1994–1996*. Washington, DC: Office of Juvenile Justice and Delinquency Prevention. Retrieved February 19, 2010, from http://ojjdp.ncjrs.org/PUBS/reform/contents.html

National Juvenile Defender Center. (2005). *2005 state juvenile justice legislation*. Washington, DC: National Juvenile Defender Center. Retrieved February 19, 2010, from http://www.njdc.info/pdf/njdc_2005_legislative_compilation.pdf

National Juvenile Defender Center. (2005). *National indigent defense and delivery oversight system.* Washington, DC: National Juvenile Defender Center. Retrieved February 19, 2010, from http://www.njdc.info/pdf/Juvenile%20Indigent%20 Defense%20Delivery%20Systems.pdf

National Juvenile Defender Center. (2007). *2006 state juvenile justice legislation.* Washington, DC: National Juvenile Defender Center. Retrieved February 19, 2010, from http://www.njdc.info/pdf/2006%20State%20JJ%20Legislation.pdf

National Juvenile Defender Center. (2008). *2007 state juvenile justice legislation.* Washington, DC: National Juvenile defender Center. Retrieved February 19, 2010, from http://www.njdc.info/pdf/2007_Legislative_Summary.pdf

Olsen, L. (2009, June 7). Juvenile transfer to adult court common. *The Houston Chronicle.*

Redding, R. (2008). *Juvenile transfer laws: An effective deterrent to delinquency?* Washington, DC: Office of Juvenile Justice and Delinquency Prevention.

Roberts, J. V. (2004). Public opinion and youth justice. In M. H. Tonry & A. N. Doob (Eds.), *Youth crime and youth justice* (pp. 495–542). Chicago: University of Chicago Press.

Sanborn, J. B., Jr. (1998). Second-class justice, first-class punishment: The use of juvenile records in sentencing adults. *Judicature, 81,* 206–213.

Schmid, M. (with Berner, K. Israel, L., Pasricha, K., & Volmert, J.). (2005). *2005 state juvenile justice legislation.* Washington, DC: National Juvenile Defender Center. Retrieved February 19, 2010, from http://www.njdc.info/pdf/njdc_2005_legislative _compilation.pdf

Schwartz, I. M., Guo, S., & Kerbs, J. J. (1993). The impact of demographic variables on public opinion regarding juvenile justice: Implications for public policy. *Crime and Delinquency, 39,* 5–28.

Scott, E., & Steinberg, L. (2008). Adolescent development and the regulation of youth crime. *Juvenile Justice, 18*(2), 15–34.

Sickmund, M. (2008). National Institute of Justice Research and Evaluation Conference Panel: Policy changes in the juvenile justice system from inside and out. Arlington, VA, July 23, 2009.

Snyder, H. N., & Sickmund, M. (1999). *Juvenile offenders and victims: 1999 national report.* Washington, DC: Office of Juvenile Justice and Delinquency Prevention.

Snyder, H. N., & Sickmund, M. (2006). *Juvenile offenders and victims: 2006 national report.* Washington, DC: Office of Juvenile Justice and Delinquency Prevention.

Sprott, J. B. (1998). Understanding public opposition to a separate youth juvenile justice system. *Crime and Delinquency, 44,* 399–411.

Steinberg, L., & Scott, E. S. (2006). Less guilty by reason of adolescence: Developmental immaturity, diminished responsibility, and the juvenile death penalty.

In T. J. Bernard (Ed). *Serious delinquency: An anthology* (pp. 129–144). Los Angeles: Roxbury Publishing Company.

Syzmanski, L. A. (2007). Juvenile delinquency purpose clauses (2007 update). *NCJJ Snapshot, 12*(11). Pittsburgh, PA: National Center for Juvenile Justice.

Syzmanski, L. A. (2007). Upper and lower age of delinquency jurisdiction (2007 update). *NCJJ Snapshot, 12*(10). Pittsburgh, PA: National Center for Juvenile Justice.

Syzmanski, L. A. (2008). Confidentiality of juvenile delinquency hearings (2008 update). *NCJJ Snapshot, 13*(5). Pittsburgh, PA: National Center for Juvenile Justice.

Torbet, P. M., Gable, R., Hurst, H., IV, Montgomery, I., Szymanski, L, & Thomas, D. (1996). *State responses to serious and violent juvenile crime.* Washington, DC: Office of Juvenile Justice and Delinquency Prevention.

Torbet, P., & Szymanski, L. (1998). *State legislative responses to violent juvenile crime: 1996–97 update.* Washington, DC: Office of Juvenile Justice and Delinquency Prevention.

Urban, L. S., St. Cyr, J. L., & Decker, S. (2003). Goal conflict in the juvenile court: The evolution of sentencing practices in the United States. *Journal of Contemporary Criminal Justice, 19,* 454–479.

Urban Institute. (2007). Past, Present and Future of Juvenile Justice Policy: Assessing Policy Options (APO). NIJ grant #2005-IJ-CX-0039.

Veneziano, C., Veneziano, L., & Gill, A. (2001). Perceptions of the juvenile justice system among adult prison inmates. *Journal of Offender Rehabilitation, 32,* 53–65.

Wool, J., & Stemen, D. (2004). *Changing fortunes or changing attitudes? Sentencing and corrections reforms in 2003.* New York: Vera Institute of Justice.

Zimring, F. E. (2005). *American juvenile justice.* New York: Oxford University Press.

## ■ Court Cases Cited

Roper v. Simmons, 543 U.S. 551 (2005).

# Policy Intersection

## PART II

CHAPTER 11   Exploring the Relationship Between Contemporary Immigration and Crime Control Policies

CHAPTER 12   Technology and Criminal Justice Policy

CHAPTER 13   White-Collar Crime and Public Policy: The Sarbanes-Oxley Act and Beyond

CHAPTER 14   Criminal Justice Policy and Transnational Crime: The Case of Anti-Human Trafficking Policy

CHAPTER 15   When Is Crime a Public Health Problem?

# Exploring the Relationship Between Contemporary Immigration and Crime Control Policies

# CHAPTER

# 11

Dana Greene

## ■ Introduction

In a past era it would have been strange to find a chapter on immigration in a criminal justice policy reader. Even today, readers of this text might approach the chapter with uncertainty and perhaps wonder why the topic is discussed alongside areas of policy that more traditionally fit under criminal justice's wide umbrella. After all, someone may ask, "Isn't immigration a labor issue or an economic issue?" Yet immigration policy, and the national debates that continue to shape it, have always incorporated elements found at the core of criminal justice discourse, including safety, danger, character, reprobation, restriction, captivity, and freedom. This association is truer today than ever before.

Social problems are socially constructed. How we understand a particular "problem" determines our response to it. Thus an examination of policy in a given arena is a window into the social processes that came to mold that issue. A brief historic overview reveals that immigration policy has developed around the precepts of order, control, and safety. The tactics and approaches to these core principles have manifested in a variety of ways across time and have targeted various immigrant groups differently. Nevertheless, immigration, like crime, has been consistently framed as a serious social problem directly related to public safety; and, like crime, the expectation has been that the immigration "problem" will be solved by centralized government action. Within this framework, immigrants are envisioned as people who must be controlled and from whom citizens and American society should be protected.

Immigration discourse, like criminal justice, does not fit the simple categories of liberal and conservative (Welch, 2002). The diverse political positions represented in immigration debates include free market proponents, nativists, civil rights advocates, protectionists, environmentalists, and humanitarians (Welch, 2002). Everyone has an opinion about immigration. Typically, people traditionally considered "conservative" advocate for worker visas and porous borders, while labor unions, typically identified as politically "liberal," call for strict limits on incoming immigrants. Conservatives clamor for large-scale enforcement (which is actually very big government), while liberals call for diminished legislation and reduced regulation (a small government stance). At another level, there are vocal English-only contingents and, in contrast, reformers who promote social services for everyone living inside our national borders. Similar to our country's unending, never-resolved, roller-coaster debate about crime and justice, many opinions are voiced from underinformed, contradictory, biased, or one-dimensional perspectives. This chapter will illustrate how immigration policy is shaped by ideology yet championed through promised outcomes.

Social processes construct criminality, illegality, and social dangerousness (Coutin, 2005). A number of scholars including Kanstroom, Miller, and Welch have recognized that the most recent era of U.S. immigration policy is replicating the structures, tactics, and rationalizations associated with getting tough on crime and the War on Drugs. American crime policy has been the guide for today's approach to immigration (Kanstroom, 2007; Miller, 2002; Welch, 2002). The current trajectory of immigration policy mirrors "policy shifts in the traditional criminal justice apparatus" (Welch, 2002, p. 7). The increasingly punitive immigration enforcement archipelago is quickly coming to resemble the nation's vast and discriminatory criminal justice network. It is no coincidence that the extensive chronicles of immigration policy have landed us in a place that looks very familiar. There is a relationship between the two systems that cannot be denied or undone: Immigration is a criminal justice matter.

## ■ U.S. Immigration Policy: Setting the Stage

The United States has an internationally celebrated history of welcoming strangers to its shores. However, a historic overview of immigration policy illustrates that criminalizing and controlling the foreign-born, who enter into the country both legally and illegally, is also a longstanding tradition. Immigration policy has evolved in haphazard patterns and does not travel a linear or progressive path. Policy direction is steered by a variety of interacting sociological, ideological, and logistical forces. These include xenophobia, labor concerns, economic undulations, social values, politics, jingoism, demographic changes in voting districts, and large-scale events such as World War II or the September 11, 2001, World Trade Center attacks.

One major historic benchmark occurred in 1913 when the Immigration and Naturalization Service (INS) was moved from the Treasury Department to the Department of Labor. The stark economic and labor landscape of the Great Depression was accompanied by "xenophobia and anti-Mexican" hysteria; immigration policy followed suit by severely restricting Mexican migration (Fernandez-Kelly & Massey, 2007). As a consequence, close to half a million Mexican nationals and Mexican-Americans were deported (Acuna, 1981; Ruíz, 1998). In 1940 the INS was moved from the Department of Labor to the Department of Justice. During World War II, contradictory immigration policies arose. On the one hand, the Bracero program was designed to import labor from south of the border (the very same population that earlier policies had maligned and excluded) as the nation endured a labor shortage. On the other hand, to quell irrational fears, Japanese-Americans were seized and held in internment camps (a policy deemed constitutional by the Supreme Court). In 2003 the entire U.S. immigration apparatus was reconfigured and placed under the auspices of the U.S. Department of Homeland Security (DHS). As the 21st century unfolds, immigration policies in the United States, and the practices associated with enacting them, are fast becoming almost exclusively a matter of crime and punishment.

Immigration debates have historically focused on *who* is entering the country, *how many* are entering, and *where* they are coming from. There have been periods in which immigrants are embraced, eras of strict exclusion, and eras that witness both postures simultaneously. There have always been classes of immigrants who were judged more desirable and those regarded as less desirable. Since its inception, U.S. immigration law has been "cyclical, turbulent and ambivalent," and is strongly linked to the country's wavering economy and its fickle perceptions of social evils (Johnson, 2007, p. 45). A constant thread amid decades upon decades of volatile immigration policy is that U.S-born citizens consistently have more civil rights than immigrants regardless of their status. Another unwavering aspect of immigration policy is that it has been, and continues to be, created and implemented through a prism of race. The historic structure of immigration policy has woven together citizenship, race, and ethnicity. People of color have been disproportionately targeted by immigration law and tend to suffer more greatly during times of heightened enforcement. Immigration policy has been consistently constructed to help the nation "maintain its whiteness," and whiteness, or "Americanness," is something that some groups—such as the various waves of European immigrants—can earn over time (Johnson, 2007). Immigrants of color can eventually realize a legal status in America, but they remain vulnerable (as the internment camps of WWII demonstrate). Moreover, they often experience social pressure to distance themselves from their culture of origin while embracing and embodying American culture in ways that are prescribed by forces outside their control. A variety of social forces police the boundaries of citizenship, and one key way in which this is accomplished is through national immigration legislation. Overall, immigration policy represents an extensive history of racializing, stigmatizing, restricting, controlling, and relocating bodies. The legacy of each initiative is generations of harm and national shame.

Daniel Kanstroom (2007), in his book *Deportation Nation*, makes a strong case for locating the earliest roots of today's approach to immigration in punishment by banishment, American Indian removal practices, and fugitive slave laws. Each of these public policies called for forcibly moving and excluding people from a given geographic location due to their legal status. Thus began the practice of officially ranking tolerability and geographically relocating the less suitable in the interest of social control. It is important to keep in mind that just as there are interrelating dynamics shaping policy, so too are there forces, beyond individual desire, that drive people to and from their native lands. For example, in the 15th and 16th centuries, Protestants were driven from England by religious persecution; in the 18th century, Georgia was a penal colony where banished convicts were sent; and for multiple centuries, the promise of the New World drew entrepreneurs and adventurers alike. Michael Welch identifies these as either "push factors," pressures that compel people to leave their nation of origin, or "pull factors," influences that draw people into the United States (Welch, 2002). Economics is a key force that pushes people from their nation of origin and pulls them toward the United States. Migration, the act of physically moving from one county to another, is not by its nature a dangerous or criminal act that must be categorically controlled. Within the United States, tens of thousands of people relocate within and outside the country every year without supervision or scrutiny. However, for a variety of reasons, and in the interest of a myriad of goals, immigration by foreigners to the United States has been and will continue to be tightly regulated.

## ■ Key Historic Benchmarks

Official governmental agencies, bureaucracies, and institutions that address immigration in the United States have transformed over time. In particular, the titular designations and the organizational location of such entities within the federal infrastructure have significantly changed throughout the nation's history. These changes are both interesting and informative, as they reveal a history of shifting landscapes, variations in policy directives, and popular thinking regarding immigration.

In 1891 the Office of Superintendent of Immigration was instituted and located within the Treasury Department. Four years later, the Bureau of Immigration was created and in 1903 relocated to the Department of Commerce and Labor. Before that decade's close, the agency was expanded to become the Bureau of Immigration and Naturalization, though in 1913 it was split into the Bureau of Naturalization and the Bureau of Immigration, both of which were located in the Department of Labor. Within the Bureau of Immigration, the U.S. Border Patrol was established, and in 1933 the two bureaus merged to form the INS. In 1940 the INS was moved to the Department of Justice. It was asserted that in order to improve safety and efficiency, the INS should be entirely reconfigured. In 2003, under the auspices of the newly created DHS, three new agencies emerged: The U.S. Citizen and Immigration Service (USCIS) was created to focus entirely on the administration of benefit applications, including the mechanisms of formal and legal immigration; enforcement and border security became the purview of Immigration and Customs Enforcement (ICE) and Customs and Border Protection (CBP).

United States immigration policy spans over 200 years. What follows is an overview of significant strategies and major historic turning points that underpin today's immigration policy. The Immigration Act of 1882 is considered the first comprehensive federal immigration law, though immigration was primarily a state issue managed through state powers until the middle of the 19th century (Kanstroom, 2007). However, the earlier Alien and Sedition Acts of 1798 began to frame the nation's approach to policing the boundaries of citizenship and fostered the sensibility that a breach of the national borders results in a vulnerable less-safe nation-state. These laws lengthened the number of years one must reside in the country to be eligible for citizenship and permitted the deportation of noncitizens if they were perceived as threatening to the nation's safety. They allowed for the criminalization and punishment of activity broadly defined as treasonable. Such legislation began the practice of legally marking a firm separation between citizens and migrants. Moreover, these legislative acts allowed for the deportation (i.e., physical removal) of noncitizens and began a tradition wherein the government defined "objectionable" migrant behavior. The groundwork put in place by these early policies—legal partitions between immigrants and citizens, expulsion, and shifting governmental definitions of unacceptable immigrant behavior—remain key elements of today's immigration practices.

By the early 20th century, U.S. immigration policy emphatically embraced a centralized federal structure (Lee, 2005). The overall strategy included policies constructed in response to race, ethnicity, and political affiliation, relying on the tactics of restriction, enforcement, and exclusion. Examples include the Immigration Act of 1917 in which literacy tests for adult immigrants were instituted to preclude entry. The Immigration Acts of 1921 and 1924 specifically restricted immigration from southern and eastern Europe. Between 1882 and 1943, a series of federal legislative acts were created to limit and exclude Asians—primarily the Chinese—from entering or staying in the country. Known as the Chinese Exclusion Acts, these laws, and the mechanisms developed to implement them, established much of the ideological and logistical framework of today's immigration system. The era of Chinese exclusion gave rise to the institutionalization of a formal and federally supported bureaucracy of immigration enforcement. Measures to formally process, admit, track, and deport immigrants were shaped and firmly established during this period. Procedures introduced during this era include physical inspection of migrating bodies, the establishment of official identity papers, and the recording and tracking of migrants' movements after having legally entered the country. A steady progression of agencies, offices, and staff emerged to fulfill the tasks now routinely associated with processing and tracking immigrants.

The first half of the 20th century is associated with the dawning of a major immigration infrastructure and gave rise to increased criminalization of the undocumented immigrant. For the first time, being in the United States illegally (i.e., not having gone through the proper administrative channels, newly generated, to enter the country) was legislatively defined as a criminal act. Thus the exclusion acts embedded the practice of punishing and deporting "illegal" immigrants by setting in motion the first border enforcement and deportation polices. These laws set the precedent for excluding migrants based on their race or class. Entry could now be, and routinely was, formally prohibited, and the prejudicial parameters for rejecting certain individuals were institutionalized. In 1924, the same year that Congress passed the National Origins Act, it also passed the Labor Appropriation Act, which established the U.S. Border Patrol as a distinct immigration enforcement agency. Perhaps the most significant legacy of this period is that the criminal justice system became formally linked to a system of immigration enforcement. According to Erika Lee, the era of Chinese exclusion is the boilerplate for our modern immigration landscape (2005).

The Palmer Raids of 1919 and 1920 are another example of framing noncitizens as dangerous criminals to justify removal. Resident aliens associated with the political left, including socialists, communists, and anarchists, were expressly targeted by government raids and subjected to arrest and mass deportation. These tactics affirmed the precariousness of a noncitizen's legal status and reinforced that immigrants were less protected than citizens regardless of how long they had been in the country or whether they were tied to the nation through family or work. Even with the official protections associated with legal resident status, resident aliens remained vulnerable.

Decades later, immigration policy was not immune to the power and momentum of the civil rights era. The 1965 Immigration Act was created with the agenda of specifically eradicating the bias and discrimination that had for too long been associated with immigration policy. With the stroke of a pen, classifying and responding to immigrants based on race, class, gender, political affiliation, and nation of origin was supposed to come to an end. This novel legislation created a new administrative structure for assessing whether to grant entry into the United States. In the wake of the 1965 Act, immigrant preference categories were based on skills and family ties. While this approach was intended to create a more equitable immigration policy, gatekeeping through patrol and enforcement remained the status quo. Immigration policy continued to operate within a framework that criminalized those who did not enter the country through proper formal channels.

Scholars differ somewhat when delineating the precise starting point of the modern era of immigration policy. By most accounts, the arrival of more than 100,000 Cubans during the Mariel boatlift of 1980 is a significant marker. There were so many Mariel Cubans arriving—so named for the harbor from which they departed—that the system was quickly overwhelmed. People were held in makeshift camps as besieged agencies attempted to process the new arrivals. Rumor, innuendo, and news reports spread the perception that these migrants were Fidel Castro's undesirables. The fact that these immigrants were darker in skin tone and less educated than previous waves of Cuban migrants fueled this speculation (Welch, 2002). Chaos ensued and the country suddenly had what came to be perceived as an "immigration emergency" (Welch, 2002). Sympathy for these alleged dissidents quickly eroded and was replaced by fear and distrust. The decade also saw an influx of immigrants from Haiti and Central America. These new immigrants aroused national anxieties associated with race and poverty, suspicions that historically surface at the slightest provocation, which in turn paved the way for a new era in immigration policy. Ronald Reagan supported the expansion of INS detention, and in 1986 this goal was realized with the passage of the Immigration Reform and Control Act (IRCA). The act criminalized the hiring of undocumented workers, significantly enlarged the border patrol budget, and embedded an overtly punitive anti-immigrant policy orientation.

According to numerous scholars, reformers, lawyers, and activists, it is during the last decade of the 20th century that the United States fully committed to a get-tough approach regarding immigration policy (Hines, 2006; Johnson, 2007; Kanstroom, 2007; Lee, 2005; Miller, 2002; Welch, 2002). Prior to 1990, immigrants could be detained for up to 6 months. In 1990 Congress enacted the Immigrant Reform Act and created an aggravated felon exception, which allowed for the extended detention of a specific class of criminal suspects. The North American Free Trade Agreement (NAFTA) was passed in 1993 and was put into effect in January 1994. Ironically, though this legislation was designed to lift trade barriers and increase the exchange of goods and capital across the U.S. border, it reinforced attention on the southern border and was accompanied by strict enforcement aimed at stopping people from entering the country. NAFTA supported trade but disallowed labor. By 1993 increased enforcement focusing exclusively on the southern border had become the status quo. Tactical initiatives designed to stop migrants from entering the United States intensified, and the strategy of prevention through force recrafted the nation's southern border into a militarized zone. During the mid-1990s the U.S. Border Patrol initiated several new campaigns including Operation Blockade, which targeted the popular El Paso–Juarez crossing point, and Operation Gatekeeper, which focused on the San Diego–Tijuana border. By decade's end, Operation Safeguard at the Arizona–Nogales border was also implemented.

In 1996 two significant pieces of legislation were passed. Considered by some to be "the most restrictive immigration bills in the history of the United States" (Hines, 2006, p. 11), the Antiterrorism and Effective Death Penalty Act (AEDPA) and the Illegal Immigration Reform and Immigration Responsibility Act of 1996 (IIRIRA) mark a major turning point in contemporary immigration policy (Hines, 2006). The new laws instituted a range of new legal practices: Judicial discretion was significantly curtailed, and the INS was granted increased powers to detain and deport. The laws mandated that those seeking asylum be detained until their cases were resolved. Not only could any immigrant, documented or not, be deported if charged with a drug offense, the laws were also applied retroactively. The legislation dramatically lengthened the list of deportable crimes by expanding the term "aggravated felony" to include nonviolent offenses and petty crimes, while simultaneously widening the category of "crimes of moral turpitude."

Together, these bills ushered in an increasingly harsh trend in immigration policy. They reframed the "immigration problem" so that it came to include not just those in the United States illegally, but all immigrants and legal residents. The language and structure of these landmark acts reinforced the idea that immigrants are involved in crime, are dangerous, and must be controlled through detention and deportation. The immigration legislation of 1996 is credited with precipitating a dramatic growth in immigration enforcement and the associated criminalization of immigration violations. The relationship between the nation's immigration infrastructure and criminal justice system is cemented by these measures.

## ■ Today's Immigration Landscape: Criminal Justice Redux

At the start of the 21st century, the forces for comprehensive immigration reform were gaining momentum. Multiple stakeholders were joining together and organizing in response to the exceedingly restrictive legislation that had emerged in the 1990s. Employers and immigrant advocates were coming together and building a pro-immigrant policy reform movement. George W. Bush openly supported reform and aimed to improve the legal status of immigrant workers by promoting guest worker legislation. However, the events of September 11, 2001, brought pro-immigrant reform momentum to an absolute standstill. The nation's hegemonic response to "9/11" was fear and xenophobia fueled by fierce nationalism. The only policy reform politicians were amenable to supporting were initiatives directed exclusively toward increased national security. Strategies designed to "protect" our borders, both the physical boundaries and the parameters of citizenship, were the only ones that could be politically justified or publicly supported.

It is tempting to blame the 9/11 attacks for the surge of harsh immigration policy and anti-immigrant fever that developed in its wake. However, the ideologies upon which the legislation and strategies of this period rest have long been a part of the nation's discourse. Even a brief historic overview demonstrates how for over 200 years such ideologies have moved through cycles of heightened popularity and latent dormancy. The climate after 9/11 did allow for a unique, wide-reaching policy to be built around longstanding fears, myths, and stereotypes. The dramatic and frightening events of September 11, 2001 also, however, reinforced underlying distrust and apprehension. Yet, the unshakable belief that safety and security could only be brought about through strict laws, large-scale enforcement, and punitive campaigns was imported directly from the nation's approach to crime. We must look to social policy in the realm of criminal justice to understand the direction of today's immigration legislation and practices (Miller, 2002; Welch, 2002).

The language we use to discuss immigrants, both in the public and private realms, places immigrants in a distinct category and in turn frames immigration as threatening. Immigration and immigrants are being constructed as a social problem through language and rhetoric (Sohoni, 2006). The two terms typically used to describe immigrants in state and federal statutes, by media outlets, in public discourse, and by political pundits are *illegal alien* and *criminal alien*. Other popular terms, often considered more neutral, include *undocumented* and *foreign-born*. These terms speak not to any action someone has undertaken, be it legal or otherwise, but to the status of their person, their disposition. The terminology thus positions immigrants as questionable, suspicious, and/or criminal beings. This standing merges easily into our structure of criminality, which has long focused on the nature of someone's character rather than on an individual act or a particular violation of the law. It becomes inevitable then to associate immigrants with danger, thereby completing their criminalization.

Once the idea that immigrants are dangerous and criminal is embedded into the nation's unconscious, our responses will serve to reinforce and bolster the initial premise. Immigration policy today is no exception. We are fully engaged in protectionism and defense by way of order, control, and enforcement. As we engage in this type of response, we also strengthen and support the argument that immigrants are dangerous, our borders are points of vulnerability, and enforcement is our only hope. It is no accident that we built the modern-day solution to our "immigration problem" on U.S. criminal justice policy of the last 30 years. Crime policy is the template for today's immigration policy (Miller, 2002). Immigration discourse has mirrored the criminal justice danger/safety ideology, and the foundation upon which immigration policy would build has been well entrenched in the national psyche. The practices, strategies, structures, and rationalizations resemble, build upon, and borrow from America's approach to crime. Since 1973, the United States has been getting tough on crime, has been committed to a War on Drugs, and has abandoned any pretense of rehabilitation. It is important to explore how these principles and tactics have shaped immigration policy.

Less than 2 months after 9/11, the USA PATRIOT Act was passed. The sweeping legislation tripled the resources for border patrol and moved immigration policy squarely and completely into a crime control mode. The PATRIOT Act focused on enhanced surveillance, domestic security, and protecting the nation's physical borders. It contained a variety of provisions specific to immigration, including mandatory detention, limitations on habeas corpus claims, the curtailment of social services for migrants, and restrictions on judicial review. Though a subject of controversy, the act was reauthorized in 2006. In 2003, the relationships between immigration and danger, the border and vulnerability, immigrant and criminal, and enforcement and safety were formally institutionalized. Early that year, all immigration agencies were moved to the newly created DHS. Under its auspices, two new agencies dedicated exclusively to investigation and enforcement were created: the ICE and the CBP. Throughout much of the 20th century, immigration discourse centered on matters of economics and labor; the trend toward a focus on physical safety as it relates to criminally dangerous immigrants emerged only in the last two decades. By the start of this century, national security, through strict immigration enforcement, became the linchpin holding all policy together. In turn, immigration enforcement has "become a fixture in the spiraling apparatus of social control" (Welch, 2002, p. 170).

The fundamental theoretical premises of deterrence and retribution that have driven recent crime policy have been transferred to a "war on terror" that has become a de facto war on immigrants. The basic idea is that people can be deterred from entering the county (Cornelius & Lewis, 2007). Deterrence, both general and specific, relies on stiff penalties, mandatory responses to transgressions, and strict enforcement. This is exactly how the DHS deploys its resources. According to the CBP Web site (2009), "Since the terrorist attacks of September 11, 2001, the focus of the Border Patrol has changed to detection, apprehension and/or deterrence of terrorists and terrorist weapons." The ICE budget ($3,556,454 in 2005) jumped to $5,581,217 in 2008 (U.S. Immigrations and Customs Enforcement, 2008a). The Border Patrol budget, a mere $94 million in 1992, was a staggering $1.94 billion in 2007. The number of agents has grown dramatically. In 1996 when AEDPA and IIRIRA were passed, there were 5863 agents; by 2005 that figure nearly doubled (Trac Reports, Inc., 2007a). The use of detention has also increased significantly. Between 2002 and 2008, the average daily population of detained people increased by 45% (U.S. Immigrations and Customs Enforcement, 2008b). The INS, now folded into the DHS, saw its staff more than double between 1994 and 2003 and its budget soar to $6.343 billion in 2003. ICE even has its own plane, which flew 367,000 illegal immigrants "home" in 2008 (Olivo, 2009).

These statistics demonstrate a growing and unrelenting commitment to enforcement. Alongside deterrence, the principal strategy of the War on Drugs is interdiction. The aim, should deterrence fail, is to stop drugs from entering the country or to seize them once they have arrived. Drugs, simply put, are prohibited and must be eliminated from society. Tactically, officers are positioned on the front lines and armed with the most sophisticated equipment available; they represent an aggressive militarized force that stands between people and the banned substance. Immigration policy and its corresponding tactical measures mirror this approach. The practice of capturing an undocumented migrant and freeing him or her on the other side of the border—a longstanding practice known as "catch and release"—has been formally abandoned and replaced by mandatory detention. In a press conference on August 23, 2006, Michael Chertoff, then Secretary of Homeland Security, announced that the new mandate for immigrants in the country illegally is "catch and detain." Every undocumented migrant is now formally processed through the system. In 2006 National Guard members were sent to the southern border to reinforce and augment border patrol efforts. At the same time, ICE engaged in sweeps and raids all over the country and became fully engaged in "policing society from within and from without" (Miller, 2002). In fact, the nation's largest workplace raid was not along the southern border but in Postville, Iowa, in the spring of 2008 (Schulte, 2008).

After 9/11, local police agencies at the state and municipal level were enlisted into immigration enforcement and interdiction as never before. There is now an understanding that national security (distinct from public safety) is a new responsibility for traditional law enforcement agencies. In the interest of increased collaboration, DHS dollars are pouring into police forces around the nation. With this influx of money, local law enforcement agencies are reorganizing by creating counterterrorism units and immigration task forces, buying new equipment, and conducting specified training (Marks & Sun, 2007). Technology has become an integral force and expensive facet of this latest "war," with hundreds of miles of fence built along the southern border, unmanned drones policing from above, and the ubiquitous presence of surveillance cameras visible across the landscape. The expansion of federal and local police agencies focused specifically on immigrants and border protection serves to bind immigration and crime while reinforcing fear and the prevailing fiction that immigrants are criminal in nature.

The rationalizations that support and legitimize the latest reorganization of the nation's immigration apparatus are reminiscent of those used to justify the last 30 years of crime policy. The primary contention is that we are in grave danger from a problem that is out of control and growing. Strict enforcement is the panacea that will reinstate order and make us safe. Central to this logic is making the border impenetrable. Indeed, the openness of a free society itself becomes a problem, justifying the curtailment of freedoms. Any response that is not harsh and muscular, anything viewed as soft, is characterized as weak, foolish, and risky. The fundamental assertion is that immigration puts national security in jeopardy and that only tough enforcement can deliver safety. In this sense, the nation's immigration problem becomes subsumed into a crime control framework.

The terms immigrant and criminal are becoming so intertwined that it is becoming difficult to tease them apart. Building on the infrastructure and logic of the prevailing approach to crime and the War on Drugs, immigration policy has contributed to an expansion of the crime control complex. A byproduct of weaving these two systems together is the growing perception that immigrants are an integral part of the drug problem. In the interest of public safety, the machinery designed to interrupt and discipline immigrants is becoming entrenched in the culture. If we need any further proof that immigration enforcement has realized the social importance of traditional criminal justice practices, we need only turn on our televisions. On January 6, 2009, alongside the legions of courtroom dramas and police procedurals, a reality program about the people and agencies associated with the DHS, *Homeland Security U.S.A.*, aired its first episode on ABC.

While contemporary immigration policy in the United States emphasizes a get-tough approach that is characteristic of the nation's approach to crime, another trend has also emerged. Across the nation, cities and counties have declared themselves "sanctuaries," locations where undocumented people can expect to be safe from heightened enforcement. Sanctuaries employ a variety of means, both formal and informal, by which to express or manifest their standing as a safe haven. Examples include executive mandates disallowing certain police actions, formal resolutions that reject federal immigration policy, a lack of cooperation with federal officers or agencies, and public condemnation of raids that occur in their locale. By August 2006, at least 31 counties or cities had declared themselves sanctuaries, and the number continues to grow (Carter, 2007). Sanctuaries can be found around the nation, including Portland, Maine; Evanston, Illinois; Takoma Park, Maryland; Fairbanks, Alaska; Chandler, Arizona; and Albuquerque, New Mexico.

A variety of arguments are used to defend or support sanctuaries. Some politicians and enforcement officials contend that empowering local law enforcement to target illegal immigration compromises public safety. They argue that such tactics weaken police–community relationships, which, in turn, compromises public safety and allows street crime to flourish. Some people view the use of local dollars toward immigration enforcement as a misuse of funds, as it is a federal initiative. There are those who dispute the morality of strict immigration enforcement. In sanctuary cities and counties, it is common for immigrants, legal and otherwise, to be viewed as valued community members, not people to be ferreted out in the interest of safety. Regardless of the reasoning, the creation of sanctuaries remains controversial (Knight & Van Derbeken, 2009; Richardson, 2009), and sanctuaries continue to challenge the sentiments underlying current immigration policy, presenting a powerful alternative for the nation to consider.

## ■ Does It Work? The Chimera of National Security

The increased scale and dimension of immigration enforcement is said to enhance national security. The expense and effort of such enforcement is justified by the safety that it purportedly yields and the threats it averts. To weigh in on the value and desirability of such an approach, it is important to assess if these objectives are realized. This is not a simple task. It is difficult to measure that which does not happen. Nonetheless, it is important to evaluate the contemporary strict enforcement regime.

The primary principle of interdiction is deterrence. Are migrants less likely to enter the United States in the wake of harsh enforcement, increased legislation, and mandatory penalties? Studies show that while patterns associated with migration can be influenced by increased enforcement, overall strict enforcement "has not stopped, nor even discouraged, unauthorized migrants" (Cornelius, 2006). Deterrence, general and specific, relies on awareness; people contemplating entering the United States would have to know about new harsh legislative strategies in order to weigh the consequences as they make decisions. Cornelius and Lewis found that most do not (2007). Moreover, the reasons that people leave their nation of origin have not been addressed by the enforcement-focused U.S. immigration policy. The push and pull factors remain, and incentives for entering the country far outweigh disincentives. People continue to migrate to the United States, and the pace is not necessarily affected by enforcement. Logic challenges the premise of deterrence, but, again, measuring that which does not occur is complicated. There is no clear consensus as to whether unauthorized migration has been reduced by increased enforcement.

Law enforcement has, at times, used an increase in arrests to support claims of efficacy (Bazar, 2008), and, on occasion, a reduction in the number of arrests has been cited as evidence of a successful strategy (Hsu, 2006). Claims become questionable when an escalation in arrests and a decline in arrests are used to support the same hypothesis. It is important to keep in mind that data about arrest rates are not data about migration behavior; rather they are information about departmental and agent activity. Thus, the consequences of strict enforcement are many, but it appears that decreased migration is not definitively one of them.

Perhaps safety is realized through other means. Apprehension rates ebb and flow yearly and differ by jurisdiction, with increased enforcement inevitably leading to higher capture rates. However, the research shows that while the cost of apprehending a single migrant has gone up, the actual number of people apprehended has hovered above or below 1 million since 1993 (Trac Reports, Inc., 2007a). The ratio of cost to capture has gone up. This means that we are actually getting less for our money because we are not capturing many more people (Cornelius, 2006). Although apprehension rates have gone up, overall they are quite low given the numbers of people entering and settling in the country illegally. The vast majority of migrants enter undetected or return after they have been deported. It is estimated that 11.9 million undocumented people live in the United States (Passel and Cohn, 2009).

The average number of unauthorized immigrants in detention on a given day has increased from 20,838 in 2005 to 31,345 in 2008, a significant leap in just 3 years (U.S. Immigrations and Customs Enforcement, 2008b). It is tempting to equate this statistic with enhanced safety, as increased incarceration is often spuriously associated with a reduction in crime (Jacobson, 2006). Incapacitation does not reflect effective deterrence, nor does it necessarily represent increased apprehension. Rather, in the case of immigration, it is a consequence of mandatory detention policies. If our concern is with what undocumented people do once they are in the country, then perhaps the growth in detention could translate into increased national security. But the liability is said to be the porous nature of the border, and detention does nothing to ameliorate this vulnerability. The criminal act, being here illegally, has already taken place once detention is exercised. Crime has not been averted, and there is no reason to correlate these data to increased public safety or a more secure nation. Of all "criminal" charges against unauthorized migrants, 86.5% involve violations of immigration law, including illegal entry or not having a valid visa (Trac Reports, Inc., 2007b). The nation's immigration detention centers are not filled with people suspected of, or charged with, doing great harm.

Since 9/11, the government has engaged in an impressive public relations campaign to sell Americans on the effectiveness of modern-day immigration policy. The rhetoric of national security is seductive; it effectively casts immigrants as dangerous and our borders as war zones. The concern that immigrants threaten our way of life is firmly cemented in the nation's collective unconscious. Getting tough on immigrants is a potent political currency. Yet the evidence does not support the assertion that strict immigration enforcement results in a safer nation. We are not detaining people suspected of violent or dangerous behavior, we are not stopping people from entering the country, and we are not catching most unauthorized immigrants. In the realm of criminal justice, shifts in policy and practice often have little to do with the volume or nature of criminal activity itself and much more to do with public perception, political expediency, and the appeasement of social forces. So it appears with immigration policy.

## ■ Consequences of a Crime Control Model

What, then, are the consequences of dedicating vast resources to an immigration policy that replicates the get-tough crime policy of the past three decades? They are considerable. Immigration enforcement is no longer about labor and economics but about "identifying and managing unruly groups" (Welch, 2002, p. 153). We can again look to crime policies and their outcomes to inform where our current immigration tactics are likely to lead. Since 1973, we have seen a 700% increase in the U.S. prison population (Clear, Cole, & Reisig, 2009). The nation's corrections budget has swelled to $70 billion, with several states spending more on criminal justice than education (Clear, Cole, & Reisig, 2009). Perhaps the most alarming outcome of this extraordinary expansion is the disproportionate impact it has had on specific populations. African Americans and Latinos are overrepresented at each stage of the criminal justice process. Studies show that this disparity is not due to differing rates of offending or differences in criminal activity across race and ethnicity (Tonry, 1996). The burden of harsh legislation, diminished judicial discretion, and strict enforcement is not distributed evenly across society (Tonry, 1996). The reasons for this are complex, systemic, and not easily isolated or dismantled. The social costs are also alarming, including a discredited criminal justice system that reinforces and perpetuates racism.

By all indications, today's immigration policy is following the same trajectory. The enormous expansion of the U.S. immigration enforcement network has been documented, with budgets and staff in some areas doubling in less than 10 years. This will in turn continue to feed the system and foster more growth. Arrests, detention, convictions, and punishment all multiply and correspond with increased policing. In the initial aftermath of 9/11, Arabs and Muslims from specific geographic regions were rounded up by the U.S. government, subjected to treatment that violated principles of due process, denied access to lawyers and family, and detained in secret locations (Hines, 2006). In subsequent years, Arabs and Muslims legally in the United States (and those perceived as Muslim or Arab) typically have found themselves under constant suspicion. Their mere identity, or suspected identity, no matter their immigration status, puts them in a precarious social position.

The national discourse on safety and border vulnerability focuses overtly on religious affiliation, often using a code phrase of "the terrorists." The overarching macro-ideology of national security is a terrorism risk-management strategy that targets Arabs and Muslims. However, the day-to-day realization of this approach, the micro-principles and -policies, are acutely most apparent at the southern border, with enforcement strategies typically focused on Mexican migrants. The weight and impact of the nation's latest wave of anti-immigrant legislation has not been distributed evenly. All immigrants of color are under suspicion, but a substantial amount of enforcement resources, particularly the most visible, are focused on immigrants from Latin America. Indeed, speaking English with an accent has become a liability that can lead to enhanced scrutiny by criminal justice officials. Certain communities and populations feel under siege as a result of harsh immigration policies, undermining claims of equality before the law. The southern border of the United States has become the ground zero of immigration enforcement.

Evidence shows that migration has been displaced, not deterred (Cornelius & Lewis, 2007; Guerette, 2007). Increased enforcement at common urban entry points, such as El Paso and San Diego, has resulted in border crossings attempted on difficult terrain. This, in turn, has generated a significant increase in migrant injuries and deaths (Cornelius & Lewis, 2007; Dunn, 2009; Guerette, 2007). The risk of physical harm also increases the likelihood that smugglers will be used to help migrants cross the border. This comes at a significant financial cost to migrants, including the possibility of exploitation and victimization by predatory "coyotes." In addition, as crossing grows more perilous and costly, migrants are more likely to stay permanently in the United States because return trips have become so much less feasible. Many migrants who would come north for seasonal work and return to their home country in the off-season are now settling in the United States without proper documentation (Fernandez-Kelly & Massey, 2007).

Once immigrants are established in the United States, strict enforcement makes life more perilous. Interacting with law enforcement or state agencies of any kind becomes risky. Accessing health care becomes risky, possibly leading to arrest, detainment, and deportation. Workers are far too frightened to report poor workplace conditions or mistreatment, making them easier to exploit. In an anti-immigrant climate fueled by dramatic media reporting and political hyperbole, hate crimes become more likely, yet seeking assistance from authorities is out of the question for the victims of such acts (Rabrenovic, 2007). Distrust of law enforcement means that communities will not share intelligence or work with local police to maintain civic order. Meanwhile, state and federal legislation has eroded many of the social services that had typically been available to undocumented residents.

Another significant byproduct of an immigration policy predicated on the crime control model is the warehousing of detainees. Strict enforcement, protracted administrative procedures, diminished discretion, and mandatory charges result in longer stays in detention and more people in detention. Several distinct problems are associated with the detention of undocumented people: Facilities are overcrowded and understaffed; cases of mistreatment abound; and detainees are often confined in remote locations far from where they were living. Their families cannot visit or are too afraid to visit. It is extremely difficult to get information about confined loved ones. Institutions often have few services in place and are unprepared for the population's various needs. There are language barriers, mental health issues, and healthcare concerns. As the volume of detainees continues to grow—and there is no end in sight—conditions will likely only further deteriorate.

Another means by which detention is growing is the new and strict mandate that those seeking asylum be detained until their cases are resolved. In the past, asylum seekers were typically free until their case was decided, but under current immigration policy that is no longer possible. Large-scale incapacitation is a tactic borrowed from the crime control model and mirrors American corrections. But a corrections framework is a poor fit with immigration. Inmate classification and strict separation of the sexes is particularly problematic in the immigrant context. Families and groups that are arrested together are quickly separated, which in turn generates a host of additional harms (Human Rights Watch, 2007).

## ■ Conclusion

The result of the developments described in this chapter is an immigration policy characterized by three trends: heightened penalties, decreased alternatives to detention, and a singular focus on managing "dangerous" populations. A comprehensive discussion of immigration must address the factors that push people from their nation of origin and must prompt an honest dialogue on how we can disentangle immigration from criminality. The time has come to shift immigration policy away from a crime control framework to a humanitarian framework that emphasizes social and economic justice.

## ■ Discussion Questions

1. What evidence can be provided to support the claim that immigration is increasingly a criminal justice matter?
2. Discuss the policies that have led some immigrant populations to feel excluded from mainstream American life at various points in history?
3. To what extent did the terrorist attacks of September 11, 2001, change the relationship between immigration and criminal justice policy?
4. Can an appropriate balance be struck between crime control and immigration enforcement? Discuss the various elements of this balance.
5. Are the concerns raised in the chapter an inevitable outcome of global migration patterns? How should policy makers think about migration in the post-9/11 world?

# ■ References ■

Acuna, R. (1981). *Occupied America: A history of Chicanos* (2nd ed.). New York: Harper & Row.

Bazar, E. (2008, October 1). Border patrol expands transportation checks. *USA Today*. Retrieved March 1, 2009, from http://www.usatoday.com/news/nation/2008-09-30-border-patrol-checks_N.htm

Carter, A. (2007, May 4). *Sanctuary cities embrace illegal immigrants*. Retrieved March 1, 2010, from Human Events Web site: http://www.humanevents.com/article.php?id=20547)

CBP.com. (2009). *FAQs: Working for Border Patrol*. Retrieved March 1, 2010, from http://www.cbp.gov/xp/cgov/careers/customs_careers/border_careers/bp_agent/faqs_working_for_the_usbp.xml

Clear, T., Cole, G., & Reisig, N. (2009). *American corrections* (8th ed.). Belmont, CA: Thomson Wadsworth.

Cornelius, W. (2006, September 26). *Impacts of border enforcement on unauthorized Mexican migration to the United States*. Retrieved March 1, 2009, from SSRC.org Web site: http://borderbattles.ssrc.org/Cornelius/

Cornelius, W., & Lewis, J. (Eds.). (2007). *Impacts of border enforcement on Mexican migration: The view from sending communities*. San Diego, CA: Center for Comparative Immigration Studies, UCSD.

Coutin, S. (2005). Contesting criminology. *Theoretical Criminology*, 9(1), 5–33.

Dunn, T. J. (2009). *Blockading the border and human rights: The El Paso operation that remade immigration enforcement* (5th ed.). Austin, TX: University of Texas Press.

Fernandez-Kelly, P., & Massey, D. (2007). Borders for whom? The role of NAFTA in Mexico-U.S. migration. *The Annals of the American Academy of Political and Social Science*, 610(1), 98–118.

Guerette, R. (2007). Immigration policy, border security, and migrant deaths: An impact evaluation of life-saving efforts under the border safety initiative. *Criminology and Public Policy*, 6(2), 245–266. Retrieved February 1, 2009, from http://www.migrationpolicy.org/pubs/Americas_Human_Rights_Challenge_1006.pdf

Hines, B. (2006). An overview of U.S. immigration law and policy since 9/11. *Texas Hispanic Journal of Law & Policy*, 12(9), 10–28.

Hsu, S. (2006, October 31). Immigration arrests down 8% for year: DHS credits deterrent effect of enforcement tactics, but analysts are skeptical. *The Washington Post*. Retrieved March 1, 2010, from http://www.washingtonpost.com/wp-dyn/content/article/2006/10/30/AR2006103001025.html?nav=email

Human Rights Watch. (2007, October 2). *US: Immigrants in detention at risk*. Retrieved March 1, 2010, from http://www.hrw.org/en/news/2007/10/02/us-immigrants-detention-risk

Jacobson, M. (2006). *Downsizing prisons: How to reduce crime and end mass incarceration.* New York: New York University Press.

Johnson, K. (2007). *Opening the floodgates: Why America needs to rethink its borders and immigration laws.* New York: New York University Press.

Kanstroom, D. (2007). *Deportation nation: Outsiders in American history.* Cambridge, MA: Harvard University Press.

Knight, H., & Van Derbeken, J. (2009, August 20). Sanctuary policy at risk, city attorney warns. *San Francisco Chronicle.* Retrieved March 1, 2010, from http://www.sfgate.com/cgi-bin/article.cgi?f=/c/a/2009/08/19/MN3S19ATB2.DTL

Lee, E. (2005). Echoes of the Chinese exclusion era in post-9/11 America. In *Chinese America: History and perspectives, January 2005.* Retrieved March 1, 2010, from http://www.chsa.org/publications/chsa_publications.php

Marks, D., & Sun, I. (2007). The impact of 9/11 on organizational development among state and local law enforcement agencies. *Journal of Contemporary Criminal Justice, 23*(2), 159–173.

Miller, T. (2002). The impact of mass incarceration on immigration policy. In M. Mauer & M. Chesney-Lind (Eds.), *Invisible punishment: The collateral consequences of mass imprisonment.* New York: The New Press.

Olivo, A. (2009, February 9). Immigration agency's airline flies tens of thousands of deportees out of U.S. *Chicago Tribune.* Retrieved March 1, 2010, from http://www.chicagotribune.com/news/nationworld/chi-deportees-09-feb09,0,5333975.story

Passel, J. and Cohn, D. (2009). A portrait of unauthorized immigrants in the United States. Washington, D.C.: Pew Hispanic Center. Retrieved March 15, 2010, from http://pewhispanic.org/reports/report.php?ReportID=107

Rabrenovic, G. (2007). When hate comes to town. *American Behavioral Scientist, 51*(2), 349–360.

Richardson, V. (2009, February 25). Lawsuits challenge sanctuary policies. *The Washington Times.* Retrieved March 1, 2010, from http://www.washingtontimes.com/news/2009/feb/25/lawsuits-challenge-sanctuary-policies/

Ruíz, V. (1998). *From out of the shadows: Mexican women in twentieth-century America.* New York: Oxford University Press.

Schulte, G. (2008, May 13). Feds say raid is nation's largest. *Des Moines Register.* Retrieved March 1, 2010, from http://www.desmoinesregister.com/apps/pbcs.dll/article?AID=/20080513/NEWS/80513022/-1/ENT06

Sohoni, D. (2006). The "immigrant problem." *Current Sociology, 54*(6), 827–850.

Tonry, M. (1996). *Malign neglect: Race, crime, and punishment in America.* New York: Oxford University Press.

Trac Reports, Inc. (2007a). *Facts and figures.* Retrieved March 1, 2010, from http://trac.syr.edu/immigration/facts/

Trac Reports, Inc. (2007b). *Immigration enforcement: The rhetoric, the reality.* Retrieved March 1, 2010, from http://trac.syr.edu/immigration/reports/178/

U.S. Immigrations and Customs Enforcement. (2008a). *Detention management.* Retrieved March 1, 2010, from http://www.ice.gov/pi/news/factsheets/detention _mgmt.htm

U.S. Immigration and Customs Enforcement. (2008b, October 23). *ICE multifaceted strategy leads to record enforcement results: Removals, criminal arrests, and worksite investigations soared in fiscal year 2008.* Retrieved March 1, 2010, from http://www.ice.gov/pi/nr/0810/081023washington.htm

Welch, M. (2002). *Detained: Immigration laws and the expanding INS jail complex.* Philadelphia: Temple University Press.

# Technology and Criminal Justice Policy

# CHAPTER
# 12

Phelan A. Wyrick

## ■ Introduction

Popular crime dramas on television and in movies portray a world of good-looking crime fighters cracking cases in darkened rooms in the glow of an assortment of seamlessly integrated crime-busting technologies. Most viewers probably recognize that these portrayals are distorted reflections of reality. After all, real crime fighters cannot possibly be that good looking! In fact, criminal justice technology has become pretty impressive. Police use stun guns to subdue hostile suspects without using lethal force; wrongly convicted prison inmates are exonerated using advanced DNA forensics; scanners detect stolen cars from among a stream of traffic and alert officials on the spot; and offenders live in communities wearing electronic monitoring devices that track their movements and alert officials when restricted zones have been entered. These applications of technology—conducted energy devices, DNA exoneration, license plate scanners, and community corrections using global positioning satellites (GPS)—might have seemed unlikely or impossible a decade ago.

To see what might be around the next bend, look at the high-priority criminal justice technology needs identified by the National Institute of Justice (NIJ) (2009). Police vehicle pursuits may be all but eliminated by remote vehicle stopping technology. High-definition surveillance cameras integrated with biometric facial recognition software and criminal intelligence databases will rapidly scan crowds and identify "suspicious" or criminal activity. Law enforcement and corrections officers will be equipped with technology to continuously monitor location and health status within structures and outside in urban and rural environments. Some have gone further and extrapolated from existing technology to portray a possible more distant future (Reed, 2008). It may become common practice for law enforcement officers to deploy unmanned aerial vehicles (UAVs) with sensor arrays to identify suspects, transmit real-time images, and track suspects in daylight or nighttime conditions. Standard-issue apparel may protect officers from bullets and chemical weapons at the same time. Augmented reality may increase situational awareness by projecting computer-generated images onto real-world vision and allowing officers to quickly spot wanted felons in crowds or coordinate multiple UAVs.

Technology promises to increase effectiveness, reduce costs, and improve the safety of officers and the public. Without a doubt, technological advances will be essential to our future system of criminal justice. However, in some cases this promise can lead to misplaced enthusiasm. In other cases, technological innovation is not met with enthusiasm because it is widely recognized that failures of technology can have negative effects ranging from inconvenience to catastrophe. We are warned not to be seduced by the allure of technology (Simoncelli & Steinhardt, 2006). Failures involving technology may stem from the limitations of the technology itself or from our own limitations. While many may think of human error in terms of mishandling or improperly operating technology, human limitations include intentional and unintentional errors of omission and commission in the policies and practices that we establish for technology.

This chapter examines criminal justice policy as it relates to technology. This policy work may be broadly characterized as a collective effort to maximize benefits and minimize the frequency and consequences of failure. However, some might consider this description as charitable, given an assortment of policy-related activities that could also be described as fragmented, inconsistent, and susceptible to bouts of irrationality.

While there is an undeniable fascination in what new technologies may await us in the future, it is more valuable for criminal justice professionals to examine the nature and origins of contemporary technology policy, thus informing our ability to make effective decisions in the future. Indeed, knowledge of future technologies may not be a prerequisite to understanding principles for improved decision making on such questions as whether to acquire a technology or what standards of practice and operation to apply.

Much of the technological innovation that we can anticipate in criminal justice will constitute incremental improvements to existing technologies, but new and unexpected technological advances will also come. As it has in the past, innovations will come to the criminal justice field from sources open to the general public; some will be specially designed for criminal justice applications; some will be adapted from military applications; and some will come from scientific fields such as biology and chemistry. The breadth of technology in criminal justice is vast. One way to characterize the scope and influence of technology in criminal justice is to review the categories of technology investment portfolios supported by the NIJ's Office of Science and Technology (see **Table 12–1**).

Thus, the term *technology* encompasses a wide range of specific applications of science and engineering to the practical challenges faced in criminal justice. Likewise, the wide range of topics included under the general term *policy* may include such things as the operational policies for specific forms of technology set by individual agencies, practical guidelines issued by professional associations, state statutes, legal precedents, federal legislation, and funding criteria. In general, criminal justice technology raises a host of policy issues and questions that must be considered at multiple levels. Does the benefit of the technology outweigh its costs? Can the technology be integrated with existing practices and other technologies? How does the technology affect the privacy of citizens? What is the general public acceptance or demand for the technology? What legal, and specifically constitutional, issues are raised? What safety and liability issues are involved? What might be the unintended consequences of the technology? Those involved in answering these questions include elected officials, the judiciary body, law enforcement and corrections professionals, researchers and scientists, legal scholars, and others. The media and public opinion can also play very important, if indirect, roles in shaping policy. Indeed, criminal justice policy has been

**TABLE 12–1** **National Institute of Justice Technology Investment Portfolios**

- Aviation
- Biometrics
- Body Armor
- Communications
- Community Corrections
- Court Technologies
- DNA Forensics
- Electronic Crime
- Explosive Device Defeat
- General Forensics
- Geospatial Technologies
- Information-Led Policing
- Institutional Corrections
- Less-Lethal Technologies
- Operations Research/Modeling and Simulation
- Personal Protective Equipment
- School Safety
- Sensors and Surveillance

Source: Reproduced from Office of Justice Programs, "High-Priority Criminal Justice Technology Needs," U.S. Department of Justice, March 2009.

strongly influenced on many occasions by salient high-profile incidents that sway public opinion.

This chapter addresses factors that influence policy decision making on technology acquisition and operation by examining the path from early technology development to widespread use in the field and by reviewing policy issues related to two particular forms of criminal justice technology: conducted energy devices and forensic DNA. For the purpose of discussing central concepts, the first part of this chapter examines a number of ideals in general terms, with few references to specific technologies. Later in the chapter, the technology profiles include discussion of the policy issues that have arisen around each of them.

## ■ From Technology Development to Saturation

Certainly there is no single path from initial technology development to widespread use in the field of criminal justice. But the NIJ, a function of the DOJ, has developed a technology research, development, testing, and evaluation (RDT&E) model that is instructive in thinking about the early stages of technological innovation in criminal justice. Beyond that, our understanding for the diffusion of innovation across networks and systems tells us much about how individual criminal justice organizations are likely to make decisions about acquisition of new technology. Finally, this section will discuss organizational approaches to improving decision making about technology policy.

### The Technology Research, Development, Testing, and Evaluation Model

The NIJ's RDT&E model (see **Figure 12–1**) includes five phases: (1) identifying technology needs; (2) developing technology program plans; (3) developing solutions; (4) demonstrating, testing, and evaluating solutions; and (5) building external capacity and conducting outreach (National Institute of Justice, 2009). In the first stage, the NIJ works with technology working groups (TWGs, pronounced twigs) for each of the technology portfolios listed in Table 12.1. TWGs are typically comprised of mid- to senior-level practitioners from around the country who have regular involvement with a given category of technology. TWG members are generally considered technology specialists in their own organizations, and many also serve as trainers or instructors in other capacities outside of the TWG. The NIJ uses members of these TWGs, and a similar group of specialists who serve on the Law Enforcement and Corrections Technology Advisory Council (LECTAC), for a number of purposes. First, TWG members identify needs in the field that can potentially be resolved through a technology solution. After identifying a problem to be addressed, they identify desired outcomes and may make initial suggestions for technology requirements. These suggestions are rolled up into scores of recommendations across technology portfolios and are prioritized by NIJ leadership with input from LECTAC. TWG and LECTAC members are also used on peer review panels at later points in the RDT&E process

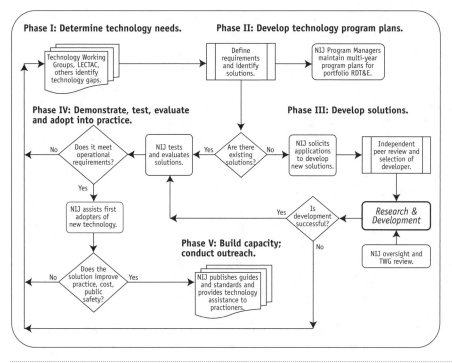

**Figure 12–1**
**Source:** Modified from Office of Justice Programs, "Research, Testing, Development and Evaluation Process Chart," U.S. Department of Justice, July 2008.

to assess potential technology solutions and are frequently involved in the testing and demonstration tasks.

If a technology need and recommended solution are determined to be high priorities, then, subject to the availability of funding, a technology program plan will be developed that includes formal definition of technology requirements and a search for existing solutions within and outside of criminal justice. Absent an existing solution, the NIJ will make one or more technology development grants through a competitive application process. NIJ program managers work with grant recipients through the development process to assess progress toward a technology that functions according to requirements.

Demonstration, testing, and evaluation involve a variety of activities intended to determine whether the technology functions under a variety of circumstances, progressing toward field settings under intended operating conditions. Functionality, of course, is only the minimal threshold at this stage. Beyond this are questions of how practical and easy it is for practitioners to use, how well its use fits with other tasks in the operational environment, and whether the technology is achieving its intended outcomes, as suggested by more specific performance measures. These initial tests may lead to refinements and redesigns to the technology based on practitioners' suggestions, or they may lead to improvements in user training. Once early versions of the technology can be made available to early users, the NIJ may launch a demonstration initiative that involves more rigorous evaluation of the impact of the technology in field settings. These demonstrations typically involve rigorous evaluation methods using counterfactuals that allow evaluators to rule out alternative explanations and inferential threats. The NIJ may publicize progress and findings at multiple points in the process, and certainly at the completion of an outcome evaluation. Publications are typically targeted to practitioners and criminal justice decision makers who are inclined toward innovative solutions.

In practice, the RDT&E process frequently does not move in a neat and linear fashion from phase I through phase V. As Figure 12.1 displays, there are multiple points at which the process may loop back based on unsuccessful models, failure to meet operational requirements, or failure to produce intended outcomes in the field. Also, the process may be overcome if new and clearly superior technologies emerge from other sources or if other developments reduce the importance of the initial identified need.

The NIJ is a federal agency with a mission to support state, local, and tribal criminal justice organizations across the United States. The RDT&E process reflects the policy of this agency to use taxpayer dollars to systematically generate technology and information about technology that can be moved into practice. The process is designed to be rigorous and to sort out ineffective or impractical technologies while identifying technologies and applications that will improve effectiveness. Through publications, conference presentations, and peer-to-peer information sharing by opinion leaders in the field (TWG members), the NIJ provides information that helps guide operational policies for technology in law enforcement, institutional and community corrections, courtrooms, forensics labs, and communities.

However, the NIJ's efforts are dwarfed by the many sources of new technology that are targeted to criminal justice practitioners. For evidence of this, pick up an industry magazine and take note of the advertisements, or visit the exhibit hall at the next annual conference for a large criminal justice organization or association. Vendors are marketing technologies that range from clever to astonishing, from mundane to highly specialized, and, undoubtedly, from very effective to completely worthless. By what standards are these technologies tested and evaluated? To what extent are independent opinions sought? What objective evidence is available to support claims of effectiveness? On the market, it is the consumer who must be informed. In the criminal justice system where decision makers are generally public officials in charge of public dollars, acquisition decisions should stand up to public scrutiny. The NIJ's RDT&E model indicates the kinds of questions that decision makers should be asking as they consider new technologies.

With support from the National Institute of Standards and Technology, Ruegg and Feller (2003) produced a toolkit for evaluating public investment in technology research and development. It is a good resource for understanding issues and methods involved in determining the merits of technology.

## Categories of Adopters

Everett Rogers was a pioneer on the topic of diffusion of innovations. As a boy, Rogers observed his father resisting new seed technology that was available to him for his work as a farmer. Rogers and his family suffered as a consequence of his father's resistance to new technology. This experience launched Rogers' interest in understanding the spread of new ideas and practices across known populations, systems, or networks of people or organizations. One of Rogers' most famous formulations was to describe five categories of innovation adopters based on their willingness to take on new practices or technologies (Rogers, 1995). Adopters typically fall across the following five categories in a normal bell-shaped curve distribution: (1) innovators, (2) early adopters, (3) early majority, (4) late majority, and (5) laggards. The distribution is such that approximately 70% split between the early and late majority and the remaining 30% split on either side of the majorities. Rogers was careful to note that these are ideal categories and that individuals or organizations may fall into one category with regard to one type of innovation and another category with a different one. Further, the characteristics or actions of individuals or organizations in any given category may be altered dramatically by outside factors. For example, because criminal justice deals with public safety, it is highly susceptible to strong movements in public acceptance, whether it be rejection of or increased demand for a given technology. Before considering each category, it is important to distinguish between a technology and an innovation. An innovation is a new practice that may or may not involve technology. But not all technology or even new technology is an innovation. New technology may simply be an incremental improvement on existing technology.

To illustrate the five categories of adopters, consider police departments and police chiefs. Those that fall in the first category—innovators—are described by Rogers as venturesome. The chiefs in these police departments are well connected to other innovators through regional, national, or even international networks. New ideas and experiences are shared across the networks with a desire to explore new possibilities and a willingness to cope with uncertainty. Such departments typically have enough control over financial resources to manage possible losses due to failed initiatives. Thus, innovators tend to be larger departments with larger budgets and wide name recognition.

True innovators are rare among police departments because of the inherent risks and costs associated with this venturesome approach. The leading edge of innovation requires not just a willingness to fail, but the expectation of a preponderance of failures. Still, successes from truly innovative approaches can be revolutionary. The Defense Advanced Research Projects Agency (DARPA) in the Department of Defense carries out the most innovative technology research and development for the military. While some military technology eventually transfers into domestic criminal justice settings (and everyday life), there is no analogous advanced technology research and development organization for criminal justice. Although the NIJ does support research on new technologies, it cannot be appropriately compared to DARPA. Thus, the risky mantle of the true innovators is left up for grabs in criminal justice, with state, local, and tribal agencies generally reluctant to take it up.

One step behind the innovators are the early adopters. The early adopter police chief is often well respected by peers and sought out by vendors and the NIJ to test products or participate in technology demonstration efforts. This chief will be interested in taking a careful look at promising technology and assessing its effectiveness in his or her own department. This chief is likely to acquire and deploy promising technology that performs well during field testing. Initial deployment within the department may occur in phases to protect against the risks of failure and build confidence within the department. Others in the field are likely to consult with early adopters for their advice before taking on a new technology or innovation of their own. Often an early adopter department will have direct access to grant writers and administrative personnel who can help secure and properly manage federal and private grant resources.

The early majority adopters tend to be more deliberative and look to the findings of the early adopters for evidence that a new technology will meet their needs. The chiefs in these departments are open to innovation and change, but they are certainly more measured in their approach and they rarely lead the way. Resources may be part of the reason. Departments that have limited access to resources through federal grants or other sources often refrain from taking the lead in technology adoption. This applies to many, but not all, smaller departments.

Late majority adopters are generally skeptical of change and tend to succumb only after they have received some amount of peer pressure or economic pressure to adopt the innovation. Compelling evidence is necessary for adopting the new technology, and even then, change must wait for financial resources. Late majority adopter chiefs tend to run smaller departments that are more local in terms of their connectivity with the rest of the field of policing.

Finally, laggards (Rogers was careful to point out that this term is not meant in a derogatory way) are characterized by extreme caution that may stem from a highly traditional orientation that is resistant to change, minimal financial resources, or a combination of the two. The laggard departments are most likely small and fairly isolated from the rest of the field. They require virtual certainty of success before they will adopt an innovative practice or technology. These are the final 10–15% of departments to accept an innovation. By the time these departments adopt a new technology, it is already well known and well established in the field.

Even though these categories of innovation adopters are ideal types that are subject to exceptions, they can tell us a lot about criminal justice policy as it relates to technology. We can make the generalization that larger organizations with more resources are more likely to consider adopting new technology earlier than smaller ones with few resources. However, this likelihood may be moderated by risk tolerance or connectivity across the professional network beyond the local level. That is, even a small organization may be an innovator or early adopter if its leadership has become widely connected or more resilient to failure for any number of individual-level, organizational-level, or contextual reasons. One contextual reason for smaller organizations to be more innovative is that they may possess specialized knowledge or attributes in their organization or jurisdiction. For example, small sheriff departments in rural areas along the southern U.S. border may have specialized knowledge about certain forms of drug and human smuggling that leads them to be innovators or early adopters of certain forms of sensor and surveillance technology.

### Deciding on a New Technology

Deciding to acquire and integrate a new technology into a criminal justice organization is no small matter, and, the previous discussion notwithstanding, this decision process is almost never driven solely by the inclinations of the chief executive. As discussed, innovators and early adopters are risk takers and experimenters. The 85–90% of criminal justice organizations behind these two groups are typically looking for much more compelling reasons and information before they are likely to make significant commitments to new technology. This is particularly true where there are sizeable financial costs or where the potential for organizational destabilization exists. In contrast, Skogan and Hartnett (2005) documented a case in which diffusion of innovative information technology was very rapid across suburban Chicago police departments. The reasons for this rapid diffusion were that the technology in question was provided initially free of cost, delivered through a strong "evangelical" outreach effort by the technology provider, and posed little threat to existing operational policies and practices. (It should be noted, however, that technology is rarely "free of cost" in practice because life cycle and training costs are often more important than up-front costs.)

The Chicago example hints at the factors that make it easy to accept innovative technology. The ideal circumstances for adopting technology involve a wider range of factors. Policy makers would have certainty that the technology in question can improve fundamental criminal justice outcomes such as reduced crime, increased officer safety, reduced costs, or reduced liability. The process for integrating the technology into existing operations would be simple, with minimal training requirements or specialized knowledge. Expenses for acquiring, implementing, and maintaining the technology would be known and reasonable relative to future benefits. No legal questions or challenges would likely arise regarding the technology. Public opinion would be favorable, and there would be no opposition to the technology from special interest groups or advocacy organizations. Finally, the technology would pose no increased safety concerns for either the public or criminal justice professionals. These ideal circumstances illustrate the difficulty and risks of making technology policy decisions.

Fox (2004) describes an increasingly common approach to addressing the numerous questions involved in assessing technology for acquisition and integration into correctional institutions. Correctional institutions are partnering at the state level to form "technology transfer" or "technology review" committees to identify, assess, educate, and build acceptance of technologies. In line with a broader movement toward using multidisciplinary teams to respond to complex challenges, the technology review committees often involve a range of practitioners, including correctional officers, engineers, attorneys, fiscal personnel, technology specialists, and wardens. Vendors seeking to market technologies to individual correctional institutions are directed to the committee to make their pitch. The committee may reject technologies outright or initiate operational pilot tests in an institution.

Technology review committees are an organizational method for increasing confidence and reducing risks in technology acquisition decisions. Engaging representatives from multiple institutions broadens the scope of interest beyond highly localized issues. Indeed, such technology review committees moderate the influence of innovation adopters at the extreme ends of the distribution. The dispositions of the freewheeling venturesome innovators on the one hand and the resistant laggards on the other are unlikely to prevail in such committees.

### But Does it Work?

The most obvious factor that should influence technology policy decisions is whether the introduction of the technology will lead to improvements in key criminal justice outcomes such as public safety, costs, and officer safety. Certainly technology vendors emphasize the potential for improved outcomes as a central theme in their marketing, but what standards of evidence should decision makers apply in judging claims of effectiveness?

Traditionally, one of the most common methods for convincing decision makers of the effectiveness of technology is through testimonials from current users who are like the intended user. Testimonials are common in criminal justice and other fields because they work. The influence of testimonials is reflected in the communication patterns described among innovators, early adopters, and the early majority. Further, criminal justice professions tend to have cohesive associations with peers due to the inherent dangers and sensitivities of criminal justice work. More cohesive groups tend to have greater amounts of trust among colleagues, increasing the likelihood that testimonials will be influential. The benefit of testimonials is that evaluative information is provided by professionals with similar responsibilities and first-hand experience using the technology in practice. However, testimonials are subject to a wide range of biases and tend to lack objective analysis or controls for other influences.

Operational pilot testing is a more demanding approach that typically involves systematic measurement of performance in a field setting or simulated environment. Data collection is often both quantitative and qualitative, with performance typically measured quantitatively and practitioner experiences and reactions captured through qualitative methods (e.g., open-ended interviews). Depending on the technology, pilot testing may be carried out over the course of hours, days, or weeks. These tests frequently reveal areas for improvement in redesign but can also be important for identifying policy implications and refining operational policies. Although pilot tests certainly are important in determining the effectiveness of technology, it is unfortunate that evaluative efforts frequently go no further. This is because pilot tests typically lack the rigor to evaluate contribution to core criminal justice outcomes in the context of real-world application and in ways that rule out alternative explanations.

Social scientists have spent decades developing methods for evaluating the effects of interventions in complex environments (e.g., Campbell & Stanley, 1963). Randomized controlled studies and quasi-experimental designs with matched comparison groups are among the most rigorous methods for generating evidence of causality between interventions and desired outcomes in applied settings. These approaches seek to isolate the influence of the intervention of interest (e.g., a criminal justice technology) and negate the influences of known and unknown factors that might also affect the intended outcomes. Deliberate efforts to be objective and systematic in establishing causality are the strengths of this approach. However, rigorous social science evaluation of criminal justice technology is uncommon because it can be expensive and time consuming, and it requires specialized knowledge. The time-consuming part can be particularly problematic. In general, criminal justice practitioners are highly motivated to receive information in a timely fashion and useful format. Social scientists typically prioritize methodological rigor and may be more motivated by peer-reviewed publications that are vital to career advancement among academic researchers. The challenges of forming and maintaining effective collaborative working relationships between criminal justice practitioners and researchers prompted the International Association of Chiefs of Police (2004) to publish recommendations for improving partnerships between law enforcement leaders and university-based researchers.

In addition to effectiveness, most decision makers are looking for objective information on costs related to benefits or, at a lower threshold of rigor, return on investment. The Washington State Institute for Public Policy, led by Steve Aos, has prepared reports on evidence-based policy options for reducing criminal justice costs that have gained widespread interest because they compare the effectiveness of prevention and treatment programs in terms of monetary savings delivered to the system (Aos, Miller, & Drake, 2006). Unfortunately, nothing similar currently exists for criminal justice technologies.

## ■ Technology Profiles

This section includes a review of two technologies that have diffused across many agencies in the criminal justice system and have attracted significant policy attention in recent years. In a sense, these very characteristics that make them worthy of discussion also make them uncommon. The vast majority of technologies entering criminal justice do not draw comparable levels of attention. The attention itself speaks to the policy challenges associated with these technologies just as the diffusion speaks to the potential benefits of these examples. The technologies that will be described are conducted energy devices and DNA forensics.

### Conducted Energy Devices

Law enforcement officers must be prepared to encounter hostile or resisting suspects during arrest situations. The legitimacy of law enforcement depends in part on its ability to handle these situations appropriately, and officers are trained on policies and tactics for effectively subduing subjects. Of course, these situations can be extremely dangerous, and officers must make split-second decisions about how to respond. In general, agencies operate under a use-of-force continuum in which officers match the lethality of the force they use to situational factors such as the level of resistance and threat. From the continuum, officers may engage suspects with verbal commands, minimal physical control tactics, hard physical strikes and grappling, intermediate weapons, and deadly weapons. Police batons, or nightsticks, fall between hard hand-to-hand tactics and the use of firearms on the continuum. Made of hard wood, plastic, or metal, these weapons inflict blunt force damage that can lead to serious injury and death, particularly when blows are to the head. In the 1980s and 1990s, interest in less lethal technologies that effectively incapacitated subjects without the negative consequences of using batons grew. One well-known, less lethal technology adopted by law enforcement was oleoresin capsicum (OC), commonly known as pepper spray. When deployed, OC may cause painful burning, eye irritation, and choking. Conducted energy devices gained interest in part because of various concerns with OC, including in-custody deaths and opposition from groups such as Amnesty International that documented incidents of misuse against peaceful, but resisting individuals.

Although the NIJ released a report in 2003 that found OC to be safe and effective when used properly, momentum had begun for finding alternatives, and in the same year, TASER International introduced the TASER X26, the most widely used conducted energy device (CED) in law enforcement today. CEDs use electromuscular disruption technology to transmit a high-voltage, low-power charge of electricity that causes involuntary muscle contractions and temporary incapacitation. CEDs can be used at distances or through direct contact. When directed at a remote target, CEDs eject two probes at high velocity for distances of up to 35 feet. When the dartlike probes strike a target, electrical pulses are transmitted through insulated wires from the handheld device to the target.

A central question for the criminal justice community is whether CEDs reduce the likelihood of injury or death to suspects, officers, or bystanders relative to other use-of-force alternatives *in actual confrontations*. Simulations and tests cannot fully substitute for all the circumstances in which officers must make use-of-force decisions. There have been a number of industry-sponsored studies and department reports of improved outcomes from using CEDs. These reports lack independence and typically follow single-group pretest/posttest study designs that fail to rule out potential alternative explanations and inferential threats. However an independent study with a more rigorous research design considered numerous factors related to use-of-force options and resulting injuries in over 1600 use-of-force incidents in two law enforcement agencies—one in South Carolina and one in Florida (Smith, Kaminski, Rojek, Alpert, & Mathis, 2007). The study found that in one agency, the odds of officer and suspect injury and the severity of injuries were reduced when a CED was deployed. The other agency used both OC and CEDs. In this case, the use of CEDs was unrelated to the odds of injury, but the use of OC was associated with reduced odds of suspect injury. Importantly, this study also found that hands-on tactics by police were associated with increased odds of injury to the officer and the suspect, while the use of canines was associated with increased odds of suspect injury. Thus, from the perspective of overall safety to officers and the public, CEDs appear to present a favorable option.

Yet deaths have occurred when CEDs have been used by criminal justice practitioners, prompting the NIJ to sponsor a medical panel to determine whether CEDs present a high risk of serious injury or death (Morgan, 2008). While the panel acknowledges that there remain many questions about the safety of CEDs that cannot yet be answered by the existing research, the current research provides no conclusive medical evidence of a high risk of serious injury or death from the direct effects of CED exposure. Morgan (2008) describes a possible explanation for some of the deaths that have occurred when CEDs were deployed. Excited delirium is a term to describe people in a highly agitated state who may be experiencing psychosis or drug intoxication. It is not a medical diagnosis, but it reflects a situation that law enforcement officers confront in the course of their duties. These individuals often have elevated body temperatures, may be violent, and may display great strength. The bodily stresses of excited delirium put individuals at risk in the first place. When CEDs are applied in this situation, the odds of death may be increased.

So in terms of reducing injury, deploying CEDs appears to be a good policy choice. From a medical perspective, the direct effects of CEDs have not been shown to cause serious injury or death. However, there are more factors to consider. In 2007, the phrase "Don't Tase me, bro!" entered our lexicon following the use of a CED on a University of Florida student. The student was attending a forum with Senator and former presidential candidate John Kerry. During a question and answer session, the student asked a series of questions in rapid succession without providing much opportunity for Senator Kerry to respond. Campus police intervened to physically remove the student, who resisted and called out to other audience members. From video footage, it appears that five officers took the student to the ground, struggled with him, issued a verbal warning that they were about to use the CED. The student's yell of "Don't Tase me, bro!" is followed by the sound of the CED being deployed and the student shouting in pain. Another student captured a video recording of the incident as it occurred. The video subsequently "went viral" across the Internet, was picked up by national and cable news stations, and was viewed by untold millions.

The exact nature of the public reaction to such an incident is unknown, but many openly criticized the use of a CED in this instance. Within the public discussion arose many questions about the propriety of this technology in any instance. Other incidents of CED use against children and the elderly have also generated widespread criticism. Within criminal justice, the response to these incidents has been to recommend improved training and policies to support proper deployment of CEDs. The International Association of Chiefs of Police (2008) issued a nine-step strategy for effective deployment that does not recommend CEDs outright, but does recommend a decision-making process for determining whether to deploy them. Step 5 of this strategy is to engage in community outreach, noting that community acceptance is essential to successful deployment. Another recommendation is to consider issues of liability. Anyone considering such issues would be interested to know that a legal analysis of 53 written opinions in U.S. courts found that most cases did not support municipal liability or liability on the part of the individual officers, particularly when the suspect was resisting (Smith, Petrocelli, & Scheer, 2007).

Policies and operating procedures for CEDs vary across departments. Connor (2006) reports on a set of CED policies developed as a result of a 2-year study by a 17-member multidisciplinary committee under the direction of the Georgia Association of Chiefs of Police. This committee recommends placing CEDs on the use-of-force continuum, properly training all officers to use CEDs, and reporting each incident of CED use. Except in exigent circumstances, the committee recommends that CEDs not be used on children, the elderly, pregnant women, or the visibly frail. Training must emphasize tactics for cuffing subjects once the CED has been applied. A medical evaluation is recommended for all who are subjected to CED applications. Finally, the committee recommends placing CEDs in front-draw holsters on the nonservice weapon side. The intent of this last recommendation is to avoid a problem that has had terrible consequences. If officers place the CED in a holster on the same side of their body as their service firearm, they may accidently draw and discharge lethal force when the intention was to use a CED. Some have resolved this by placing the CED on the opposite side in a reverse position to allow a rapid cross-draw motion. This approach reduces the chance of drawing the wrong weapon, but the reverse positioning may increase the probability for weapon takeaways and nondeployment in close quarters. Therefore a forward position on the off-hand side is preferred.

While these policy recommendations appear reasonable on the surface, there is no consensus on them. Staton (2008) objects to lists of people who should be excluded from CED use, arguing that officer decision making during a confrontation should not be clouded by consideration of a suspect's eligibility for a specific force option. Further, he advocates a case-by-case approach and argues against policies requiring medical evaluation for every CED use. He notes that research indicates a reduced number of injuries using CEDs and that other use-of-force options do not have this requirement. Finally, he takes minor issue with recommendations for front-draw positioning of the CED. The cross-draw is more natural than using the off-hand to forward draw and then either deploying the weapon with the off-hand or switching it to the other hand. However, he acknowledges the critiques of the cross-draw holster and reiterates the need for rigorous training as the best way to assure proper performance in the field.

In summary, CED policy is by no means resolved. The contributions of independent social scientists, medical professionals, and legal researchers suggest that CEDs have demonstrated effectiveness in reducing injury in practice, are unlikely to be directly related to serious injury or death, and are unlikely to lead to liability claims. But high-profile incidents continue to occur that contribute to negative public opinion of this technology, and leading criminal justice organizations recognize the importance of public acceptance to any decision to deploy CEDs across an agency. Even the professionals who have accepted this technology have varying recommendations regarding appropriate operating policies. But given developments to date, it seems likely that there will continue to be a place in criminal justice practice for CEDs or similar devices for years to come.

### Forensic DNA

Deoxyribonucleic acid (DNA) is the genetic material in the cells of all living organisms that helps define each individual's characteristics. Each organism has a unique DNA sequence (with the exception of identical twins, who share the same sequence). The DNA sequence is the same in every cell within an organism. This allows scientists to extract unique DNA profiles from biological samples of cells (e.g., blood, semen, skin cells, perspiration) and compare those to other DNA profiles. In criminal justice, DNA has been used as an investigatory tool most frequently in instances of homicide and sexual assault. These crimes are more likely to involve blood and semen that can be used to extract DNA profiles. The level of certainty that comes from DNA analysis serves as strong evidence in prosecution. It has also been used to exonerate wrongfully convicted prisoners, sometimes decades after a guilty verdict.

For many in the public, initial exposure to use of forensic DNA came during the highly publicized murder trial of O. J. Simpson in 1995. In that trial, DNA evidence was presented that connected Simpson to the murder scene. The defense scrutinized the handling of the DNA evidence and cast doubt on the practices of the technicians and detectives who handled the case. Questioning police procedure was a good strategy on the part of defense attorneys because the actual science behind DNA analysis is hard to assail. Many will recall that Simpson was found not guilty.

Just what exactly qualifies as admissible scientific forensic evidence in criminal proceedings has changed over time with two major legal decisions (Bradley, 2004). For many years, courts were guided by what is known as the Frye standard that emerged from a 1923 appeals court ruling that ended unfavorably for defendant James Frye. In the appeal, the defense attempted to introduce a blood pressure test as a method for determining whether someone was telling the truth. The court ruled that the technology from which the evidence comes must be generally accepted in the particular field to which it belongs. Since the blood pressure test did not meet this standard, the conviction was upheld. In subsequent years, the standard of general acceptance in the applicable scientific field was the guiding legal principle. That changed in 1993 with the case of Daubert v. Merrell Dow Pharmaceuticals, Inc. The new Daubert standard requires that "any and all scientific evidence and testimony is not only relevant, but reliable" (National Academy of Sciences, 2009). Among the factors to be considered under Daubert are whether the technique (1) has been tested, (2) has been subjected to peer review, (3) has a known error rate, (4) has standards of operation, and (5) is accepted in the scientific community. This standard has contributed to increasing scrutiny of the scientific foundations of forensic activities with some coming under strong criticism (e.g., fingerprint analysis), but with DNA forensics better positioned to withstand this higher standard of evidence (National Academy of Sciences, 2009).

While the science itself is sound, DNA collection and handling practices continue to receive scrutiny. Bond and Hammond (2008) completed a study of the processing of approximately 1500 biological samples recovered from crime scenes in the United Kingdom to assess the factors related to converting them into DNA profiles and matching them to other samples to identify a suspect. One of the more important factors was the nature of the sample, with a DNA match most likely when the sample was blood. But the strongest factor in determining whether a sample led to a match was whether it was handled by an accredited crime scene examiner. Accreditation involves training and testing based on a set of standard knowledge and skill areas. In the United States there are no standards for mandatory accreditation of forensic science laboratories or forensic scientists (National Academy of Sciences, 2009). Interestingly, a recent U.S. study on the use of DNA evidence in property crimes found that forensic technicians were no more effective than patrol officers in collecting biological evidence (Roman et al., 2008).

Of the practical steps involved in using DNA in a criminal investigation, it is the use of DNA databases that has generated the most controversy. Jacobs and Crepet (2008) provide a review of policy decisions on DNA databases that underscores how expectations and practices shifted dramatically through a series of incremental policy changes. In the late 1980s, states first began collecting DNA from convicted sex offenders and put them into databases that could be searched for matches with evidence from other crime scenes. The frequent existence of DNA evidence combined with patterns of serial sex offending and the horrific nature of the offense itself were all advanced in the initial justifications for these databases. In 1990, the federal government launched a pilot project that linked several state databases using software called the Combined DNA Index System (CODIS). In 1994, Congress authorized the FBI to build the National DNA Index System (NDIS) that serves as the national database for DNA records and is still linked to the state databases through CODIS. In 2000, Congress passed the DNA Backlog Elimination Act that requires DNA testing for persons convicted of certain federal offenses and provided funding to states to collect DNA from convicted sex offenders. In 2006, federal legislation under the Violence Against Women Act authorized the federal government to collect DNA evidence from all federal arrestees. Non-U.S. citizens detained for immigration law violations may also be required to submit DNA. Forty-four states have expanded DNA collection from violent and sex offenders to all convicted felons. At least 11 states now require all arrestees to provide DNA samples. In June 2001, NDIS included over 600,000 offender DNA profiles. By October 2007, the number was over 5 million.

What began as a state-by-state effort to retain DNA profiles of sex offenders grew into a national system of DNA databases that not only collects DNA from all felony offender types but also from all felony arrestees at the federal level and in multiple states. Is there any harm in that? After all, increasing the size of DNA databases should increase the probability of making a DNA match and identifying criminal suspects. This does not apply only to cases of homicide or sexual assault. An NIJ experiment found that when property crime cases like burglary were randomly assigned for DNA analysis and investigation versus non-DNA investigation, more than twice as many suspects were identified, twice as many suspects were arrested, and more than twice as many cases were accepted for prosecution (Ritter, 2008).

One of the main critiques of current policies for DNA databases is the inclusion of samples from all arrestees. Simoncelli and Steinhardt (2006) argue that the principle of the presumption of innocence is undermined by this policy. One may be arrested, compelled to provide a DNA sample, then released with no trial, or the arrestee could go to trial and be found not guilty by a jury of his or her peers. In either case, although the arrestee was never convicted of any crime, he or she has provided a DNA sample that will reside in the database for comparison in every subsequent search unless something is done to remove it. Current policies around removal of DNA profiles put the burden on the individual to appeal to what may be multiple systems at the state and federal level.

Another concern about DNA databases has to do with the potential for future reuse of DNA samples for purposes other than those for which they were collected. DNA is sometimes referred to as a "genetic fingerprint"; however a fingerprint is a two-dimensional image that can be used to identify someone, whereas DNA samples include an individual's entire genetic code. Potential foreseeable uses of the DNA data banks include research on predispositions for disease, physical traits, or behavioral dispositions. Critics have noted multiple possible misuses including research seeking to identify genetic markers for aggression, substance addiction, criminal tendencies, and sexual orientation (Simoncelli & Steinhardt, 2006). Such inquiries are viewed as a slippery slope toward discriminatory policies based on genetic makeup. To reduce the potential for misuse, some have proposed policies for regulating access to these databases for purposes other than identification. One such proposal is to institute hardwired constraints that destroy all informational elements that are not critical to the core purpose of criminal investigation (Lazer & Mayer-Schonberger, 2006).

Familial searching is an example of a controversial use of DNA databases that has already been implemented in the United Kingdom. Familial searching is a procedure used when a DNA sample is not matched to an existing profile on the database. This procedure searches for similar DNA profiles that have a direct genetic relationship to the sample (i.e., parent, child, or sibling). In 2002, officials in the United Kingdom used familial searching to help identify the son of a man who was suspected of having committed multiple rapes and murders 27 years earlier (Williams & Johnson, 2006). The primary suspect had passed away, but familial searching and the collection of additional DNA samples from other living family members led to a sufficiently credible case that police were given permission to exhume the body of the deceased primary suspect and extract a DNA sample. The sample from the deceased suspect matched the crime scene samples, and the case was solved. But familial searching also raises ethical questions about the use of family members who have had no involvement in the crime. Through participating in this type of investigation, such family members may learn of a genetic link between individuals that was previously unknown by one or both of the parities. Similarly, an investigation may reveal the absence of genetic links where participants had assumed them to exist. The use of familial searching means that the implications for the storage of DNA extend well beyond even those who are in the database.

# Conclusion

Technology contributions are vital to the progress of criminal justice. Yet, we must recognize the limitations of technology and our own limitations in applying technology. Through a discussion of basic principles such as categories of innovation adopters and of the use of CEDs and DNA forensics, this chapter has considered many of the influential factors that contribute to technology policy in criminal justice. Through the NIJ, practitioners help identify technology needs and possible solutions. Federal and private investments in technology development lead to pilot testing and coordination with innovators and early adopters in the criminal justice community. These early adopters spread the word across their professional networks and promote the early findings that help identify technologies as promising. Technology review committees and professional associations help set standards for review and thresholds for acceptance prior to acquisition and deployment of technology among early and late majority adopters. Researchers conduct studies and evaluations on the effectiveness of technologies. Professionals from related sciences such as medicine or biology comment on the contributions of their fields to the application of the technology. Practitioners weigh in on operational policies to maximize effectiveness. Legal analysts and ethicists frame arguments that address regulation, liability, and propriety issues. State and federal legislators pass laws that authorize and provide resources for technology. Throughout this process, public interest periodically spikes—typically in reaction to a high-profile event.

The combined effect of all of these influences makes for a complicated and uncertain process for technology policy in criminal justice. However, the better criminal justice professionals understand this process, the better they will be able to use it as a system of checks and balances that lead to the most effective and efficient technology in advancing criminal justice.

# Discussion Questions

1. Discuss the various policy issues that must be considered before a new criminal justice technology is widely adopted.
2. Describe the process used by the NIJ to assess the viability of a particular technology in criminal justice.
3. What is the role of criminal justice practitioners in the development and implementation of technology in the criminal justice system?

4. Outline the dilemmas and challenges encountered by policy makers charged with making recommendations on the use of CEDs in law enforcement.

5. Describe some of the controversial issues policy makers face when considering an expansion in the use of DNA in criminal justice.

## ■ References

Aos, S., Miller, M., & Drake, E. (2006). *Evidence-based public policy options to reduce future prison construction, criminal justice costs, and crime rates.* Olympia, WA: Washington State Institute for Public Policy.

Bond, J. W., & Hammond, C. (2008). The value of DNA material recovered from crime scenes. *Journal of Forensic Sciences, 53*(4), 797–801.

Bradley, R. C. (2004). *Science, technology, & criminal justice.* New York: Peter Lang Publishing.

Campbell, D. T., & Stanley, J. C. (1963). *Experimental and quasi-experimental designs for research.* Chicago: Rand McNally.

Connor, G. (2006). Essential elements in TASER policy and procedure. *Law & Order, 54*(8), 87–90.

Fox, A. (2004). Collaborating for technological success. *Corrections Today, 66*(4), 66–70.

International Association of Chiefs of Police. (2004). *Unresolved problems & powerful potentials: Improving partnerships between law enforcement leaders and university based researchers.* Alexandria, VA: Author.

International Association of Chiefs of Police. (2008). *Electro-muscular disruption technology: A nine-step strategy for effective deployment.* Alexandria, VA: Author.

Jacobs, J., & Crepet, T. (2008). The expanding scope, use, and availability of criminal records. *New York University Journal of Legislation and Public Policy, 11*(2), 177–213.

Lazer, D., & Mayer-Schonberger, V. (2006). Statutory frameworks for regulating information flows: Drawing lessons for the DNA data banks from other government data systems. *Journal of Law, Medicine & Ethics, 34*(2), 366–374.

Morgan, J. (2008). Medical panel issues interim findings on stun gun safety. *NIJ Journal, 261,* 20–23.

National Academy of Sciences (2009). *Strengthening forensic science in the United States: A path forward.* Washington, DC: The National Academies Press.

National Institute of Justice. (2003). *The effectiveness and safety of pepper spray: Research for practice* (NCJ 195739). Washington, DC: Office of Justice Programs.

National Institute of Justice. (2009). *High priority criminal justice technology needs* (NCJ 225375). Washington, DC: Office of Justice Programs.

Reed, B. Jr. (2008). Future technology in law enforcement. *FBI Law Enforcement Bulletin, 77*(5), 15–21.

Ritter, N. (2008). DNA solves property crimes (But are we ready for that?). *NIJ Journal, 261*, 2–12.

Rogers, E. M. (1995). *Diffusion of innovations* (4th ed.). New York: The Free Press.

Roman, J. K., Reid, S., Reid, J., Chalfin, A., Adams, W., & Knight, C. (2008). *The DNA field experiment: Cost-effectiveness analysis of the use of DNA in the investigation of high-volume crimes.* Washington, DC: Urban Institute.

Ruegg, R., & Feller, I. (2003). *A toolkit for evaluating public R&D investment: Models, methods, and findings from ATP's first decade* (NIST GCR 03-857). Gaithersburg, MD: National Institute of Standards and Technology.

Simoncelli, T., & Steinhardt, B. (2006). California's Proposition 69: A dangerous precedent for criminal DNA databases. *Journal of Law, Medicine & Ethics, 34*(2), 199–208.

Skogan, W. G., & Hartnett, S. M. (2005). The diffusion of information technology in policing. *Police Practice and Research, 6*(5), 401–417.

Smith, M. R., Kaminski, R. J., Rojek, J., Alpert, G. P., & Mathis, J. (2007). The impact of conducted energy devices and other types of force and resistance on officer and suspect injuries. *Policing: An International Journal of Police Strategies & Management, 30*(3), 423–446.

Smith, M. R., Petrocelli, M., & Scheer, C. (2007). Excessive force, civil liability, and the Taser in the nation's courts. *Policing: An International Journal of Police Strategies & Management, 30*(3), 398–422.

Staton, J. (2008). A TASER policy that works. *Law & Order, 56*(3), 93–97.

Williams, R., & Johnson, P. (2006). Inclusiveness, effectiveness and intrusiveness: Issues in the developing uses of DNA profiling in support of criminal investigations. *Journal of Law, Medicine & Ethics, 34*(2), 234–247.

## ■ Court Cases Cited

*Daubert v. Merrill Dow Pharmaceuticals, Inc.*, 509 U.S. 597 (1993).

# White-Collar Crime and Public Policy: The Sarbanes-Oxley Act and Beyond

# CHAPTER
# 13

David O. Friedrichs (with Sarah Youshock)

## ■ Introduction

Crime and its control have been a major focus of domestic public policy for much of our recent history, especially since the 1960s. Opinion polls over the years have consistently listed crime at or near the top of citizen concerns, and addressing "the crime problem" has been high on the list of political party platforms (Beckett, 1997; Gest, 2001; Macionis 2005, 136). Historically and into the present, the concern with crime has focused principally upon conventional, or "street" crime, and its control. Assault, rape, murder, burglary, and auto theft are among the classic examples of conventional crime. Unsurprisingly, those conventional crimes that involve physical harm to other persons have been accorded the highest priority. In addition to such crime and its control being the focus of politicians and public officials, criminologists have also largely adopted this focus. For a recent special issue of the journal *Criminology and Public Policy* on "informing public policy," 27 respected criminologists responded to an invitation to set forth crime-related public policy proposals that could be derived from criminological research. These proposals, on matters ranging from juvenile curfews to police gang units to probation in relation to murder, addressed conventional crime (and terrorism), and its control, quite exclusively (Clear & Frost, 2007).

The media—in both its entertainment and news dimensions—has featured conventional crime, particularly in its most violent and sensational forms (Ferrell, Hayward, & Young, 2008). In a content analysis of the highest rated prime-time news shows on the major cable news channels, reported in 2007, it was found that 21% (approximately one-fifth) of the shows' news segments were devoted to crime (Phillips & Frost, 2007). Earlier studies documented that this media news coverage of crime was very disproportionately devoted to conventional crime (Friedrichs 2007a, pp. 16–18). The directness of the threat and experience of conventional crime means that public fear of it is especially pronounced. Any public policy initiatives that are perceived as reducing this threat (e.g., more police, tougher sentences) are widely supported. Accordingly, if only as a matter of expediency, it has been attractive for politicians and public officials to endorse and promote such initiatives.

Since the 1960s in particular, broadly diffused public concern over drug use has led to the adoption of a series of laws and policies that collectively make up the War on Drugs (Gordon, 1994). Again, whether on balance this war has achieved more good than harm continues to be debated. Altogether, this war, and the broader war on crime that intensified from the 1970s on, produced record numbers of prison inmates in the recent era (Mauer & Chesney-Lind, 2002; Thompson, 2008). These inmates are disproportionately younger, male, minority, inner-city, lower-class conventional crime offenders and drug dealers. The driving forces behind this get-tough approach have been thoroughly examined (Beckett, 1997; Elikann, 1996; Waller, 2006). Although significant declines in many forms of conventional crime have occurred in recent years, it is far less clear that such declines are principally attributable to the get-tough policies (e.g., see Karmen, 2000; Zimring, 2007). Furthermore, these policies have had myriad harmful consequences, such as those seen in inner city communities (e.g., see Clear, 2007; Thompson, 2008). Public policy initiatives such as "three strikes and you're out" and "Megan's law" are especially vivid illustrations of the phenomenon of crime policies that are adopted in the wake of heinous crimes, with politicians responding directly to public outrage. But the effectiveness of these specific legal initiatives as public policy has also been questioned (Gest, 2001; Worrall, 2008). The distinguished criminologist Gilbert Geis (1996), in fact, argued a three strikes policy might be most appropriately applied to white-collar crime offenders. In sum, public policy pertaining to crime has been a substantial endeavor in the recent era, but it has focused principally on conventional crime and its control, and both its effectiveness and fairness in promoting social justice has been questioned.

## ■ Conventional and White-Collar Offenders (and the Control of Conventional and White-Collar Crime) Compared

The profile of the typical white-collar offender differs from that of the typical conventional crime offender. White-collar offenders tend to be older, better-off financially, and more educated, and—virtually by definition—they are more likely to be gainfully employed in a legitimate occupation (Friedrichs, 2007a; Piquero & Weisburd, 2009). Further, they are less likely to have a criminal record and are more likely to be part of an organization; indeed, some of the most significant white-collar crime is committed by (or through) an organization. Although some studies have suggested areas of overlap and intersection between white-collar offenders and conventional crime offenders (e.g., parallels in records of past offenses), such findings may be a function of how these studies choose to define white-collar crime and the type of data they use. However, the generally significant differences between white-collar crime offenders and conventional crime offenders mean, among other things, that public policy that might be appropriate and effective in response to conventional crime and delinquency is not necessarily relevant for addressing the challenges of white-collar crime, especially in its more complex and sophisticated forms.

It is well understood that conventional crime is addressed primarily by criminal law and by the principal divisions of the criminal justice system: the police, the prosecutor's office, the criminal courts, and the correctional agencies. For white-collar crime, the situation is quite different (Friedrichs, 2007a). Much of the harmful conduct of corporations, small businesses, professionals, and those engaged in legitimate occupations is addressed by civil and regulatory law, as well as criminal law. Federal enforcement agencies respond to much white-collar crime, as do some state enforcement agencies; for local police, white-collar crime is typically a very small portion of their workload. Furthermore, the Internal Revenue Service and the U. S. Postal Inspection Service are among the various other federal enforcement agencies often involved in addressing significant forms of white-collar crime, as are the inspectors general associated with federal departments such as the U.S. Department of Defense and the Housing and Urban Development Department. A large network of federal regulatory agencies, including the Environmental Protection Agency, the Federal Trade Commission, and the Securities and Exchange Commission, also play a central role in responding to white-collar crime, especially that of corporations and financial institutions. Self-policing—for example, by professional associations such as the American Medical Association—is also much more of a factor with white-collar crime than with conventional crime. Professionals, such as accountants, have certain "policing" responsibilities in relation to audits. We depend upon corporations and financial institutions to undertake much self-regulation, through the activities of corporate boards and compliance officers, for example. Altogether, then, the policing of white-collar crime involves a much broader range of entities than is typically engaged in the policing of conventional crime, and, accordingly, it calls for a far more complex range of public policy responses.

## ■ White-Collar Crime as a Focus of Public Policy

White-collar crime has been far less of a focus of public policy, and of popular concern, than has been the case with conventional crime. Although the recognition that powerful and elite parties have engaged in actions with profoundly harmful social consequences is hardly new, the concept of white-collar crime itself is fairly recent. It was introduced by Edwin H. Sutherland (1940), often described as the single most important criminologist of the 20th century, in his American Sociological Society presidential address in 1939. Ten years later, in 1949, Sutherland's (1949a) classic contribution to the literature, *White Collar Crime*, was published. Sutherland (1949b, p. 511) set forth a number of definitions of white-collar crime and the white-collar criminal, including the assertion that "the white-collar criminal is defined as a person with high socioeconomic status who violates the laws designed to regulate his occupational activities." Yet, in part due to Sutherland's own failure to focus more upon developing a single, wholly consistent and coherent definition of white-collar crime, there has been some confusion ever since on its definition, and the term *white-collar crime* has been applied in quite different ways. We will not address these definitional disputes here (but see Friedrichs, 2007a; Geis, 2007; Shover & Cullen, 2008). Rather, for present purposes, we will regard "white-collar crime" as a generic term for the whole range of illegal, prohibited, and demonstrably harmful activities involving a violation of a private or public trust; committed by organizations and individuals occupying a legitimate, respectable status; and directed toward financial advantage or the maintenance and extension of power and privilege.

In a typological scheme, the following principal forms of white-collar crime can be delineated: First, there are the two *core* forms, corporate crime and occupational crime. Second, there is a principal *cognate* form of white-collar crime, governmental crime (crimes of states and political white-collar crime). Third, there are *hybrid* and *marginal* forms of white-collar crime, including state-corporate crime, crimes of globalization, finance crime, enterprise crime, contrepreneurial crime, technocrime, and avocational crime. Readers are referred to *Trusted Criminals* (Friedrichs, 2007a) for an elaboration on these forms of white-collar crime. In terms of public policy, it should be self-evident that there can be no single, comprehensive public policy that addresses all the different forms of white-collar crime. The concluding chapter of *Trusted Criminals* addresses the whole range of public policy options that can be adopted in relation to white-collar crime, with at least indications of the specific public policy challenges arising in relation to the different specific forms. In this chapter, we will focus selectively upon dimensions of corporate crime and a high-profile form of white-collar crime, as well as the public policy responses.

Why has white-collar crime been the focus of far less public policy concern than conventional crime? The harms associated with conventional forms of crime—from homicide to rape to burglary to robbery—tend to be more direct and visible. The media has overwhelmingly focused upon such crime because it is easier to dramatize and represent, is more "colorful" than white-collar crime, and is more "sensational." A long history promotes images of criminals as "the other," different from respectable middle and upper class members of society in terms of social class membership, ethnicity, appearance, and so forth. White-collar offenders are typically much closer to the social norm, and accordingly there is a deeply rooted resistance to viewing them as criminals. The media owners, as well as politicians themselves, have many ties (including important forms of financial dependence for commercial sponsorship and political campaign donations) with the corporate, business, and professional classes, and accordingly have reason to avoid associating members of these classes with criminal activities. Moreover, as a practical matter, it is far more time consuming, complex, and costly to successfully prosecute most white-collar crime cases than most conventional crime cases.

For all kinds of reasons, then, it has commonly been assumed that the general public is more concerned with conventional forms of crime than with white-collar crime. It has followed from this perception that the priorities of those who produce public policy have reflected such public priorities. Nevertheless, in the recent era—and especially since the early 1970s—various forces have helped increase attention to white-collar crime, broadly defined. Emerging consumer and environmental movements, for example, drew attention to the serious harms perpetrated by large corporations. The Vietnam War and the Watergate case helped initiate a broader public appreciation of wrong-headed or corrupt activities of powerful leaders. And relatively independent prosecutors began to discern some benefits to pursuing white-collar crime cases. The section that follows addresses a shift in public perceptions of white-collar crime.

## ■ The Perceived Seriousness of White-Collar Crime

Historically, from the time of Sutherland on, the typical assumption has indeed been that the general public regards white-collar crime as less serious than conventional crime. However, surveys as early as the 1950s have indicated some significant public recognition of the seriousness of such crime (Friedrichs, 2007a, pp. 40–41). Although public attitudes have been found to vary according to such variables as gender, race, socioeconomic status, occupation, and the like, those forms of white-collar (or corporate) crime that have physically harmful consequences in particular have been rated as serious in public polls. Furthermore, some very recent surveys have documented an increasing perception of seriousness of white-collar crime generally. One such survey found that while the majority of those polled rated violent conventional criminals as more serious than white-collar offenders, at least one-third had the opposite view (and two-thirds called for more government resources for addressing white-collar crime) (Holtfreter, Van Slyke, Bratton, & Gertz, 2008). Still another recent survey found—perhaps as a consequence of the "corporate scandals" involving Enron, WorldCom, and other corporations—that public opinion has now been transformed qualitatively, with high-profile white-collar offenders viewed as "bad guys" (Cullen, Hartman, & Jonson, 2009). Americans, especially African Americans, seem to increasingly favor getting tough on corporate illegality and punishing it accordingly (Unnever, Benson, & Cullen, 2008).

In light of these recent findings, it seems quite reasonable to hypothesize that the immensely high-profile wrongdoing on Wall Street brought to light during the recent ongoing financial and economic crisis, with its devastating consequences for many millions of Americans as citizens, savers, workers, homeowners, and taxpayers, may well produce an even more broadly held view of white-collar crime as serious and meriting tough responses. And a measurable increase in public perception of the seriousness of white-collar crime would almost certainly translate into more public policy attention. Political officials are highly likely to be responsive to such heightened public concern.

### ■ The Harms Caused by White-Collar Crime: Why It Merits More Public Policy Attention

The harms caused by conventional or street crime are substantial by any measure, and certainly an immense amount of pain is associated with such crime. The survivors of homicide victims and the victims of rape often endure a lifetime of intense emotional pain associated with these criminal events. Property losses from all forms of conventional crime total billions of dollars annually in the United States alone. But the claim here is that by at least some measures the harms caused by white-collar crime—especially in its corporate form—greatly exceed those of conventional crimes (Friedrichs, 2007a; Pontell & Geis, 2007). The financial losses are exponentially greater; there have been cases in which a single corporate or financial entity, or individual investment manager, has caused greater financial losses than the total losses associated with bank robbery in a single year.

Furthermore, it has been common to think of white-collar crime exclusively in terms of financial losses, and it is true that this is the way in which Sutherland originally represented white-collar crime. However, since Sutherland's pioneering work, the expansion of the concept of white-collar crime to include a wider range of corporate activities, as well as the recognition of crimes of state as a cognate form of white-collar crime, has transformed this perception. Accordingly, corporate environmental practices, work-related conditions, and the production of unsafe products, are associated with a vast range of physical harms, from premature deaths to debilitating illnesses to paralysis. Even if this violence is not direct in the sense of a specific corporate intent to cause harm, it is very real, emanating from the pursuit of profit above all other considerations. With that in mind, however, this chapter focuses on forms of white-collar crime with devastating consequences that are primarily, but not exclusively, in the corporate and financial realm.

## ■ Financier Crime and Public Policy

For the purposes of this chapter we focus principally upon an especially significant form of white-collar crime: *financier crime*, or the illegal and unethical activities of corporate and financial institution elites (i.e., CEOs and high-level executives), primarily for their own benefit. It is worthwhile to distinguish this form of white-collar crime from the dominant, core types—corporate crime, or crime committed on behalf of the corporation, and occupational crime, or crime committed within the context of a legitimate occupation. Such crime differs from typical manifestations of corporate crime in terms of its focus on individual benefit and from occupational crime in terms of the immense scope of its consequences. Financier crime, then, is a further refinement of *finance crime*, or crime that occurs within the context of high-finance and financially large-scale enterprises (Friedrichs, 2007a). A dictionary definition of a financier is "One who deals with finance and investment on a large scale" (Merriam-Webster, 2007, p. 469). While there is inevitably some arbitrariness in distinguishing this type of crime from the others described, the distinction seems useful nevertheless. The two highest profile "waves" of white-collar crime that have occurred in the United States in the first decade of the 21st century are best classified as this type. They involved high-level executives of major corporations and financial institutions whose primary focus seemed to be advancing their own financial well-being rather than that of their company, and whose policies and practices had devastating, systemic effects, vastly transcending the effects of most corporate and occupational crimes that cause economic losses.

## ■ The Enron Case—And the Cases That Followed It

The collapse of the Enron Corporation, amid allegations of massive financial misrepresentations and accounting fraud, was the first widely publicized major white-collar crime of the 21st century (Fox, 2003; Friedrichs, 2007b; McLean & Elkind, 2003). In the wake of the Enron case, similar large-scale misrepresentations of the finances of various other major American corporations—including Adelphia Communications, Computer Associates, Dynergy, Global Crossing, Qwest, Rite Aid, Tyco International, WorldCom, and Xerox—were also exposed (Berenson, 2002). Although such misrepresentations by corporations were hardly new, the scope and scale of the subsequent losses in these most recent cases were so great that they attracted major attention from the American public and from various political entities. Investors and pensioners were estimated to have lost tens of billions of dollars, and many other adverse consequences for specific communities and the American economy also occurred.

At the time of its collapse, Enron was listed as the seventh largest Fortune 500 company and claimed to be doing some $100 billion a year in business. Between the beginning of 1998 and the end of 2000, the stock price of Enron had experienced a phenomenal rise, and Enron was celebrated on Wall Street as an extraordinary investment opportunity. At the beginning of 2001, Enron was still valued at some $60 billion. A year later, the value of its stock had declined by some 99% and was virtually worthless. The corporation itself had declared the largest bankruptcy in American history up to that time.

The frauds carried out at Enron were complex. A major accounting fraud was involved, with Enron's massive debts shifted to off-the-books partnerships (controlled by Enron insiders) that allowed Enron to misrepresent its income flow and profits grossly. Enron also disguised its immense debt by listing billions of dollars of loans as trades. It also evolved into a classic "pump-and-dump" operation. This enduring form of investment fraud is characterized by misrepresentations that drive stock prices up, with insiders then dumping their own stock at the top of the market, leaving the many misled investors to suffer huge losses when the stock inevitably plummets as misrepresentations come to light. Although such pump-and-dump misrepresentations are clearly illegal, some of the aggressive accounting practices adopted by Enron were legal, at least at that time. For example, Enron was allowed to immediately declare profits on highly speculative, long-term energy deals by a technique known as "mark-to-market" valuation. Such techniques produced a wildly inflated notion of Enron's profitability, pumping up its stock price. Existing laws too often sanction practices that are inherently deceptive and unethical.

Although the Enron case (and the other cases of that period) were widely referred to as corporate crime cases, they were characterized principally by top executives and their professional associates focused upon their own financial gain, and in this sense are financier crime. The top executives of Enron were key to the massive misrepresentations that brought about the collapse of the company and the billions of dollars of losses, but many other parties were involved as well. The company's board of directors—including some prominent executives at other major corporations—failed in one of the basic responsibilities of any corporate board by not ensuring that the corporation was conducting itself honestly and honorably. As with many such boards, the board members all too often had inherent conflicts of interest, sometimes because they were well-compensated consultants or contractors for Enron, sometimes because they had too many other responsibilities to focus properly on how Enron worked. Investment banking houses and stock analysts were profiting immensely from Enron's dramatic rise in stock price and did not adequately question the basis of Enron's claims about its financial performance. Lawyers, including a prominent Houston law firm, either assisted in structuring some of Enron's dubious partnerships and other misleading practices or failed to clearly identify the illegalities involved in such activities. Credit-rating agencies failed to scrutinize Enron's financial claims with sufficient skepticism. Regulatory bodies were often intimidated by Enron's political clout.

The party most directly complicit in Enron's massive financial misrepresentations was the Arthur Andersen accounting firm. This long-established and prestigious major firm was Enron's auditor, and it should have uncovered and exposed the massive forms of accounting fraud. But Andersen was collecting over $50 million a year in fees from Enron, with more of that amount coming from consulting services than for auditing. Ultimately, as Enron began to collapse, Andersen employees shredded many of their auditing documents and tampered with e-mail messages. The Andersen firm was indicted on federal obstruction-of-justice charges and collapsed following its conviction on those charges. This prosecutorial action by the federal government was highly controversial, as tens of thousands of Andersen employees who had nothing to do with the Enron matter lost their jobs with the collapse of their firm. The U.S. Supreme Court ultimately overturned the conviction, on the basis of faulty charges the judge in the case made to the jury, but by that time the Andersen firm no longer existed. Andersen had previously served as auditor for a number of other major corporations, including Sunbeam and Waste Management, which were subsequently found to have fraudulently misrepresented their finances on a massive scale. This was one of the reasons federal prosecutors decided to pursue criminal charges against Andersen in the Enron case. Furthermore, some former Andersen employees claimed after the company's collapse that in its final era the production of high revenue became the overwhelming goal of Andersen (Squires, Smith, McDougall, & Yeack, 2003). A "culture of greed" supposedly became dominant, where concerns about the quality of audits and about the unacceptable accounting practices of major clients were wholly overshadowed by this focus on revenue production.

## ■ The Sarbanes-Oxley Act as a Public Policy Response to Enron

The revelations of massive financial misrepresentations at Enron, WorldCom, and other major American corporations, with their attendant bankruptcies and devastating financial consequences, generated much public concern and anger during the final months of 2001 and the first half of 2002. Although the September 11, 2001 attack and the ongoing concern about another such terrorist attack, as well as the decision to take military action against Afghanistan, was the biggest story of this period, the corporate scandals were arguably the second biggest story. Accordingly, the political leadership in Washington DC was under considerable public pressure at this time to address what had happened at these corporations. Congressional hearings were held, with top executives of the failed corporations subpoenaed to testify.

Such congressional hearings can always be viewed in conflicting ways. Are they undertaken by dedicated public servants focused on uncovering and addressing corporate practices that are harmful to the American public, or are they opportunities for grandstanding by politicians principally focused on their own political careers and personal interests? Surely elements of both dimensions are involved in such hearings, and this was especially the case involving the corporate scandals hearings. Commentators noted that some of the same legislators who had solicited and accepted large campaign contributions from these corporations and their executives, and who had pushed through legislation that had in various ways facilitated the fraudulent misrepresentations undertaken by the executives of these corporations, used the hearings to excoriate them for the benefit of a large television audience (McLean & Elkind, 2003). At least some of the Congress members involved in these hearings were hypocritical in this regard.

The Sarbanes-Oxley Act, officially named the Public Company Accounting Reform and Investor Protection Act of 2002, came into law on July 30, 2002 (Recine, 2002). The act was named after its sponsors, Senator Paul Sarbanes, a Maryland Democrat, and Republican Representative Michael G. Oxley from Ohio. The Sarbanes-Oxley Act (widely known as "SOX") was enacted directly in response to Enron and the other high-profile cases of massive financial misrepresentations. One of the act's main goals is ensuring that corporate directors are accountable for the financial condition of their companies. This can be clearly seen in the act's requirement that companies' board of directors have an auditing committee. The company's auditing committee must police the auditing firm of that company, and in turn the auditors must directly report to the auditing committee. Furthermore, members of the committee cannot be employees of the company they are auditing, and the auditing firms are required to make known which members of the committee are actual financial experts. Also, the audit committee is required to address complaints made about the accounting practices of that company.

One can question whether this new auditing policy goes far enough in addressing the conflicts of interest that rendered audits of Enron so misleading. Perhaps a more effective public policy approach would require companies that by law must produce public audits to contribute an amount to finance the auditing of their company (commensurate with the size and complexity of the company and so forth) to a fund maintained by a government-run auditing council. Auditors would then be assigned from a pool of qualified auditing companies to undertake the auditing of companies, based upon their specific expertise and resources. But the auditing companies would be wholly independent of the audited companies, with their findings reported to the auditing council. Needless to say, accounting and auditing specialists might raise a range of questions about any such structure, but the key here is that if higher-level white-collar crimes of the type that occurred in the Enron and Andersen case are to be prevented, policies that address fundamental conflicts of interest must be adopted.

The Sarbanes-Oxley Act includes other provisions directed toward producing accurate financial statements. The CEO of every company that is subject to the Sarbanes-Oxley Act must certify in writing that the company's financial disclosures comply with the law and actually reveal the company's financial status, without any misrepresentations. More specifically, section 906 requires that CEOs and CFOs certify their company's financial reports periodically to comply with sections 13(a) and 15(d) of the Securities Exchange Act of 1934 and "fairly [present], in all material respects, the financial condition and results of operations of the issuer" (18 U.S.C. 1350, 2002). If a knowing violation occurs, the punishment can be as severe as 10 years in prison and a $1 million fine (18 U.S.C. 1350, 2002). The White Collar Crime Penalty Enhancement Act is the part of the larger Sarbanes-Oxley Bill that spells out punishments attached to violations of the bill's provisions (Katz, 2007). These provisions have been controversial.

The Sarbanes-Oxley Act also instructs corporations on which records need to be stored and for how long (Salinger, 2007). This legislation states that all business records must be saved for at least 5 years. In addition, the act includes specific prohibitions on the destruction of such records. The consequences for not complying include fines and/or imprisonment.

Overall, the Sarbanes-Oxley Act raised financial standards in the areas of corporate governance, securities analysis, and the performance of audit work. The ultimate goal of the act was to restore public confidence by making corporations produce truly accurate financial statements, which would in turn encourage the public to continue to invest in corporate stocks.

## ■ The Backlash to the Sarbanes-Oxley Act

Several years after passage of the Sarbanes-Oxley Act, there was evidence of a favorable impact on corporate governance practices and on financial accountability for corporations; many corporations had to file financial restatements, but no cases equivalent to Enron and WorldCom had emerged (Armour & McCahery, 2006; Labaton, 2006; Nocera, 2005). But perhaps inevitably, once the immediate public and political fury in response to the corporate scandals had died down somewhat, a significant backlash against the Sarbanes-Oxley Act and some of its specific provisions began to emerge. Some commentators suggested that the haste with which the Sarbanes-Oxley Act was adopted led to the inclusion of deficient corporate governance provisions (Romano, 2005). Representatives of corporations and businesses covered by the Sarbanes-Oxley Act claimed that its provisions are complex and difficult to understand (Lerner & Yahya, 2007), making it difficult for CEOs to know when they are misrepresenting financial statements. The costs of compliance can total millions of dollars for larger corporations.

Altogether, the Sarbanes-Oxley Act has been attacked as an unwarranted violation of the core laissez-faire principle of capitalism, and a compromising of the free hand of the marketplace that is detrimental on many grounds. Congress member Ron Paul was among the critics who claimed that the Sarbanes-Oxley Act was an unnecessary and costly governmental interference with corporations that put American corporations at a competitive weakness with foreign corporations, thus driving potential business out of the United States. In many other countries, corporations are not required to devote time and resources to complying with the kinds of regulations spelled out in the Sarbanes-Oxley Act and would seem to have a competitive advantage. Furthermore, these foreign corporations may feel more inclined to exclude the United States in business transactions and instead conduct business with corporations in other countries that have less strict policies concerning corporations' operations.

## ■ Department of Justice Policy on Corporate Crime Cases

During the period leading up to the Enron case (and the related cases), the U. S. Department of Justice adopted a series of new policy initiatives for addressing corporate crime. On June 16, 1999, during the Clinton administration, the U.S. Deputy Attorney General Eric Holder issued a formal memo, which became known as the "Holder memo," to prosecutors in relation to prosecuting corporations (Holder was sworn in as Attorney General of the United States in 2009, following his appointment to that position by President Barack Obama). The memo identified nine criteria for prosecutors to consider when deciding whether to prosecute corporations, including the seriousness and the nature of the offense, the pervasiveness of the wrongdoing, and the corporation's history of similar conduct (Bohrer & Trencher, 2007, p. 1484). But the issues raised in this memo did not receive much attention until after the Enron case and the other cases of massive corporate misrepresentation of finances surfaced in 2001–2002.

On January 20, 2003, Deputy Attorney General Larry Thompson issued a revision of the Holder memo, entitled "Principles of Federal Prosecution of Business Organizations" (otherwise known as the "Thompson memo") (Bohrer & Trencher, 2007). This memo called for prosecutors considering the pursuit of an indictment of a corporation to take into account whether the corporation had a compliance program, as well as the adequacy of any such program; it authorized the use of pretrial diversion in corporate crime cases; and it called for careful consideration of the authenticity of any cooperation offered by the corporation in connection with allegations of criminal conduct. Controversially, it allowed prosecutors to pressure corporations to waive their defense privileges and to refrain from paying legal fees for corporate employees under investigation in the criminal matter (Arkin, Pope, & Prinz, 2008). Furthermore, unlike the Holder memo, this memo was characterized as binding on federal prosecutors.

Subsequently, Deputy Attorney General Paul McNulty released another set of guidelines (the "McNulty memo") on December 12, 2006. This memo imposed a requirement that prosecutors obtain authorization from the Deputy Attorney General before seeking privileged information waivers from corporations, and it made some distinctions between different levels of privileged information (Arkin,

Pope, & Prinz, 2008; Bohrer & Trencher, 2007). But the Thompson and McNulty memos taken together were harshly criticized by the corporate defense bar, corporate spokespeople, former government officials, and current political office holders with pro-business biases. These criticisms led to various legislative and judicial responses that compelled the Department of Justice—via Deputy Attorney General Mark Filip—to issue some new guidelines in August 2008, claiming that federal prosecutors would no longer require corporate waivers of privileges or hold against a corporation its payment of employees' legal fees. Whether these new guidelines resolved the concerns of critics of the earlier memos remained to be seen. Some commentators argued that these guidelines continued to impose substantial pressure on corporations to provide evidence against their own employees and to waive privileges under various circumstances (Arkin, Pope, & Prinz, 2008). Prohibiting the payment of employees' legal fees had already been struck down in a judicial ruling.

The new approach to corporate crime pursued by federal prosecutors during the post-Enron period has also been characterized by a proliferation of deferred prosecution agreements and nonprosecution agreements (Spivack & Raman, 2008). In essence, if the corporation being investigated for possible criminal conduct signs off on a set of conditions stipulated by the prosecutor, then the prosecutor will not proceed with a criminal case against the corporation. This prosecutorial trend, however, has been criticized from two opposing directions. On the one hand, some argue that corporations are being pressured to waive their rights or face what could be dire consequences in a criminal prosecution. Accordingly, these agreements have been characterized as a form of prosecutorial bullying. On the other hand, some argue that such agreements allow culpable corporations to avoid the full consequences of their criminal actions. In this view, the proliferation of deferred prosecution agreements reflects a prosecutorial retreat from aggressively pursuing criminal cases against corporate offenders.

## ■ The Worst Economic Crisis Since the Great Depression and White-Collar Crime

For those hoping that the series of corporate scandals beginning with the Enron case in 2001, and the strong public and political reaction to these scandals, would be the last major white-collar crime wave for some time, the events of 2007–2008 and beyond were profoundly disheartening. In 2008, the American economy and the world of high finance was in a state of crisis, with billions lost through the subprime mortgage market collapse, millions of homeowners in foreclosure or threatened by such, major investment banks destroyed by massive losses, oil prices spiraling upward, and a recession in effect, or at least pending (Morris, 2008; Sloan, 2008). In the fall of 2008, following the collapse of Lehman Brothers, the bailout of the insurance giant AIG, the government takeover of mortgage giants Freddie Mac and Fannie Mae, the passage by a divided Congress of a contentious $750 billion "bailout" bill, and dramatic drops in the stock market, the financial crisis was widely characterized as the most serious since the Great Depression of the 1930s (Cresswell & White, 2008).

This crisis occurred in the wake of several decades of high finance playing a central role in the production of vast wealth—with a disproportionate distribution to a relatively small population at the center of this system. Deregulation, globalization, and technological innovation were three key factors in all of these developments (Lahart, 2008). Some dimensions of this crisis reflected global economic forces and inevitable economic cycles. But in what one commentator has characterized as a "Tinker Bell financial market," many people expected their house prices to keep rising, borrowed vast sums of money, and bought securities they did not understand (Sloan, 2008, p. 80). Fraudulent representations of various kinds played a significant role in all of this.

A criminogenic structure at the heart of high finances generates immense financial rewards for those who produce and promote investments and financial instruments on the rise; other parties very disproportionately pay the price when these investments decline or collapse. According to Sloan (2008), in the world of increasingly unregulated high finance,

> If you take big, even reckless, bets and win, you have a great year and you get a great bonus—or in the case of hedge funds, 20% of the profits. If you lose money the following year, you lose your investors' money rather than your own—and you don't have to give back last year's bonus. Heads you win; tails, you lose someone else's money. (p. 82)

Tens of millions of investors, taxpayers, homeowners, and consumers were suffering during the course of the current economic crisis period, at least in part as victims of misrepresentations and manipulations in the world of high finance—as victims, in other words, of white-collar crime.

The massive financial crisis spiraling out of control in late 2008 was in important ways precipitated by the collapse of the subprime mortgage loan market (Morris, 2008; Sloan, 2008). Fraud on many different levels occurred. Banks and other lenders have often taken advantage of generally unsophisticated borrowers with modest incomes who need to pay their bills or want to buy a house (Moss, 2004). Lenders discovered that "subprime" loans—or loans to financially marginal people—could be very profitable. If not entirely new, the pursuit of the subprime mortgage market ramped up dramatically during the early years of the 21st century. Mortgage borrowers often discovered they were misled on escalating interest rates and fees and ended up with monthly payments they could not afford. In some cases, these borrowers attempted to consolidate their debts with new, high-interest mortgages (Moss, 2004). Many elderly homeowners were persuaded to borrow against their homes (sometimes with "reverse mortgages"), ultimately being saddled with high fees and debts they could not pay (Duhigg,

2008a). Some of these people lost their homes. Many of those who went into foreclosure then found themselves charged with exorbitant fees in that process (Morgenson, 2007). Misrepresentations on many different levels occurred.

In *Confessions of a Subprime Lender*, Richard Bitner (2008) writes of his disenchantment with a business that became progressively more greedy and fraudulent. In his book, he exposes how unscrupulous brokers tricked both lenders and gullible borrowers and turned unqualified applicants for mortgage loans into "qualified borrowers" by fraudulent misrepresentations. In his estimation, a staggering three out of four subprime mortgage loans were fraudulent. The blame for this fiasco, which he estimates will eventually result in losses in the trillions, is widespread. Beyond borrowers who made fraudulent representations, blame can be attributed to Wall Street investment banks who transformed tens of thousands of these mortgages into complex securities that were sold to investors who often did not understand the risks built into them; the credit rating agencies such as Moody's and Standard & Poor that failed to rate these securities properly; and the Federal Reserve, consumers, retail lenders, homebuilders, and realtors.

Specific crimes involved in subprime lending practices include wire and mail fraud, securities fraud, bank fraud, and violations of the Continuing Financial Crimes Enterprise Act (Seltzer & Ryan, 2008). By the middle of 2008, criminal investigations of the mortgage industry were intensifying (Browning, 2008). During the fall of 2008, the main focus was on restoring confidence in the financial system rather than prosecuting criminal wrongdoing, but it seemed likely that eventually more attention would be directed toward the criminal cases.

In 2003 and 2004, Freddie Mac and Fannie Mae—as the two huge mortgage giants are known—were investigated for various forms of accounting fraud (O'Brien & Lee, 2004). These entities have traditionally generated billions of dollars of mortgages from commercial banks to enable them to make further loans; accordingly, they have played a central role in American home ownership. Their top executives are extremely well compensated—with salaries exceeding $20 million a year in the case of Fannie Mae (Duhigg, 2008b). In 2008, Franklin Raines and other former Fannie Mae executives were required to donate $2 million to charity (and give up worthless stock options) to settle charges relating to violations of accounting rules (Hagerty, 2008). But Raines was allowed to retain some $90 million for 5 years as CEO of Fannie Mae.

In September 2008, in a costly bailout, the U.S. government took over Freddie Mac and Fannie Mae in the wake of the subprime mortgage market collapse (Labaton & Andrews, 2008). They had been pressured by both Wall Street and Congress to buy up hundreds of billions of dollars of mortgage loans to risky borrowers. Evidence surfaced in August 2008 that the CEO of Freddie Mac rejected internal warnings about these risks (Duhigg, 2008b). In 2008, it was reported that the accounting fraud problems were not successfully addressed, and Freddie Mac greatly overstated the size of its capital base (Morgenson & Duhigg, 2008). So once again, major accounting fraud within these entities was being investigated.

The subprime mortgage lending frauds have ultimately been one of the root causes of the massive financial crisis of 2008 and beyond, with countless victims. The victims were disproportionately poor people, minorities, and the elderly (Wright, 2008). The victims include the millions of homeowners facing foreclosure (at least some of whom are themselves blameworthy), communities and neighborhoods with high rates of foreclosures, investors and retirement account holders, laid-off employees, and taxpayers (for bailout costs) (Bajaj, 2008; Morris, 2008). Altogether, the subprime mortgage crisis may come to be recognized as a central part of the monumental white-collar crime wave of this era.

## ■ Public Policy Responses to the Financier Crime at the Center of the Economic Crisis

The financier crimes so central to the economic crisis at the end of the first decade of the 21st century have certain parallels with the corporate scandals at the outset of that decade. In both cases, the pursuit of exorbitant compensation in some form was one fundamental criminogenic factor driving the massive financial misrepresentations. The absence of appropriate regulation and public policy to ensure transparency and accountability in the realm of major corporations and investment banks was, importantly, a reflection of the immense influence of major corporations and Wall Street on public policy. It is also important to recognize that public policy responses to crime in the corporate realm and within the world of Wall Street confront a fundamentally different challenge from that directed toward conventional crime, in that productive activity is deeply intertwined with destructive and harmful activity. The enduring challenge is to formulate public policy that addresses the harmful and destructive dimensions without fundamentally impacting the productive activities. Much is at stake in getting this balance right, with the promotion of renewed public trust in corporations and the financial institutions essential to the economic well-being of society.

## ■ Conclusion

Going forward in the 21st century, crime is likely to maintain its status as a major focus of public policy. Although at least some recent public policies are regarded as having achieved a measure of success in addressing conventional forms of crime (such as some novel policing policies), other policies are widely regarded as relative failures, and in some cases sources of more harm than they alleviate. In particular, the predominantly retributive and punitive thrust of public policy toward conventional crime in the recent era has been challenged. It remains to be seen, for example, whether the restorative justice alternative to the retributive and punitive model will gain momentum. It also remains to be seen whether the major economic crisis, along with other emerging social conditions, will lead to a significant resurgence of conventional forms of crime, along with attendant increases in public concern. However, it is a core premise of this chapter that as a society we are long overdue for much broader public policy initiatives addressing white-collar crime, broadly defined.

Many factors, from relative age distributions within the population to entirely new structures of opportunity through technological breakthroughs, suggest that the scope of the white-collar crime problem may well expand rather than contract as we move forward. In the first decade of the 21st century, in addition to significant new cases for every form of white-collar crime, we have also witnessed two major white-collar crime waves, the so-called corporate scandals at the outset of the decade and the Wall Street subprime mortgage–related meltdown toward the end. Each of these waves had a broad range of devastating consequences, including massive losses of jobs, homes, and savings. Although the crimes involved were complex, were intertwined with legitimate, productive activity, and involved many different parties on many different levels, financier elites played a central role. In this chapter, we have attempted to place white-collar crime and public policy issues within the broader context of enduring crime and public policy issues, and we have focused on the two major American white-collar crime waves of the new century and the public policy response to them.

## ■ Discussion Questions

1. Why has white-collar crime received so little attention from policy makers when compared to street crime and drug-related crime?

2. Discuss the recent developments that have generated widespread concern for white-collar crime in the United States.

3. What is financier crime? Discuss the challenges such crimes present for criminal justice policy makers.

4. Discuss the origins and implications of the Sarbanes-Oxley Act. Does the act represent an effective response to financier crime?

5. What additional steps should policy makers in the Department of Justice take to prevent the recurrence of the white-collar crimes and corporate scandals of the past decade?

## ■ References

Arkin, S. S., Pope, P. B., & Prinz, B. N. (2008). New DOJ guidelines on prosecuting businesses. *Business Crime Bulletin, 16*(1), 5–7.

Armour, J., & McCahery, J. A. (2006). *After Enron: Improving corporate law and modernising securities regulation in Europe and the US*. Portland, OR: Hart Publishing.

Bajaj, V. (2008, March 27). Inquiry assails accounting firm on lender's fall. *New York Times,* p. A1.

Beckett, K. (1997). *Making crime pay: Law and order in contemporary American politics.* New York: Oxford University Press.

Berenson, A. (2002, February 24). Three-decade-old echoes, awakened by Enron. *New York Times,* p. C1.

Bitner, R. (2008). *Confessions of a subprime lender.* New York: John Wiley.

Bohrer, B. A., & Trencher, B. L. (2007). Prosecution deferred: Exploring the unintended consequences and future of corporate cooperation. *American Criminal Law Review, 44,* 1481–1502.

Browning, L. (2008, June 6). Inquiry into a guarded world. *New York Times,* p. C1.

Clear, T. R. (2007). *Imprisoning communities: How mass incarceration makes disadvantaged neighborhoods worse.* New York: Oxford University Press.

Clear, T. R, & Frost, N. A. (2007). Informing public policy. *Criminology & Public Policy, 6,* 633–640.

Cresswell, J., & White, B. (2008, September 28). Wall Street R.I.P. *New York Times* p. Week 1.

Cullen, F. T., Hartman, G., & Jonson, C. L. (2009). Bad guys: Why the public supports punishing white-collar offenders. *Crime, Law & Social Change, 51,* 31–44.

Duhigg, C. (2008a, May 20). Bilking the elderly, with a corporate assist. *New York Times,* p. A1.

Duhigg, C. (2008b, August 5). At Freddie Mac, chief discarded worry signs. *New York Times,* p. A1.

Elikann, P. T. (1996). *The tough-on-crime myth: Real solutions to cut crime.* New York: Plenum Press.

Ferrell, J., Hayward, K., & Young, J. (2008) *Cultural criminology: An invitation.* Los Angeles: Sage.

Fox, L. (2003). *Enron: The rise and fall.* Hoboken, NJ: Wiley.

Friedrichs, D. O. (2007a). *Trusted criminals: White collar crime in contemporary society* (3rd ed.). Belmont, CA: Thomson Wadsworth.

Friedrichs, D. O. (2007b). Enron. In J. Gerber & E. Jensen (Eds.), *Encylopedia of white-collar crime* (pp. 85–89). Westport, CT: Greenwood Press.

Geis, G. (1996). A base on balls for white-collar criminals. In D. Shichor & D. Sechrest (Eds.). *Three strikes and you're out: Vengeance and public policy* (pp. 244–264). Thousand Oaks, CA: Sage.

Geis, G. (2007). *White-collar and corporate crime.* Upper Saddle River, NJ: Pearson.

Gest, T. (2001). *Crime and politics: Big government's erratic campaign for law & order.* New York: Oxford University Press.

Gordon, D. R. (1994). *The return of the dangerous classes.* New York: Norton.

Hagerty, F. R. (2008, April 19). Fannie Mae ex-officials settle. *Wall Street Journal,* p. 3.

Holtfreter, K., Van Slyke, S., Bratton, J., & Gertz, M. (2008). Public perceptions of white-collar crime and punishment. *Journal of Criminal Justice, 36*, 50–60.

Karmen, A. (2000). *New York murder mystery: The true story behind the crime crash of the 1990s.* New York: New York University Press.

Katz, R. (2007). United States. In L. Salinger (Ed.). *Encyclopedia of white-collar crime and corporate crime* (2nd ed., pp. 838–841). Thousand Oaks, CA: Sage.

Labaton, S. (2006, January 5). Crime and consequences still weigh on corporate world. *New York Times,* p. C1.

Labaton, S., & Andrews, E. L. (2008, September 8). Mortgage giant taken over by U.S.: A costly bailout. *New York Times,* p. A1.

Lahart, J. (2008, April 28). Has the financial industry's heyday come and gone? *Wall Street Journal,* p. A2.

Lerner, C. S. & Yahya, M. A. (2007). "Left behind" after Sarbanes-Oxley. *American Criminal Law Review, 44*, 1383–1405.

Macionis, J. (2005). *Social problems* (2nd ed.). Upper Saddle River, NJ: Pearson.

Mauer, M., & Chesney-Lind, M. (2002). *Invisible punishment: The collateral consequences of mass imprisonment.* New York: The New Press.

McLean, B., & Elkind, P. (2003). *Smartest guys in the room: The amazing rise and scandalous fall of Enron.* New York: Penguin.

Merriam-Webster, Inc. (2007). *Merriam-Webster's Collegiate Dictionary* (11th ed.). Springfield, MA: Author.

Morgenson, G. (2007, November 6). Dubious fees hit borrowers in foreclosures. *New York Times,* p. A1.

Morgenson, G., & Duhigg, C. (2008, September 7). Mortgage giant overstates size of capital base. *New York Times,* A1.

Morris, C. (2008). *The trillion dollar meltdown.* New York: Public Affairs.

Moss, M. (2004, October 10). Erase debt now. Close our home later. *New York Times,* p. C1.

Nocera, J. (2005, December 3). For all its cost, Sarbanes law is working. *New York Times,* p. C1.

O'Brien, T., & Lee, J. (2004, October 3). A seismic shift under the house of Fannie Mae. *New York Times,* p. 3/1.

Phillips, N. P., & Frost, N. A. (2007). Talking heads: Crime reporting in the media. A Paper Presented at the Academy of Criminal Justice Sciences Meeting, Seattle (April).

Piquero, N. L., & Weisburd, D. (2009). Developmental trajectories of white-collar crime. In S. Simpson & D. Weisburd (Eds.), *The criminology of white-collar crime* (pp. 153–171). New York: Springer.

Pontell, H. N., & Geis, G. (Eds.). (2007). *International handbook of white-collar and corporate crime.* New York: Springer.

Recine, T. S. (2002). Examination of the white-collar crime enhancements in the Sarbanes-Oxley Act. *American Criminal Law Review, 39*, 1536–1570.

Romano, R. (2005). The Sarbanes-Oxley Act and the making of quack corporate governance. *Yale Law Review, 114,* 1521–1603.

Salinger, L. M. (2007). Obstruction of justice. In L. M. Salinger (Ed.), *Encyclopedia of white-collar and corporate crime* (2nd ed., pp. 573–575). Thousand Oaks, CA: Sage.

Seltzer, M. D., & Ryan, D. M. (2008, April). Potential criminal liability for subprime lending practices. *Business Crime Bulletin,* pp. 3–4.

Shover, N., & Cullen, F. T. (2008). Studying and teaching white-collar crime: Populist and patrician perspectives. *Journal of Criminal Justice Education, 19,* 155–174.

Sloan, A. (2008, April). On the brink of disaster. *Fortune,* pp. 78–84.

Spivack, P., & Raman, S. (2008). Regulating the 'new regulators': Current trends in deferred prosecution agreements. *American Criminal Law Review, 45,* 159–193.

Squires, S. E., Smith, C. J., McDougall, L., & Yeack, W. R. (2003). *Inside Arthur Andersen.* Upper Saddle River, NJ: Prentice Hall.

Sutherland, E. H. (1940). White collar criminality. *American Sociological Review, 5,* 1–12.

Sutherland, E. H. (1949a). *White collar crime.* New York: Holt, Rinehart & Winston.

Sutherland, E. H. (1949b). The white collar criminal. In V. C. Branham & S. B. Kutash (Eds.), *Encyclopedia of criminology* (pp. 511–515). New York: Philosophical Library.

Thompson, A. C. (2008). *Releasing prisoners, redeeming communities.* New York: New York University Press.

Unnever, J. D., Benson, M. L., & Cullen, F. T. (2008). Public support for getting tough on corporate crime: Racial and political divides. *Journal of Research in Crime and Delinquency, 45,* 163–190.

Waller, I. (2006). *Less law, more order: The truth about reducing crime.* Westport, CT: Praeger.

Worrall, J. L. (2008). *Crime control in America: What works?* Boston: Pearson.

Wright, K. (2008, July 14). The subprime swindle. *The Nation,* pp. 11–20.

Zimring, F. E. (2007). *The great American crime decline.* New York: Oxford University Press.

# Criminal Justice Policy and Transnational Crime: The Case of Anti-Human Trafficking Policy

CHAPTER 14

Barbara Ann Stolz[1]

## ■ Introduction

In the last decade of the 20th century, globalization of transportation and communication supported the extension of criminal networks and activities across national boundaries, thereby presenting new challenges to law enforcement. One such transnational challenge was the trafficking of human beings. Although not a new problem (Bales, 1999; Barry, 1979; Outshoorn, 2004), during the 1990s human trafficking became a policy concern of many governments of the world, including the United States, and international organizations. This concern spread with the recognition of the links between trafficking, transnational crime, and government corruption in emerging democracies and the threat these factors posed to national security. What was new in the 1990s was not the behavior, but the governmental recognition that human trafficking was a problem that governments needed to address.

In the United States, criminal justice policy was traditionally the responsibility of local and state governments; the federal government played only a limited role (e.g., investigating and prosecuting crimes on the high seas, treason, or offenses committed on Indian reservations). With congressional enactment of legislation to combat a range of crimes, including narcotics control, organized crime, and white-collar crimes, usually involving interstate commerce, the role of the federal government expanded during the 20th century (Oliver & Marion, 2008; Stolz, 2002a). Federal criminal justice agencies were established and grew to accommodate the enforcement of federal criminal justice policy. With the emergence of new transnational crimes, such as human trafficking, the need arose for additional federal laws to respond to these acts.

Prior to 2001, human trafficking cases were prosecuted under a number of federal criminal statutes, including involuntary servitude statutes, the Mann Act, and labor laws concerning workplace conditions and compensation (U.S. Department of Justice, 2006, p. 17). In response to growing concerns both within the United States and internationally, the U.S. Congress passed, and President Bill Clinton signed into law, the Trafficking Victims Protection Act of 2000. In accordance with the act, the legislation was reauthorized in 2003, 2005, and again in 2008 (Stolz, 2005, pp. 408–409). With multiple reauthorizations within an 8-year period, the development, implementation, and reauthorizations of the criminal justice provisions of U.S. trafficking in persons legislation provides an excellent case study of the complexities of the criminal justice policy-making process.

Whether the policy area is education, health care, labor law, or the focus of this article—criminal justice—the policy process consists of several stages: agenda setting, legislation, implementation, and reauthorization (Stolz, 2002b). Each stage of the process provides points of access to decision making where policy can be influenced. That is, the process provides more than one bite of the apple. To understand a particular criminal justice policy, it is necessary to analyze the full spectrum of decision points in the policy-making process.

This chapter lays a framework for analyzing the various stages of the U.S. federal criminal justice policy-making process. The framework is then applied to a case study of U.S. antitrafficking policy. Although a variety of factors that may affect policy outputs can be examined employing the framework, this article uses it to analyze the role that nongovernmental interest groups and the bureaucracy played in the development of U.S. human trafficking policy. Since the early days of the United States, interest groups have participated as informal actors in U.S. politics, "filling in the gaps" in the formal structures and processes and seeking to influence the policy-making process to serve the ends of the groups (Stolz, 2002b, pp. 52–53). In addition, the bureaucracy contributes to policy making by interpreting legislation through the implementation process and may also seek to influence policy through the legislative process, especially reauthorization. The analysis of U.S. human trafficking policy using this framework helps to shed light on how this policy evolved through the stages of the criminal justice policy-making process. In so doing, it enhances general understanding of how criminal laws and criminal justice policy are made, including where and when such policy can be influenced.

## ■ The Framework

U.S. policy making, including criminal justice policy making, is a continuous and iterative process. The political science and public policy literature has, however, tended to focus on the legislative process (Greenwald, 1977; Key, 1964; Truman, 1951). While a critical stage in the policy-making process, studying only the legislative process and resulting policy outputs does not explain, for example, why the same legislatively created program may be implemented differently; why some problems come to be defined as issues needing governmental attention and others do not; or why a problem is not viewed as a policy concern at one point in time but is at another. To answer these questions, it is necessary to identify all decision points in the policy-making process where participants may influence the content of that policy. **Figure 14–1** depicts the four fundamental stages of the criminal justice policy-making process—agenda setting, legislation, implementation, and reauthorization (Stolz, 2007b, p. 171).

At each stage, participants may access the process and attempt to influence policy outputs. Each stage may involve a number of substages that are points of access.

### Agenda-Setting Stage

Why do certain problems become matters of public policy and others do not? Considering the agenda-setting stage as part of the policy-making process emphasizes the ability of interest groups to draw attention to an issue, thereby precipitating legislative action, or, at times, preventing issues from being considered by legislatures (Stolz, 2002b, p. 60). Kingdon (1984, pp. 1, 53) identified this predecision stage in the policy areas of health and transportation policy. He found interest groups advocating certain proposals or blocking issues to keep them off the public policy agenda (Kingdon, 1984, p. 52). A study of federal efforts to enact domestic violence legislation suggested the potential for such a stage in

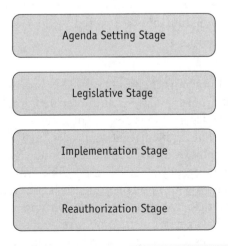

**Figure 14–1** Stages of the Criminal Justice Policy Making Process

the development of criminal justice policy (Stolz, 1999). During this stage, interest groups seek to educate policy makers about the problem and the need for them to address it—that is, they reconceptualize the problem from a concern to an issue on the public policy agenda. To communicate this message, groups use a variety of techniques and venues, including letters or visits to policy makers; public forums, such as conferences or public hearings (this venue overlaps with the legislative process described next); and the media.

### Legislative Stage

The description of the legislative stage of the policy-making process has received much attention, generally, although the relevant criminal justice literature is somewhat sparse (e.g., Baker & Meyers, 1980; Fairchild, 1981; Marion 1994, 1995, & 1997; Melone & Slatger, 1983; Oliver & Marion, 2008; Roby, 1969; Stolz, 1999, 2002a, 2002b, 2005, 2007a, & 2007b). This stage can be broken down into substages—drafting legislation, public hearings, debate and compromise, congressional passing of legislation, and presidential signing into law. Each substage provides an access point or points for participants. For example, interest group representatives or agency officials may provide proposals for draft legislation or discuss provisions with congressional staff. Usually congressional hearings are held to discuss draft legislation; sometimes several bills are introduced. Occasionally, hearings about a problem may be held for informational or educational purposes, with legislation subsequently introduced. Draft legislation is debated within congressional subcommittees, committees, and the floor of both the House and Senate. Groups may seek to influence legislative provisions by testifying at hearings, meeting with members and/or staff to discuss provisions of interest, or by engaging in letter-writing or other communication campaigns to communicate their positions to members and staff. Agency officials may testify at public hearings or meet informally with members and/or staff. Most legislation dies on the vine, but some measures are passed by the Congress and signed into law by the President.

### Implementation Stage

Policies and programs must be implemented, a process that in and of itself may require the creation of regulations and interpretation by bureaucrats (Lipsky, 1980). Once federal legislation is enacted, it must be implemented by one or more federal agencies. This implementation process may involve direction from the White House through executive orders, the promulgation of regulations, and program planning by the agency/agencies responsible for carrying out the policy. In addition, those charged with implementing the legislation may confront challenges, including unanticipated gaps in legislation and implementation issues that must be addressed to carry out the policy or program. In doing so, agency officials influence policies, although this role is often not acknowledged. The implementation process also provides a point of access for interest groups seeking to influence policy. For example, they may seek to influence the regulations or design of the approach taken in the programs that carry out the policy. Thus, implementing policy is not a neutral process; rather, it involves decisions that affect the final policy output as the legislative provisions are interpreted, fleshed out, and put into practice.

### Reauthorization Stage

Depending on provisions in the law, congressionally established policies and programs may be subject to legislative reauthorization and amendment. In recent years, Congress has frequently required the reauthorization of policies and programs at regular intervals. During the reauthorization process, interest groups have the opportunity to seek changes in the law, providing another point of access. Reauthorization also provides the opportunity for bureaucracies to request and obtain changes in the law to fill in gaps, clarify existing provisions, or add new provisions.

To analyze the role of participants in the policy-making process, at each stage of the policy process it is necessary to ask the following questions:

- Who sought to influence the policy?
- What were the goals and objectives of each participant?
- What techniques were used to influence policy outcomes?
- What were the policy outcomes?

## ■ Methods

Two primary methods were used to gather the information analyzed: (1) the case method and (2) a content analysis of the 2000, 2003, 2005, and 2008 anti-trafficking legislation, related executive orders, and other implementation documents. The case study method has been used by social scientists, particularly political scientists, to examine political decision making in urban settings or legislatures (e.g., Banfield, 1961; Banfield & Meyerson, 1955; Dahl, 1961; Redman, 1973). The content analysis of legislation and related executive branch documents provides evidence of policy outputs that can be analyzed in terms of group interests and bureaucratic policy making.

Frequently, researchers who study interest groups develop case studies using legislative documents, newspaper accounts, and interviews to identify and describe who participated in a particular policy decision. By expanding the case study to the pre-legislative/agenda-setting stage, through interviews and Web searches, groups participating in agenda setting can also be identified (Stolz, 2002b, pp. 59–60.) In doing so, additional roles of these participants, including raising awareness of problems, can be identified. Similarly, further extending the case study to the implementation and reauthorization stages can identify interest group participation, as well as the techniques used by groups to influence policy outputs as they are carried out or amended. It also provides the means to examine the implementing agency's contributions to policy outputs. The participants and their roles during the pre- and post-legislative stages can be compared and contrasted with those involved in the legislative process to develop a fuller understanding of who affected the final policy, when, and how.

The case study method was used to determine which groups helped to set the U.S. human trafficking policy agenda and which influenced the policy outcomes. Through a Google search on human trafficking, analysis of congressional hearings (Congressional Information Services, 2000, 2003), and a review of congressional hearings and written documents, interviewees were identified. Additional interviewees, some who were not as publicly acknowledged as those who appeared at hearings, were identified through a snowball technique—that is, initial interviewees were asked to identify other participants who were then contacted for interviews. These individuals included executive branch and congressional staff, as well as representatives of various interest groups. The different roles played and policy positions articulated by diverse groups at the various stages in the policy-making process were compared and contrasted. In addition, the interviews with policy makers and representatives of interest groups provided information on the variety of techniques used by the groups at different access points in the process to try to influence policy outputs. The steps taken by agencies to implement policies were also identified through this process. An effort was made to corroborate interview results with written documentation to validate the description of the human trafficking policy-making process.

A content analysis of proposed and enacted U.S. trafficking legislation and related executive branch documents was undertaken. The policies in these documents were compared with the positions of the various groups that sought to influence the legislation and implementation of the policy. The language of specific provisions was tracked across documents to ascertain changes in wording in order to identify possible changes in policy, and to determine who influenced the provisions successfully.

## ■ U.S. Anti–Human Trafficking Legislation

### Chronology of U.S. Anti–Human Trafficking Legislation

U.S. anti–human trafficking policy evolved between the late 1990s and the first 8 years of the 21st century. **Figure 14–2** presents the chronology of that evolution.

Human trafficking emerged as a policy issue almost a decade before the enactment of U.S. anti–human trafficking legislation in 2000, as various groups became aware of the problem. Interviews, speeches given by policy makers, and documents indicated that during the 1990s, White House and executive branch officials and staff in the Clinton administration, as well as First Lady Hillary Clin-

Figure 14–2 U.S. Human Trafficking Legislation Chronology

ton and her staff, became aware of human trafficking. That awareness developed through a number of sources (Stolz, 2005, p. 413).

One venue used to raise this awareness was international and national conferences. For example, in 1995, in Beijing, China, during remarks before the United Nations Fourth World Conference on Women, First Lady Hillary Clinton identified the selling of women and girls into the slavery of prostitution as one of the adverse conditions experienced by women (Clinton, 1995). While not specifically employing the word "trafficking," the speech, which had been vetted with the National Security Council and other experts, substantively marked the appearance of human trafficking "on the radar screen" of the Clinton administration. In April 1997, nongovernmental organizations and interest groups concerned about human trafficking interacted with White House and executive branch policy makers at a conference on criminal justice issues and the international exploitation of women and children, sponsored by the U.S. Departments of State and Justice and the Federal Judicial Center. During the conference, the Global Survival Network's[2] Codirector Gillian Caldwell presented a documentary video of its 2-year undercover investigation of sex trafficking in Russia— *Bought and Sold*. In July of 1997, the First Lady, with her chief of staff Melanne Verveer, attended a conference sponsored by the global partnership, Vital Voices,[3] in Vienna, Austria. During a comment period, Oksana Horbunova of La Strada[4] in Ukraine asked for international help for women being trafficked from Ukraine (Stolz, 2005, pp. 413–414). These conferences brought together academics, interest groups, bureaucrats, and elected officials, and provided the means to raise awareness of the problem through formal and informal interactions and discussions.

In addition to conferences, interest groups engaged in other types of activities. They helped prepare federal government-issued pamphlets on trafficking. Representatives of various groups met with the House and/or Senate members and their staffs to inform them about the trafficking problem. The groups usually met with policy makers who held general policy views similar to those of the group—representatives from conservative groups usually met with conservative policy makers and representatives from liberal groups usually met with liberal policy makers. At the same time, parallel efforts to educate policy makers and develop antitrafficking policies were taking place on the international level; for example, the United Nations developed a protocol on human trafficking. By the end of the 20th century, human trafficking had a place on the U.S. and international policy agendas, having been identified as a problem in need of a response by governments and international organizations.

The legislative policy-making stage was relatively short, beginning with hearings during the summer of 1999 and culminating with the passage of the 2000 Trafficking Victims Protection Act (TVPA) in the fall of 2000 (*Congressional Almanac*, 2000). Both the House and Senate conducted hearings. In June 1999, the Commission on Security and Cooperation in Europe (CSCE), chaired by Representative Chris Smith (Republican), held the first anti–human trafficking hearing, focusing on sex trafficking (CSCE, 1999). The Commission, however, did not have the authority to process legislation. In September, Smith held further hearings on sex trafficking in his role as the chair of the U.S. House International Relations Committee's Subcommittee on International Operations and Human Rights (1999). The Senate began consideration of anti–human trafficking legislation, with hearings held by the Committee on Foreign Relations, Subcommittee on Near Eastern and South Asian Affairs (U.S. Congress, Senate, 2000), in February and April 2000. These hearings were chaired by Senator Sam Brownback (Republican) with Ranking Minority Member Paul Wellstone (Democrat).

A review of congressional hearing records shows that diverse nongovernmental groups with different political perspectives participated in the formal legislative process that led to TVPA by testifying at human trafficking hearings held during 1999 and 2000 (CSCE, 1999; U.S. Congress, House, 1999; U.S. Congress, Senate, 2000). They included human rights, legal rights, religious, and feminist organizations, as well as service providers (e.g., nongovernmental organizations assisting refugees). The witnesses generally agreed that a three-pronged approach—prevention, protection, and prosecution—was needed. In addition, there was a consensus that the trafficked should be treated as victims even if they had engaged in criminal behavior, such as prostitution. Nongovernmental organization hearing witnesses, however, differed in their views as to how trafficking should be addressed in the initial legislation (Stolz, 2005). Specifically, some organization witnesses stated that the initial legislation should focus only on sex trafficking, which they believed to be the most serious form of trafficking; other witnesses testified that the legislation should include labor and sex trafficking, as they believed that trafficking included a range of types of unfree labor (U.S. Congress, House 1999, pp. 41–43; U.S. Congress, Senate, 2000, pp. 44–45).

Federal departments also sought to influence U.S. antitrafficking legislation during hearings. Government witnesses included officials from the Departments of State and Justice. Specifically, in April 2000 hearings before the Committee on Foreign Relations, Subcommittee on Near Eastern and South Asian Affairs (U.S. Congress, Senate, 2000, pp. 76–86), Justice Department witnesses described current prosecutorial efforts against human trafficking, expressed the need for legislation to strengthen prosecutorial tools, and stated the department's support for a special visa for victims of trafficking.

According to interest group representatives and congressional and executive branch staff interviewed, some nongovernmental organizations participated behind the scenes during the legislative process. Some groups provided an educational forum for policy makers but did not testify. Still other organizations used the press to communicate their views on trafficking (Stolz, 2007b, pp. 318–319). Agency officials also met informally with congressional staff to discuss provisions in the legislation.

The House and Senate each passed antitrafficking bills in July 2000. Because the bills differed, a conference committee was appointed to resolve the differences and develop a single bill. The conference committee report was agreed to by both the House (371 to 1) and Senate (95 to 0) in October. President Clinton signed the passed legislation into law on October 28, 2000 (Stolz, 2007a, pp. 178–179; 2007b, p. 319).

The implementation of the 2000 TVPA fell to the Bush administration, who identified the eradication of human trafficking as a priority. In February 2002, the President issued Executive Order 13257, establishing a cabinet-level interagency task force to monitor and combat trafficking. The task force was to direct the implementation of the act in a coordinated fashion throughout the federal government. Also in February 2002, the President issued National Security Presidential Directive 22 (NSPD-22). NSPD-22 identified human trafficking as an important national security issue and instructed federal agencies to strengthen their collective efforts, capabilities, and coordination to support the President's goal of abolishing human trafficking. NSPD-22 also asserted that the abolition of prostitution, the driving force behind sex trafficking, was integral to the abolition of human trafficking (U.S. Department of Justice, 2006, pp. 5–6).

Between 2001 and 2005, a number of congressional committees held hearings on trafficking in persons for the purposes of oversight and reauthorization, as the 2000 TVPA required the reauthorization of the statute every 2 years (U.S. Congress, House, 2001, 2002, 2003a, 2003b, 2004a, 2004b, 2005a, 2005b, 2005c,

2005d; U.S. Congress, Senate, 2002, 2003, 2004a, 2004b; CSCE, 2005). The majority of those testifying at these hearings focused on issues involving trafficking for sex (Stolz, 2007b, p. 319). On occasion, federal agency officials testified. For example, in 2007 during House Judiciary Committee hearings on the reauthorization of the trafficking legislation, representatives from the Departments of Justice (Rothenberg, 2007) and Homeland Security (Forman, 2007) reported on their respective agency's antitrafficking efforts.

The 2000 anti–human trafficking legislation has been reauthorized three times. In 2003, Congress passed the Trafficking Victims Protection Reauthorization Act (TVPRA); in 2005 it again reauthorized the legislation. Most recently, in December 2008, Congress passed the William Wilberforce Trafficking Victims Protection Reauthorization Act. In each instance, Congress authorized the appropriation of funds to support antitrafficking programs, as well as amendments of the legislation. After each reauthorization, the designated federal agency/agencies implemented the provisions.

### Participants: Interest Groups

Numerous interest groups have sought to influence U.S. anti–human trafficking policy. An analysis of the views of the groups suggests a clustering of these groups into two spheres of interest—the "antiprostitution" sphere and the "human trafficking" sphere (Stolz, 2007b). The clusters differ in their perspectives on what human trafficking is, what antitrafficking policies should emphasize, and how these policies should be implemented. **Figure 14–3** depicts the two cluster spheres. With regard to specific groups, although there is some evidence of migration by individual groups across clusters, primarily from the antiprostitution to the human trafficking sphere, the core group participants and the respective views on human trafficking have remained the same over time.

The two spheres differ in their views of whom to focus on as a trafficking victim. The antiprostitution sphere has focused on women and girls, as its concern is trafficking for sex. In contrast, the human trafficking sphere is concerned about any victim of unfree labor—man, woman, boy, or girl.

In regard to defining trafficking as a crime, the two spheres differ. The antiprostitution sphere conflates trafficking and prostitution. Whether the individual consented to committing illegal behavior or whether he or she was subject to "force, fraud, and coercion" is considered irrelevant to their status as a victim of human trafficking. In contrast, the human trafficking sphere views trafficking

**Figure 14–3** U.S. Human Trafficking Legislation Interest Groups

as a crime in and of itself. Accordingly, it considers "fraud, force, and coercion" to be a key element of the legal definition of human trafficking, as it is this language that distinguishes trafficking from related criminal behaviors, such as alien smuggling and prostitution.

In addition, to some extent the groups differ in their focus on immigrant and domestic victims of trafficking. The 2000 trafficking legislation primarily addressed issues related to immigrant victims of human trafficking and located the primary structures for overseeing the implementation of the programs enacted in the State Department. Although the act did not include crossing the U.S. border as part of the definition of trafficking in persons, the victims' assistance provisions were primarily directed toward non–U.S. victims. For example, the act established a special visa to allow trafficking victims to remain in the United States as long as they met certain conditions. These visas are available to victims of trafficking for sex or labor. This perspective generally reflects the views of the human trafficking sphere. In contrast, focusing on trafficking for sex and conflating sex trafficking with prostitution, the antiprostitution sphere has sought to direct federal attention to the problem of juvenile prostitution. Groups within this sphere consider juvenile prostitutes to be victims of human trafficking. This population includes not only immigrant but also domestic victims. By defining juvenile prostitution as human trafficking, the antiprostitution sphere broadened the concept of a human trafficking victim. They have sought an interpretation of existing human trafficking policies and advocate new policies that combat crimes against and provide assistance to this population.

These basic differences in the perspectives and goals of the groups within each of these spheres affected the types of policies each sought to promulgate.

### Bureaucrats as Participants

Federal agencies sought to influence antitrafficking policy during the legislative process. The Justice Department had primary responsibility for enforcing statutes used to combat human trafficking prior to 2000. In 2000 congressional testimony, a department representative enumerated the following limitations of existing laws: (1) They criminalized only a narrow range of trafficking acts (e.g., the sex trade); (2) they failed to reach all who profit by forced labor of persons (e.g., farm laborer contractors); and (3) they required that the defendant be shown to have used actual force, threat of force, or illegal coercion to enslave the victim,[5] thereby limiting the ability of federal law enforcement to reach more subtle forms of trafficking (U.S. Congress, Senate, 2000, pp. 77–78).

The implementation of the U.S. antitrafficking program is carried out by numerous federal agencies, including components of the Departments of State, Justice, Labor, Health and Human Services, and Homeland Security.[6] **Figure 14–4** depicts the key agencies and their respective responsibilities related to the investigation and prosecution of trafficking in persons crimes (U.S. Government

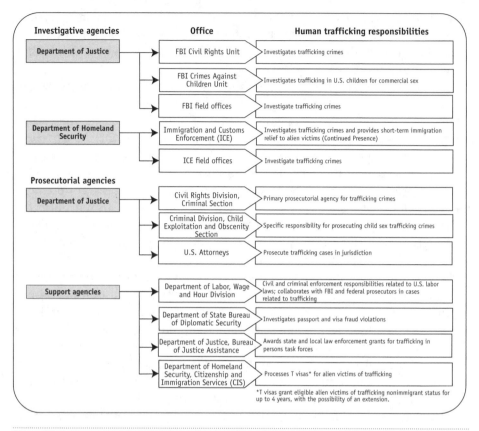

**Figure 14–4** Federal Agencies with Primary Responsibilities for Investigating and Prosecuting Trafficking in Persons Crimes
**Source:** Reproduced from "Human Trafficking: A Strategic Framework Could Help Enhance the Interagency Collaboration Needed to Effectively Combat Trafficking Crimes," U.S. Government Accountability Office Report to Congressional Requesters, July 2007.

Accountability Office, 2007, p. 10). These federal agencies developed programs and initiatives, and provided training to staff to implement their various responsibilities under the act.

Between 2001 and 2007, agency officials occasionally testified on the enforcement of antitrafficking criminal justice provisions during oversight and reauthorization hearings. For example, during 2007 House Judiciary Committee hearings on the reauthorization of the antitrafficking legislation, a representative of the Justice Department reported on the results of the multifaceted approach to trafficking, including the number of prosecutions, the number of victims assisted, and the steps taken to prevent trafficking. He stated that the Justice Department strongly supported the reauthorization of the antitrafficking act and expressed the department's desire to work with Congress on specific provisions in the draft legislation, although the provisions were not identified (Rothenberg, 2007).

## ■ Analysis: Stages in the Criminal Justice Policy-Making Process— The Case of Human Trafficking

Underlying the 2000 TVPA was a three-pronged approach to human trafficking—protection, prevention, and prosecution. Namely, the approach advanced a range of new protections and assistance for victims of trafficking who cooperate with law enforcement investigations, expanded U.S. activities internationally to prevent victims from being trafficked, and enhanced the penalties available to federal investigators and prosecutors pursing traffickers (U.S. Department of Justice, 2006, p. 17). Between 2001 and 2008, the federal agencies charged with enforcing the new law took steps to implement the provisions of the act and in doing so interpreted and fleshed out U.S. anti–human trafficking policies. Through the reauthorization process, provisions were added and some changes were made to the 2000 legislation in 2003, 2005, and 2008.

Analyzing U.S. antitrafficking policy in terms of the four-stage policy-making process (see Figure 14–1) sheds light on how interest groups and federal agencies affected this policy. Specifically, the next section of this chapter analyzes the evolution of three aspects of U.S. antitrafficking policy—criminal law provisions, restrictions on organizations eligible for grants to provide assistance to victims of trafficking, and state and local law enforcement anti–human trafficking task forces. These three examples, although not mutually exclusive, illustrate how interest groups and bureaucracies affected human trafficking policy outputs.

### Criminal Law Provisions

During the agenda-setting stage, the identification of the limitations of existing laws and the education of policy makers about those limitations fell primarily to Justice Department staff. They found that existing statutes did not always treat the workers involved as victims and failed to address modern manifestations of human trafficking, such as the use of psychological coercion (United States Department of Justice, 2006, p. 19). Representatives of the Justice Department interacted with other executive branch officials, representatives of nongovernmental organizations, and congressional staff at conferences. During this stage, members of interest groups from both spheres articulated their respective views on how human trafficking should be defined, at least in broad terms. The antiprostitution sphere emphasized the importance of addressing crimes related to trafficking for sex. The human trafficking sphere sought a broader definition of human trafficking encompassing both sex and labor. They communicated these views at conferences, through the media, and in meetings with officials (Stolz, 2005, 2007b).

During the legislative phase, representatives of the Justice Department met with congressional staff and testified at public hearings regarding the limitations of current law (Stolz, 2007a, pp. 177–178). Provisions in the 2000 U.S. anti–human trafficking legislation addressed concerns raised during the agenda-setting phase by Justice Department officials and articulated by them during hearings and discussions with congressional staff. Specifically, the act did so through the following measures, among others (U.S. Department of Justice, 2006, p. 18; U.S. Government Accountability Office, 2007, p. 46):

- Criminalizing the obtaining or maintaining of persons for commercial sexual activity
- Criminalizing the use of certain kinds of force, fraud or coercion to provide or obtain persons for any labor or services (e.g., work in farms, factories, and households)
- Including nonviolent coercion and threats of harm to third persons in federal involuntary servitude laws
- Making attempted trafficking crimes punishable
- Increasing the maximum penalty of slavery and involuntary servitude offenses from 10 to 20 years or to a life sentence if the offense involved factors like death, kidnapping, or aggravated sexual abuse
- Providing legal status and special benefits to aliens certified as trafficking victims in the United States who are willing to assist law enforcement

By addressing both trafficking for sex and for labor, the act encompassed the positions of both the antiprostitution and human trafficking spheres. The act defined trafficking in terms of "fraud, force, and coercion." The inclusion of this language was opposed by the antiprostitution sphere, as they viewed the consent of the victim to be irrelevant, and it undermined the support of some groups for the final legislation. In contrast, the human trafficking sphere supported the inclusion of this language to distinguish trafficking from related criminal behaviors. A vital consideration in the development and implementation of criminal law is that the law be enforceable. Legal experts interviewed underscored that the "force, fraud, and coercion" language was necessary to distinguish trafficking from related crimes (e.g., prostitution or alien smuggling), took into account jurisdictions in Nevada where prostitution was legal, and averted potential conflicts with other federal criminal laws (Stolz, 2005, 2007b, pp. 319, 322–324, 333). The debate over the inclusion of this language has repeatedly resurfaced during the reauthorization stage.

With the initial implementation of the 2000 anti–human trafficking legislation, agencies confronted a number of challenges. Underlying some of these challenges was the fact that implementing trafficking laws conflicts with the traditional law enforcement paradigm. For example, unlike typical crimes, such as robbery, where the criminal behavior precipitates the investigation, identifying traffickers usually involves proactive efforts by law enforcement to identify the victims. Furthermore, victims of trafficking may be engaged in illegal behaviors that would typically lead law enforcement to treat them as offenders or coconspirators rather than victims. Accordingly, treating individuals involved in criminal behavior as victims requires a change in the law enforcement paradigm. The prosecution of trafficking cases also presents a range of evidentiary challenges, such as the need to rely primarily on victim testimony, obtain corroborating evidence from sources located beyond U.S. borders, and establish a causal connection between subtle forms of coercion and a victim's servitude. Department of Justice officials addressed some of these challenges by establishing state and local human trafficking law enforcement task forces, discussed next.

Subsequent reauthorizations of the 2000 antitrafficking legislation included additional prosecutorial tools. The 2003 reauthorization (1) added new tools to combat child sex tourism by raising the statutory maximum sentences for violations involving child sex tourism; (2) included trafficking as a predicate for money laundering and racketeering under the RICO Act[7]; (3) codified the requirement that victims cooperate with law enforcement, while relieving victims younger than 18 years of age from this requirement; and (4) created a new civil action that allowed trafficking victims to sue their traffickers in federal court (U.S. Department of Justice, 2006, pp. 19–20). The 2005 reauthorization added a new violation to enhance the prosecution of trafficking in persons offenses committed by persons employed by or accompanying the federal government overseas and allowed for the application of forfeitures. The 2008 legislation, among other things, (1) established a new trafficking-specific conspiracy statute and enhanced penalties for conspiring to commit trafficking crimes; (2) added provisions penalizing those who knowingly benefit financially from ventures engaged in trafficking; (3) clarified the application of the forced labor provision; (4) broadened the crime of sex trafficking by force, fraud, or coercion by including reckless disregard, as well as knowledge of the fact that such means would be used; (5) expanded the crime of sex trafficking of minors by eliminating the knowledge-of-age requirement in certain instances; (6) added new crimes of obstructing human trafficking enforcement and labor contracting crime; and (7) clarified the definitions of "serious harm" and "abuse or threatened abuse of the legal process," which are among the prohibited means of coercion under the forced labor and sex trafficking statutes.

Many of the new criminal provisions reflected the technical needs of federal investigators and prosecutors charged with pursing these crimes. For example, by including forms of human trafficking as predicates to money laundering and racketeering offenses, the 2003 reauthorization legislation provided new enforcement tools that prosecutors could use to dismantle trafficking syndicates (U.S. Department of Justice, 2006, p. 20). The 2008 reauthorization provisions were to enhance the ability of federal prosecutors to charge, convict, and punish human traffickers and vindicate the rights and dignity of trafficking victims (Axam, 2009).

Some provisions in the reauthorization legislation suggest the influence of interest groups. For example, child sex tourism was a concern of the antiprostitution sphere. However, not all criminal provisions supported by the various interests were enacted. Accordingly, examining provisions not implemented also sheds light on interests that may be pursued through other access points. For example, a provision that would have in effect federalized the prosecution of pandering-, pimping-, and prostitution-related offenses was considered during the discussions over the 2008 reauthorization. This proposal reflected the positions of the antiprostitution sphere, specifically its opposition to prostitution and belief that consent is irrelevant to a victim of trafficking. The proposed provision was opposed by the Justice Department and was not included in the final bill; however, the fact that it was raised and considered demonstrates how groups may continually try to influence policy to reflect their views.

Overall, the breadth of U.S laws to combat human trafficking expanded between 2001 and 2008. This expansion reflected both the concerns of components of the federal bureaucracy charged with implementing the statutes and the concerns of interest groups seeking to influence U.S. policy on human trafficking.

### Victim Assistance Programs

During the agenda-setting phase, to convince policy makers that human trafficking was a public policy issue that needed to be addressed, it was necessary to reconceptualize policy makers' perceptions of the problem. Key to the reformulation of their perception of trafficking was changing the characterization of the trafficked person from criminal to victim. For example, rather than the traditional depiction of women in the sex trade as criminals or coconspirators, interest groups sought to characterize trafficked women and girls as victims. Similarly, persons trafficked for labor purposes were also depicted as victims (Stolz, 2005, pp. 422–423). Once those who were trafficked were viewed as victims, attention could be focused on the needs of the victims and the establishment of policies and programs to address those needs.

As part of the protection prong of the victim-centered approach embodied in the 2000 legislation, the act mandated the establishment of a grants program to assist victims of trafficking. Benefits were to be provided to trafficking victims meeting certain conditions, without regard to their immigration status. Specifically, the act authorized the Attorney General to make grants to states, Indian tribes, units of government, and nonprofit, nongovernmental victim services organizations to develop, expand, or strengthen victim service programs for victims of trafficking. Both the antiprostitution and human trafficking spheres viewed trafficked persons as victims, although, as previously discussed, they focused on different populations.

To implement the 2000 antitrafficking legislation, the White House, under President Bush, promulgated NSPD-22. The directive stated that U.S. policy opposed the legalization of prostitution. To further this policy, NSPD-22 directed that federal funds to assist victims of trafficking were not to be allocated to groups that supported the legalization of prostitution. While the directive came from the White House, the position on prostitution found therein was consistent with the goals of the antiprostitution sphere. The practical effect of the policy was that some organizations that had provided assistance to victims of trafficking were no longer eligible to receive funds, while some groups that were new to the trafficking issue were better positioned to apply for these funds because of their stance on prostitution. Through the implementation process, the antiprostitution sphere achieved what it had not achieved through the initial legislation process.

These restrictions were codified in the 2003 antitrafficking reauthorization legislation. Specifically, the act restricted the awarding of funds and required a grantee to state in the grant application or agreement that it would not promote, support, or advocate the legalization of prostitution. Thus, the antiprostitution sphere was able to solidify through legislative action the policy initially promulgated by executive branch action during the implementation stage.

The provisions restricting the awarding of grants for victim assistance were not changed in the 2008 trafficking reauthorization legislation and remained in effect as of May 2009. Because appointees of the Obama administration are likely to disagree with these restrictions, as they are likely more closely aligned with the positions of the human trafficking sphere, it remains to be seen what steps will be taken to change this policy.

## State and Local Law Enforcement Task Forces

As discussed, during the legislative stage, emphasis on the victim-centered approach to human trafficking led to the establishment of a grants program to assist victims of trafficking. As federal prosecutors sought to prosecute traffickers under the legislation, however, the challenge of identifying trafficking victims, which was necessary in order to identify the traffickers, became evident. Only a small number of victims were officially identified as trafficking victims and received federal assistance. Federal prosecutors recognized that support from state and local law enforcement was necessary to locate victims. Responding to this challenge, prosecutors in the Justice Department's Civil Rights Division, which had primary responsibility for prosecuting trafficking cases, developed a concept paper outlining a partnering of federal prosecutors and Department of Health and Human Services' staff. This partnership would oversee the assistance of trafficking victims and enhance federal capabilities to identify trafficking victims.

To implement this partnership, the Justice Department sponsored a conference in Tampa, Florida, in 2004, during which it launched a grants program to support the establishment of state and local human trafficking law enforcement task forces. Communities were identified where federal officials believed such task forces could be developed. The conference was to introduce the model of victim-centered investigations and the concept of human trafficking prosecutions through the creation of local task forces. It brought together about 500 participants, including 21 teams, each consisting of about 20 state, local, and federal officials. After the conference, the teams were expected to work together on human trafficking in their respective communities. To further the task force initiative, the Bureau of Justice Assistance (BJA)—the Justice Department agency responsible for supporting local, state, and tribal efforts to achieve safer communities—developed and implemented a human trafficking competitive grants program. The grants were to be awarded to state or local police agencies to work with the local U.S. Attorney's Office, federal law enforcement entities, and nongovernmental organizations that may come into contact with victims of trafficking (U.S. Department of Justice, 2006, pp. 35–40; U.S. Government Accountability Office, 2007, p. 11).

This state and local human trafficking initiative provides an example of how bureaucracies can make policy through the implementation process. Confronted by the difficulties of identifying victims and recognizing the need to obtain the assistance of state and local law enforcement agencies to help identify victims, federal agencies developed and implemented the task force initiative. In so doing, the bureaucracy promulgated trafficking policy. The initiative was expanded by the 2005 antitrafficking reauthorization legislation, which authorized for appropriation additional funding to support the program. By 2006, the BJA had awarded grants up to $450,000 for a three-year period to each of 42 task forces in communities across the country. The Justice Department sponsored a second trafficking conference in New Orleans in 2006 to help train task force members. In plenary and breakout sessions, federal agency officials from Justice, Immigration and Customs Enforcement, and Labor along with representatives of nongovernmental organizations and successful trafficking task forces presented information on investigative strategies, victim services, interviewing witnesses, and other topics (U.S. Government Accountability Office, 2007, p. 28). These task forces were viewed by federal investigators and prosecutors as a key component of U.S. antitrafficking efforts.

## ■ Conclusion

This chapter began with the assertion that U.S. federal criminal justice policy making is an iterative and continuous process. It laid a framework for analyzing the criminal justice policy-making process, including four stages—agenda setting, legislation, implementation, and reauthorization. The framework was then used to analyze the development of U.S. antitrafficking policy from 2000 through 2008. This case study points to different access points and means through which policy can be influenced. In addition to legislation establishing a policy or programs, executive orders, directives, and practice through agency efforts provide means to affect policy during the implementation stage. The three examples of specific aspects of U.S. antitrafficking policy analyzed illustrate how interest groups sought to influence human trafficking policy at the various stages, suggesting the motto, "If at first you don't enact, try another access point at another stage in the process." The examples also demonstrate how the bureaucracy can affect policy through the implementation and legislative processes.

Both the antiprostitution and human trafficking spheres sought access at different stages in the policy-making process. The evolution of the restrictions on organizations' eligibility to receive grants to assist trafficking victims illustrates how interest groups can successfully achieve policy goals over time by using different access points. The restrictions were not included in the 2000 antitrafficking legislation; rather, they became policy for the administration through a presidential directive during the implementation stage and became law—policy for future administrations—through the 2003 reauthorization legislation. In part, the successes of the antiprostitution sphere between 2001 and 2008 may be attributed to its goals being similar to those of the Bush administration.

The conceptualization and establishment of the state and local human trafficking law enforcement task forces illustrate how the bureaucracy can affect policy. The task forces were a response by the bureaucracy to challenges confronted in identifying trafficking victims, a necessary step in the pursuit of the perpetrators of trafficking crimes. Begun as an implementation initiative, the policy was solidified in reauthorization legislation.

The evolution of the human trafficking criminal law provisions illustrates not only how such policies evolve incrementally but how issues that seem to be resolved at one point may continue to resurface. Provisions in the original legislation responded to gaps and limitations in existing statutes that made it difficult for prosecutors to pursue trafficking crimes. The inclusion of provisions—for example those that enable prosecutors to pursue RICO, conspiracy, or money laundering in the context of human trafficking—further enhanced their arsenal.

In addition, the case study demonstrates how groups may continue to raise the same issues even when it was thought that the compromises reached would lay the issue to rest. One such example is the "force, fraud, and coercion language." Although a provision to federalize commercial sex acts in or affecting interstate commerce with no requirement of force, fraud, and coercion considered as part of the debate on the 2008 trafficking legislation was not included in the final legislation, its consideration demonstrates that groups opposing the "fraud, force, and coercion" language continue to contest to this language. Such proposals and others seeking to expand the scope of trafficking to domestic prostitution—especially juvenile prostitution—show how groups continue to take bites, even the same bites, at the apple.

These examples illustrate that policy making is a continuous, iterative process, albeit not necessarily linear. With a change in administration, one seemingly more amenable to the perspective of the "human trafficking" sphere, U.S. human trafficking policy may change direction, at least to a degree. One can expect some policies, such as the restrictions on groups eligible to receive grants to provide victims assistance, to be changed or modified. Other aspects of trafficking policy, such as the state and local law enforcement task forces and some of the more technical criminal law positions, will more than likely continue and may even be enhanced. All this remains to be seen.

As a transnational crime, the study of human trafficking policy making provides insight into the federal criminal justice policy-making process, generally. That is, interest groups and agencies seek to influence criminal justice policy through a process that is continuous and iterative. Because the basic structures and processes of state and local criminal justice policy making are similar to federal structures and process, the framework can also be applied to analyze criminal justice policy at these levels of government. In addition, for those seeking to influence criminal justice policy, the framework provides a tool for analyzing where best to attempt influencing the process to produce the desired policy output. In short, the framework provides a tool for both analyzing and influencing criminal justice policy making at all levels of government.

## ■ Discussion Questions

1. Describe the fundamental stages of the criminal justice policy-making process. Are there any elements that can be added to the framework presented in this chapter?

2. Drawing on the case of human trafficking, discuss the relationship between U.S. criminal justice policy and international criminal justice policy.

3. Using the framework presented in Chapter 1, discuss the policy community that has evolved to address the problem of human trafficking.

4. Discuss the relationship between politics and policy in the case of human trafficking.

5. Has the policy-making process described in the chapter generated an effective response to the problem of human trafficking?

## ■ Endnote

1. The views expressed in this article are solely those of the author and are not intended to reflect those of the U.S. Government Accountability Office.

2. The Global Survival Network (GSN) is a human rights and environmental (endangered species) nonprofit organization based in Washington DC. GSN works as an international organization with consultants in many different

countries. GSN stumbled upon human trafficking while investigating the illegal trade in wildlife in the Russian Far East. A Russian "mafia group" that was trading tiger bones to China and tiger skins to Japan was also trading Russian women to Japan.

3. Vital Voices grew out of the U.S. government's successful Vital Voices Democracy Initiative, which was established after the United Nations Fourth World Conference on Women in Beijing, to promote the advancement of women as a U.S. foreign policy goal.

4. La Strada is a nongovernmental organization in Ukraine with a bidirectional goal: to prevent trafficking of women and to help the victims of trafficking.

5. In *United States v Kozminski*, 487 US 931 (1988), the Court held that involuntary servitude exists only when the master subjects the servant to (1) threatened or actual physical force, (2) threatened or actual state-imposed legal coercion, or (3) fraud or deceit when the servant is a minor or an immigrant or mentally incompetent.

6. The Department of Homeland Security, specifically Immigration and Customs Enforcement, was established in March 2003.

7. Racketeer Influenced and Corrupt Organization Act, 18 U.S.C 1961–1968.

## ■ References

Axam, H. (2009, January 12). Human trafficking email. Washington, DC: Department of Justice, Human Trafficking Prosecution Unit, Civil Rights Division, Criminal Section.

Baker, R., & Meyers, F., Jr. (1980). *The criminal justice games: Politics and players.* Belmont, CA: Duxbury Press.

Banfield, E. C. (1961). *Political influence: A new theory of urban politics.* New York: The Free Press.

Banfield, E. C., & Meyerson, M. (1955). *Politics, planning, and the public interest.* New York: The Free Press.

Bales, K. (1999). *Disposable people.* Los Angeles: University of California Press.

Barry, K. (1979). *Female sexual slavery.* New York: New York University Press.

Clinton, H. R. (1995, September 5). Remarks for the United Nations Fourth World Conference on Women. Beijing, China.

Commission on Security and Cooperation in Europe. (1999, June 28). *The sex trade: Trafficking of women and children in Europe and the United States: Hearings.* 106th Congress, 1st Session.

Commission on Security and Cooperation in Europe (2005, June 7) *Exploiting Americans on American soil: Domestic trafficking exposed: Hearings.* 109th Congress, 1st Session.

*Congressional Almanac.* (2000). *Anti-crime package seeks to protect women and children from violence* (Vol. LVI, pp. 15–26). 106th Congress, 2nd Session. Washington, DC: CQ Press.

*Congressional Information Service/Index to legislative histories.* (2000, 2003). Bethesda, MD: Congressional Information Service.

Dahl, R. (1961). *Who governs? Democracy and power in an American city.* New Haven, CT: Yale University Press.

Fairchild, E. S. (1981). Interest groups in the criminal justice process. *Journal of Criminal Justice, 9,* 181–194.

Forman, M. M. (2007, October 31). Statement concerning combating modern slavery: Reauthorization of the anti-trafficking programs before the U.S. House of Representatives, Committee on the Judiciary.

Greenwald, C. S. (1977). *Group power: Lobbying and public policy.* New York: Praeger.

Key, V. O. (1964). *Politics, parties, and pressure groups* (5th ed.). New York: Thomas Y. Crowell.

Kingdon, J. W. (1984). *Agendas, alternatives, and public policies.* United States: HarperCollins.

Lipsky, M. (1980). *Street-level bureaucracy: Dilemmas of the individual in public services.* New York: Russell Sage Foundation.

Marion, N. E. (1994). *A history of federal crime control initiatives, 1960–1993.* Praeger Series in Criminology and Crime Control Policy. Westport, CT: Praeger.

Marion, N. E. (1995). *A primer in the politics of criminal justice.* Albany, NY: Harrow and Heston.

Marion, N. E. (1997). Symbolic policies in Clinton's crime control agenda. *Buffalo Criminal Law Review, 1,* 67–108.

Melone, A., & Slatger, R. (1983). Interest group politics and the reform of the Federal Criminal Code. In S. Nagle, E. Fairchild, & A. Champagne (Eds.), *The political science of criminal justice* (pp. 41–55). Springfield, IL: Charles C. Thomas Publishers.

Oliver, W., & Marion, N. E. (2008). *The making of criminal justice policy in the United States.* New York: Lewiston.

Outshoorn, J. (2004). *The politics of prostitution: Women's movements, democratic states and the globalisation of sex commerce.* Cambridge, UK: Cambridge University Press.

Redman, E. (1973). *The dance of legislation.* New York: Simon & Schuster.

Roby, P. (1969). Politics and criminal law: Revision of the New York State penal law on prostitution. *Social Problems, 17*(Summer), 83–109.

Rothenberg, L. E. (2007, October 31). Statement concerning combating modern slavery: Reauthorization of the anti-trafficking programs before the U.S. House of Representatives, Committee on the Judiciary.

Stolz, B. A. (1999). Congress, symbolic politics and the evolution of the 1994 Violence against Women Act. *Criminal Justice Policy Review, 10,* 319–341.

Stolz, B. A. (2002a). Criminal justice policy making: Federal roles and process. Westport, CT: Praeger Publishers.

Stolz, B. (2002b). The roles of interest groups in U.S. criminal justice policymaking: Who, when, and how. *Criminal Justice, 2*(1), 51–69.

Stolz, B. A. (2005).The Victims of Trafficking and Violence Act of 2000: Interest groups and setting the criminal justice policy making agenda. *Criminal Justice, 5,* 407–430.

Stolz, B. A. (2007a). Interest groups and the development of the U.S. Congress's response to human trafficking. *International Journal of Comparative and Applied Criminal Justice, 31*(2), 167–190.

Stolz, B. A. (2007b). Interpreting the U.S. human trafficking debate through the lens of symbolic politics. *Law & Policy, 29*(2), 311–338.

Truman, D. B. (1951). *The governmental process.* New York: Knopf.

U.S. Congress, House. (1999, September 14). *Trafficking of women and children in the international sex trade: Hearings before the Subcommittee on International Operations and Human Rights Committee on International Relations.* 106th Congress, 1st Session.

U.S. Congress, House. (2001, November 29). *Implementation of the Victims of Trafficking Protection Act: Hearings before the Committee on International Relations.* 107th Congress, 1st Session.

U.S. Congress, House. (2002, June 19). *Foreign government complicity in human trafficking: A review of the state department's "2002 Trafficking in Persons Report" hearings before the Committee on International Relations.* 107th Congress, 2nd Session.

U.S. Congress, House. (2003a, June 25). *Global trends in trafficking and the "Trafficking in Persons Report": Hearings before the Subcommittee on International Terrorism, Nonproliferation and Human Rights, Committee on International Relations.* 108th Congress, 1st Session.

U.S. Congress, House. (2003b, October 29). *The ongoing tragedy of international slavery and human trafficking: An overview: Hearings before the Subcommittee on Human Rights and Wellness, Committee on Government Reform.* 108th Congress, 1st Session.

U.S. Congress, House. (2004a, June 24). *Trafficking in persons: A global review: Hearings before the Subcommittee on International Terrorism, Nonproliferation and Human Rights, Committee on International Relations.* 108th Congress, 2nd Session.

U.S. Congress, House. (2004b, July 8). *Trafficking in persons: The federal government's approach to eradicate this worldwide problem: Hearings before Subcommittee on Human Rights and Wellness, Committee on Government Reform.* 108th Congress, 2nd Session.

U.S. Congress, House. (2005a, March 1). *United Nations organization mission in the Democratic Republic of Congo: A case for peacekeeping reform: Briefing and hearings before the Subcommittee on Africa, Global Human Rights and International Operations, Committee on International Relations.* 109th Congress, 1st Session.

U.S. Congress, House. (2005b, March 9). *Combating human trafficking: Achieving zero tolerance: Hearings before the Subcommittee on Africa, Global Human Rights and International Operations, Committee on International Relations.* 109th Congress, 1st Session.

U.S. Congress, House. (2005c, April 28). *Combating trafficking in persons: Status report on domestic and international developments: Hearings before the Subcommittee on Domestic and International Monetary Policy, Trade, and Technology, Committee on Financial Services.* 109th Congress, 1st Session.

U.S. Congress, House. (2005d, June 22). *Combating trafficking in persons: An international perspective: Hearings before the Subcommittee on Domestic and International Monetary Policy, Trade, and Technology, Committee on Financial Services.* 109th Congress, 1st Session.

U.S. Congress, Senate. (2000, February 22 and April 4). *International trafficking in women and children: Hearings before the Subcommittee on Near Eastern and South Asian Affairs, Committee on Foreign Relations.* 106th Congress, 2nd Session.

U.S. Congress, Senate. (2002, March 7). *Monitoring and combating trafficking in persons: How are we doing? Hearings before the Subcommittee on Near Eastern and South Asian Affairs, Committee on Foreign Relations.* 107th Congress, 2nd Session.

U.S. Congress, Senate. (2003, April 9). *Trafficking in women and children in East Asia and beyond: A review of U.S. policy: Hearings before the Subcommittee on East Asian and Pacific Affairs, Committee on Foreign Relations.* 108th Congress, 1st Session.

U.S. Congress, Senate. (2004a, July 7). *Examining U.S. efforts to combat human trafficking and slavery: Hearings before the Subcommittee on Constitution, Civil Rights and Property Rights, Committee on Judiciary.* 108th Congress, 2nd Session.

U.S. Congress, Senate. (2004b, July 13). *Human trafficking and mail order bride abuses: Hearing before the Committee on Foreign Relations.* 108th Congress, 2nd Session.

U.S. Department of Justice. (2006). *Report on activities to combat human trafficking fiscal years 2001–2005.* Washington, DC: Author.

U.S. Government Accountability Office. (2007). *Human trafficking: A strategic framework could help enhance the interagency collaboration need to effectively combat trafficking crimes.* Washington, DC: Author.

## ■ Laws Cited

Mann Act, ch. 395, 36 Stat 825 (1910) (codified as amended at 18 USC §§2421–2424).

Racketeer Influenced and Corrupt Organization Act, 18 USC 1961–1968.

Trafficking Victims Protection Reauthorization Act, Pub L No 108–193, 117 Stat 2875 (2003).

Trafficking Victims Protection Reauthorization Act of 2005, Pub L No 109–164, 119 Stat 3558 (2006).

The William Wilberforce Trafficking Victims Protection Reauthorization Act of 2008 Pub L No 110–457 122 Stat 5044 (2008).

Victims of Trafficking and Violence Protection Act of 2000 (TVPA), Pub L No 106–386, 114 Stat 1464 (2000).

## ■ Court Cases Cited

*United States v. Kozminski*, 487 U.S. 931 (1988).

# When Is Crime a Public Health Problem?

# CHAPTER 15

Gregory J. DeLone and Miriam A. DeLone

## ■ Introduction

As a social science discipline, criminology strives to answer three key questions: (1) Why do we have the laws we have? (2) Why do we use these laws the way we do? and (3) Why do people commit crime? However, after decades of research and academic development, the answers to these questions remain elusive. Moreover, our discipline's ability to influence policy formation and implementation is inconsistent. How might we more effectively reduce crime? We suggest a reinvigoration of the impact of our discipline from the point of view reflected in this new question: When is crime a public health problem?

This chapter proposes that the study of crime causation and crime prevention will be enhanced by extending its historic tendency to be interdisciplinary to include the priorities and research strategies of public health. Most often the connection between crime and public health centers on violent behavior (e.g., Sampson, Morenoff, & Raudenbush, 2005); however, we propose that nearly every issue in criminology can benefit from the development of research and policy agendas that emerge from viewing crime from a public health perspective.

What is public health? Mariner (2003) contends that "public health began as a social reform movement and current public health recognizes the influence of political and economic conditions and human rights on health" (p. 529). The American Public Health Association (APHA; www.apha.org) states that public health is prevention, policy development, and population health surveillance. Specifically, according to the APHA mission statement, public health convenes constituencies, champions prevention, provides evidence-based policy and practice, and advocates for healthy people and communities.

Frenk (1993) asserts that "as a field of knowledge and as a social practice, public health has historically been one of the vital forces that have led to reflection on and collective action for health and well-being" (p. 469). Among the advantages of the public health approach to addressing problematic social issues is that public health "promotes the welfare of individuals by fostering healthy, strong and safe families, communities, and workplaces. It views the individual within a social milieu and explores the influence of cultural, family, and community values on behavior" (Korn & Shaffer, 1999, p. 306).

## ■ When Is Crime a Public Health Problem?

As described, crime is a public health concern in nearly every instance. The health and well-being of communities and populations have been evident in criminology policy and theory for hundreds of years, in both private and in public settings. The imagery of crime epidemics and of contagion prevention can be linked to the earliest of criminological perspectives and is still evident in modern "moral panics" about crime (e.g., drugs, immigration, and terrorism).

### Historic Convergence of Theory and Policy

The historic emergence of positivism and the birth of sociology impacted the priorities of criminological research and policies. However, a historic discussion of crime control policies started even before Lombroso, with such policies as banishment and containment of contagions (e.g., witchcraft trials) of the pre-enlightenment Western societies. Strategies of purging the sinner's body of evil strengthened the populace through participation in public condemnation, fitting loosely into a public health agenda (Foucault, 1979). Lombroso's research led to the categorization of criminals as atavistic anomalies and the suggestion of eugenics as a policy of crime control (Akers, 2000; Vold & Bernard, 1986). This movement was followed by the classification of the mentally ill as a dangerous group threatening the health of communities (Foucault, 1973) and the resulting "age of the asylum" (Rothman, 1971).

Western history is full of policies of forced assimilation to resocialize unsuitable populations (Aguirre & Turner, 2007; McNamara & Burns, 2009). American examples from the late 1800s to the mid-1900s include Indian boarding schools (Jenkins, 2007), orphan trains (O'Connor, 2001), and the Child Savers (Platt, 1977).

The Orphan Trains began in 1854 and lasted until 1929 (O'Connor, 2001). This "public health" initiative was run by the private Children's Aid Society as an alternative to orphan asylums. Orphaned children of the subordinated classes in New York City were forced into resettlement plans and relocated to a "better place" to assimilate them to the appropriate middle class moral values of the time. Orphan Trains were organized by private welfare groups that sent children to families in the Midwest, to conditions that often resembled indentured servitude disguised as foster care and/or adoption.

By the turn of the century in Chicago (1899), a more enduring institution had emerged: the Juvenile Court. Platt's (1977) analysis of the emergence of the Juvenile Court in the United States serves as an exemplar of the influence of moral entrepreneurs, using the medical model of social diseases and emerging social work strategies to influence the creation of government institutions for the control of children of crime-prone populations. The moral entrepreneurs in this case were private citizens and organizations referred to as "the Child Savers." The legal foundation for the new courts was the common law tradition of *parens patriae*, which allowed the state to act as parent in situations of neglect, abuse, and delinquency. These courts were governed by the policy of doing what was in the best interest of the child. In contrast to the Orphan Trains, which were organized by private and religious rescue societies, the Child Savers were private citizens influencing the construction of a legal bureaucracy. The juvenile court as a legal entity has spread to all 50 states and endures into the 21st century.

### Contemporary Convergence of Theory and Policy

The health and well-being of communities prevalent in social ecology, neo-Marxist theory, and routine activities influenced contemporary policies to foster healthier neighborhoods. Subsequent policies involved improving gun control, increasing target guardianship (neighborhood watches), and decreasing potential targets (vulnerable populations). Confederations of public and private groups attempted to "weed out" elements of an unhealthy neighborhood and "seed in" institutions that fostered social organization.

The imagery of the crime "epidemic" and the classification of dangerous populations emerged in the mid- to late 20th century from neo-Marxist theories explaining the adverse responses by subpopulations to conditions of alienation. Spitzer (1975) presented the classifications of social dynamite and social junk as socially constructed definitions created for subordinate populations to express their potential for disruption to the relations of production. Subordinate populations such as criminals were most often seen as direct threats (through violence or disruption of the relationships of private property), and their actions were criminalized, with punishment leading to institutionalization. The mentally ill and criminally insane were also seen as surplus populations, but as less of a threat to the relations of production and more as a drain on resources, they were categorized as social junk, to be dealt with through policies of exclusion and extermination.

Drug control strategies are often dominated by the War on Drugs imagery, but they do include population prevention strategies (e.g., Drug Abuse Resistance Education [DARE]), addiction treatment and recovery programs (Kennedy, 2006), and drug courts as an alternative to traditional crime policy and practices. Also diverging from traditional practices is the increasing use of offender reentry programs—often with the recognition of the mental health problems faced by former inmates (Sample, 2008).

Current criminology prevention practices focus on vulnerable populations: keeping school children from being bullied and keeping the elderly from being abused. Rescue strategies at the earliest possible point of intervention are routinely available for abused children and victims of domestic violence. These strategies are not always guided by theory, but focus on desistence strategies. Criminology's commitment to the quality-of-life issues that impact citizens, neighborhoods, and communities are seen in fear-of-crime research, the prevention of hate crimes, and zero-tolerance policing.

Efforts to prevent environmental racism and human rights violations are among some of the newest issues gaining attention in the field of criminology but are easily absorbed into the neo-Marxist theory developed by Spitzer (1975). Critics of environmental racism and proponents of environmental justice argue that individuals and communities cannot enjoy the social and economic benefits of prosperity and industrialization without their individual and community health (McNamara & Burns, 2009, p. 15). Similarly, the issues of terrorism, genocide, and human trafficking are seen as not just threats to national security and the health of our democracy, but to global security.

## ■ Integration of Two Intellectual Traditions

There are a number of crime-based public health concerns that have attracted the attention of public and private agencies for decades. These strategies have often been short lived or have transformed significantly over time to survive. Recently, more voices have echoed the need in criminology for evidence-based practices and real interaction with policy makers (Liberman, 2009).

Frenk, a public health researcher, identifies five potential barriers between researchers and decision makers (1993). We contend that criminology as a discipline can substantively advance its impact on public policy by integrating with the more mature public health perspective. Frenk's five barriers are priorities, time management, language and accessibility of results, perceptions about the final product of research (discovery vs. decision), and integration of different findings on the same problem.

Priorities in criminological research, funding, and policy making are often the result of "crisis" situations that are framed by "moral entrepreneurs" and limited information. For example, the crack cocaine crisis and resulting concerns about crack babies (Chiricos, 1996) led to the passing of federal sentencing policies that were not based on the pharmacology of the drug or systematic information about its effects. Similar historic examples include alcohol prohibition and sex offender regulation. Moreover, criminology's priorities for research are often channeled by sources of funding rather than the objective identification of risk. The emergence of the Chicago School, which resulted in the "social pathologists" (Mills, 1943) of the social ecology school in the 1920s, was spurred by the private funding (the Rockefeller Foundation) of a small private college to focus research on the ills of the urban milieu.

The priorities of criminological research are also influenced by the U.S. Congress and federal agencies through such legislation as the Juvenile Justice and Delinquency Prevention Act (1974, with subsequent amendments). This act created the government office, within the National Institute of Justice, to oversee the program and research initiatives by the states to (1) deinstitutionalize status offenders, (2) separate juveniles from adults during incarceration, and (3) monitor disproportionate minority contact with law enforcement and the juvenile court system. Later, under the Clinton administration, crime prevention policy and academic research was strongly influenced by the COPS initiative (Community Oriented Policing Services, created through the Violent Crime Control and Law Enforcement Act of 1994). This influence came in the form of funding for 100,000 new police officers on the street (despite research suggesting this would not reduce crimes rates) and for research projects assessing the impact of community policing. These initiatives were the response to a sense of crisis that had emerged about the existence and causes of crime in America and were not necessarily based on objective, evidence-based research findings.

A more successful, less crisis-driven, integration of criminology and public health comes from the work of Robert Sampson and colleagues in Chicago neighborhoods. In the recent essay, "Social Anatomy of Racial and Ethnic Disparities in Violence," Sampson, Morenoff, and Raudenbush (2005) assert that "the public health of the United States has long been compromised by inequality in the burden of personal violence" (p. 224). In their research, they analyze "key individual, family and neighborhood factors to assess competing hypotheses regarding racial/ethnic gaps in perpetrating violence" (p. 224). Their multilevel (individual constitution, family, neighborhood), multivariate analysis of nearly 3000 participants indicates that neighborhood context is the most important course of reducing the gap between white and black levels of violence; conversely, the least significant source of the explanatory factors are the constitutional factors of the individual respondent. Moreover, Sampson et al. find that the relationships observed between neighborhood level predictors of violence and individual level predictors of violence are robust (not changing) across racial/ethnic groups (p. 231). Thus, Sampson et al. argue that their results suggest

> that generic interventions to improve neighborhood conditions may reduce the racial gap in violence. Policies such as housing vouchers to aid the poor in securing residence in middle-class neighborhoods may achieve the most effective results in bringing down the long-standing racial disparities in violence. Policies to increase home ownership and hence stability of residence may also reduce disparities. (p. 231)

As for family indicators, their findings support the relevance of the positive "social influence of marriage and calls for renewed attention to the labor-market contexts that support stable marriages among the poor" (p. 231; see also Kiecolt-

Glaser & Newton, 2001; Wilson, 1996). In conclusion, Sampson et al. contend that the "large racial/ethnic disparities in violence found in American cities are not immutable. Indeed, they are largely social in nature and therefore amenable to change" (p. 231).

Dowdy (1994) argues that the influence of public funding since the 1970s has led criminological researchers to focus more on "individualistic as opposed to structural explanations of crime" (p. 77). However, during this period, private foundations such as the H. A. Guggenheim Foundation, the Ford Foundation, and the Rockefeller Foundation have been more focused on funding projects that address more comprehensive peace and social justice issues rather than simply crime.

Frenk suggests the best solutions for setting priorities for funding and policy generation involve integrating educated consumers of research within the governing legislative bodies and executive agencies. Criminologists must also be more proactive with the diagnosis and surveillance of potential crisis situations to allow for the timely data collection, analysis, and transmission of findings to decision makers. In a 2008 issue of the journal *Criminology and Public Policy*, scholars outline the need for such improvements as (1) the use of advanced technology to keep researchers and policy makers well informed about current crime patterns, rather than current calendar year reporting and (2) automatic sunset provisions in new legislation to monitor the impact of haphazardly implemented policies and mediate the impact of hastily and poorly conceived policies.

## ■ Conceptual Overlap of Two Disciplines

Korn and Shaffer (1999) present the unique elements of public health in a description that complements the goals of criminology. They assert that public health is a

> vantage point [that] encourages the application of a conceptual continuum to the range of risk, resiliency, and protective factors that can influence the development and maintenance of related problems. A public health perspective also offers an integrated dynamic approach that emphasizes a "systems" view rather than a primary focus solely on individuals or isolated events. (p. 306)

Given this encouragement, how can criminology advance toward greater effectiveness as a discipline? We suggest that the synthesis of the disciplines of crime and public health will lead to a stronger foundation to overcome the barriers we face in accomplishing meaningful and lasting social change.

While the conceptual overlap seems evident between the disciplines of criminology and public health, this chapter uses the work of Bourdieu (1993) on defining intellectual fields as a basis for integration of criminology and public health. Bourdieu (1993) refers to four key elements to anchor the understanding of an intellectual field: conceptual base, production base, reproduction base, and utilization base. This section offers an understanding of criminology and public health in this framework of distinct intellectual fields and concludes with an argument that merging these two fields creates a solid foundation for the strengthening of criminology as a discipline.

Applying Bourdieu's (1993) perspective on the key elements of intellectual traditions to the field of criminology, a number of weaknesses in the foundation of criminology as an intellectual tradition emerge.

*Conceptual Base.* The conceptual base establishes the limits of the specific areas of research, teaching, and action for the discipline/field. Criminology as a social science strives to answer the following questions: Why do we have the laws we do? Why do we use these laws the way we do? And, why do people commit crime? Criminologists address these three questions, but seldom do we address the entire picture of these questions in our research. As a social science, criminologists focus on understanding, explanation, prediction, and control, but seldom do we accomplish the full range of these goals, as we are most effective at understanding, less effective at explanation, and even less successful with prediction and control. Our research is most often microlevel or macrolevel and seldom combined across levels of analysis. Our theory development is ideologically based, thus our research and policy initiatives are ideological as well. Often policy initiative from criminal justice agencies have no theoretical foundation but promote a more philosophical foundation for policy (e.g., conservatism vs. liberalism).

*Production Base.* Institutions, having a critical mass of researchers necessary to generate a body of knowledge, serve as the production base. Criminology researchers abound in academia and private research groups (e.g., RAND, Urban Institute). Knowledge is generated through theory testing to achieve the traditional social science goals of understanding and explanation, with less attention to prediction and control. Program evaluations occur in public and private settings with attempts to assess the implementation and impact of new policies.

*Reproduction Base.* The reproduction base is seen in the educational associations, publications, and general associations that reflect an aggregation of interest and allow for the exchange of ideas. Publications in criminology have subtitles such as "interdisciplinary" but fail to reach the full potential, as research is predominantly theory testing that is mostly sociological and occasionally psychological in nature. In the last few decades, criminology as a discipline has progressed along the traditional path of scientific development that rewards fragmentation, and our publications reflect an increasing number of divisions: critical criminology, gender and crime, minorities and crime, international criminology, and policing as specialized journals. The main professional organization in criminology has no dedicated code of ethics but refers Web site visitors to codes in allied fields (e.g., the Academy of Criminal Justice Sciences, the American Sociological Association).

*Utilization Base.* The utilization base is highlighted when policy institutions translate knowledge into evidence-based decision making. The primary professional organization for criminology (the American Society of Criminology), created in 1941, has in its more than 60-year history issued only two policy statements: (1) in 1989 condemning the racial bias present in the use of the death penalty in the United States, and (2) in 2007 specifying that the use of Uniform Crime Report data are not suitable for establishing a ranking of most dangerous and safest cities. Note the inconsistency in these statements: the first is for the benefit of a subpopulation of citizens for a goal of social justice, while the second is a rather benign statement about the misinterpretation of data. In a similar vein, the current statements on the American Society of Criminology Web site on how to effectively influence policy are limited to instructions on how to send letters to your elected representatives. More recently, evidence-based knowledge has received a more central focus in criminology with the emergence of the *Criminology and Public Policy* journal, but has failed to achieve its projected potential for influencing policy makers.

A recent essay in *The Criminologist: The Official Newsletter of the American Society of Criminology* highlights the importance of the evidence-based approach and is notably written by a government-based researcher rather than an academic-based researcher (Akiva M. Liberman, 2009). In this essay, Liberman outlines opportunities for evidence-generating policy and four key elements for the useful planning of evidence-generating policy. She argues that our policy makers move beyond our too-often "ill-informed, and even faddish policy-making" (p. 5). To do so we must reconcile the traditional standard of nonpartisanship with issue advocacy to ensure our field has a "constructive agenda to incrementally improve the formulation, implementation, and ultimately the content of criminal justice policy" (p. 5).

Two conditions help explain this imbalance between research and policy in criminology. First is our lack of consensus on the foundations of prediction. Second is our reluctance to advance control of our research priorities beyond those set by governmental funding sources.

Frenk (1993) offers a description of public health using Bourdieu's (1993) framework that offers fertile ground for integration.

*Conceptual Base.* Public health research is described as having two main scientific study objectives: (1) epidemiological research into health conditions and (2) health systems research into organized social responses at the micro- and macrolevels (Frenk, 1993, pp. 472–475). Epidemiology refers to research that establishes the frequency, distribution, and determinants of the health needs of populations and subpopulations, with an emphasis on identifying multiple determinants. Health system research includes the research done on (1) microlevel and intraorganizational health system organization (both services and resources) and (2) macrolevel and interorganizational health policy research. Frenk describes these studies as investigating "social, political, and economic processes that determine specific forms adopted by the organized social response. Therefore, it studies the determinants, design, implementation and consequences of health policies" (p. 475).

*Production Base.* Frenk (1993) contends that the dominant model for production of scientific knowledge at the population and subpopulation levels with an emphasis on interdisciplinary input puts public health at odds with traditional biological health research, thus restricting researchers from conducting studies and disseminating results. In Frenk's words, the "conventional image of scientific progress necessarily describes a growing fragmentation of the objects of study and the consolidation of independent disciplines" (p. 480). By contrast, public health as a discipline seems to promote growth through synthesis. Public health research is fundamentally juxtaposed with biomedical research and clinical research in that they have competing levels of analysis. However, public health departments, schools, and colleges are emerging throughout Western academia to ease the integration of numerous disciplines and their seemingly competing levels of analysis as considered reflection leads to the conviction that biological and sociobehavioral sciences can find common ground and improve the search for answers to modern social problems.

*Reproduction Base.* Public health education has emerged from the crisis of ambiguous direction that, according to Frenk, describes the early 1990s, to a discipline with increased focus, stronger refereed publications, more interaction with key public and private stakeholders (resulting in increased research funding and more effective policy implementation), and a strong code of ethics. More education departments are focusing on public health with an interdisciplinary mission (e.g., the University of Nebraska Medical Center's College of Public Health).

*Utilization Base.* Public health is based on advocacy and is developing avenues of communication with policy makers and policy implementers to use academic knowledge by policy makers and key stakeholders. In addition, it is crucial to have avenues of information to detect emerging needs that should receive appropriate prevention at all levels: primary, secondary, and tertiary (Callahan & Jennings, 2002). The quality and quantity of research for evidence-based (mission-oriented) policy initiatives is more readily accepted as relevant and useful to policy makers and program implementers. Callahan and Jennings note that the public health discipline has increasingly focused on public and legislative arenas (p. 169).

When criminology more fully draws on the more mature aspects of the public health perspective, we become stronger researchers with more comprehensive research questions, more extensive levels of analysis, and a stronger grounding in policy relevance. Drawing from the strengths of each of these intellectual traditions, we find that research needs to be anchored in the ability to use results in a constructive way. Policy makers need to feel like stakeholders in the products of academia. We find that policy should be enacted and improved with evidence-based research, not simply the result of moral panic and philosophical justifications. Moreover, we become more effective advocates, more fully embrace social justice, and increase our commitment to ethical standards.

Considering the impact of policies to increase individual, community, and population health, a number of American crime control strategies come to mind that can be improved:

- What are the benefits of Head Start programming for creating healthier individuals and communities? What is its contribution to saving "crack babies" and fetal alcohol syndrome babies?
- Have methadone maintenance programs decreased addiction? Are they more effective as community-based programs or incarceration-based programs? What are some of the unintended consequences of this programming (positive and negative)? What is the cost–benefit ratio in terms of crime reduction and health improvement?
- What are the short-term and long-term benefits of domestic violence shelters? Have they decreased the domestic violence victimization recurrence risk? Decreased the instance of child abuse? Decreased the domestic violence homicide rate for abusers or victims? Improved educational attainment for adult victims and their children?
- What have neighborhood watches and weed and seed programs brought to community health? Have the impacts been short term, long term or both? Have they decreased fear of crime? What are the unintended consequences?
- Has the organizational advocacy of Mothers Against Drunk Driving decreased vehicular injuries and homicides?
- Have public service announcements decreased the incidence of shaken baby syndrome? Hate crimes? Elder abuse? Drug use?
- What are the short-term and long-term impacts of corporate and government environmental racism at the community and population level?
- Have university extension programs for elderly caregivers decreased the instance of elder abuse? Have hospice programs decreased the occurrence of physician-assisted suicide?
- Have school programs decreased bullying? School shootings? How can these programs be improved? Expanded?

## ■ Conclusion

The public health perspective offers a number of advantages in understanding crime patterns and the causes of criminality. This perspective allows for consideration of macrolevel features, individual-level factors, and the interactions between these causes within the framework of public policy.

As Frenk (1993) contends:

> Health is a crossroad. It is where biological and social factors, the individual and the community, and social and economic policy converge. In addition to its intrinsic value, health is a means of personal and collective advancement. It is, therefore, an indicator of the success achieved by a society and its institutions of government in promoting well-being. (p. 469)

## Potential for Growth in Criminology: Advantages and Disadvantages

The potential for growth in criminology, with the expansion of research, policy, and treatment to include the public health perspective, offers many potential advantages, and perhaps a few disadvantages. First, advantages start with a more comprehensive focus on crime detection and prevention. A public health perspective leads researchers to focus on vulnerable populations in terms of potential offenders and potential victims. Criminal justice interests have expanded beyond deterrence theory to include detection and treatment, thus benefiting a myriad of vulnerable populations and increasing the points of intervention beyond reactive policing. Additionally, we know that both individual characteristics and community/population characteristics are correlated with crime. Criminologists recognize several influences that are correlated with crime at an individual and community level, thus preventing these correlates (e.g., lack of individual self-control, relative deprivation, community-level poverty) can reduce crime, even if the precise microlevel and macrolevel mechanisms of causation are not known. Moreover, the interest in modeling the often recursive nature of crime and victimization relationships—such as the potential for the abused becoming the abusers later in life, or the crime victim the criminal offender—becomes an advantage for crime detection prevention and treatment. In short, a more comprehensive focus on prevention, detection, and treatment in tandem will be more successful than current fragmented strategies.

Additionally, a dynamic approach to crime detection and prevention would allow for quicker adaptation to new types of crime (e.g., terrorism, new drug threats, cyber-bullying) that seem to expand rapidly once they emerge. This approach can move crime detection and prevention beyond the rhetoric of moral entrepreneurs and moral panic to lead to faster action at multiple prevention levels (individual and community) with multiple goals (prevention, detection, and treatment), rather than the unsubstantiated War on Crime imagery that relies on a deterrence approach (harsh, swift, and certain punishment) to crime, with the primary focus on punishment of the individual.

Integration of scientific approaches outside of the traditional criminological approach to crime detection and prevention will move beyond the deterrence approach to embrace and include the sociopsychological and the sociobiological. Such broadening of focus will lead to a more comprehensive understanding of the phenomena of addiction, the emergence of terrorism, and the creation of serial killers. Such strategies will include population-, community-, and individual-level approaches to crime detection and prevention, contributing to the understanding of crime and criminals/victims and victimization from multiple levels of analysis. An ethical foundation for future crime detection and prevention will lead criminologists to focus on the need for risk identification (for victims and offenders) that promotes community health, without the negative impact of profiling.

Finally, an aim toward social justice will lead criminologists to greater focus on fighting such human rights violations as genocide and environmental racism. Critics of environmental racism and proponents of environmental justice argue that individuals and communities cannot enjoy the social and economic benefits of prosperity and modernization without individual and community health (McNamara & Burns, 2009, p. 15). Similarly, the issues of terrorism, genocide, and human trafficking are seen not just as threats to national security and the health of our democracy, but as threats to global security.

The potential negative impact of this broader focus of preventing crime from a public health perspective is threefold. First, the data smog created by the expanded field and increased levels and units of analysis may result in policies that decrease crime and victimization without understanding it. In short, the broader detection and policy strategies that emphasize correlates to crime rather than waiting for the causes of crime may be a quicker response to restore quality of life, but at the detriment of causation research. Criminologists generally argue for creating sound theory-based policies rather than remaining content with the reduction of crime victimization and criminal offending. Second, the communication barriers with regard to both research findings and policy priorities between a broad array of scientists (criminologists, psychologists, social workers, etc.), multiple levels of policy makers (legislative and bureaucratic), and policy implementers (police, social workers, probation officers, etc.) are significant to policy success (Frenk, 1993). Third, the generation, implementation, and continuation of policies for detection, prevention, and treatment of crime and victimization are costly with regard to time and financial resources (Frenk, 1993).

Knowing the potential disadvantages of "preventing crime from a public health perspective" could lead to pessimism and cynicism about the success of this strategy. However, we argue the opposite is true—that the advantages of a more comprehensive and dynamic strategy to crime prevention, detection, and treatment can be realized by attending to and minimizing the disadvantages. An expanded and integrated field should lead to a more comprehensive focus on research and policy, reflecting integrated and dynamic cultures.

The disciplines of criminology and public health have individually succeeded in these areas, but integrating the efforts of these disciplines and addressing crime from a public health perspective will lead to an even stronger alliance.

Arguably, a higher quality of life for individuals, communities, and populations is a result worth consideration by researchers, policy makers, and justice professionals. While the often stated and implied goal of research and policy on crime control may be egalitarian in nature, the operative reality is often one of negative multiculturalism. The potential for positive multiculturalism will be seen not only in accepting the public health tenet that age and gender matter but also by addressing whether race, ethnicity, and culture matter in terms of criminological research and policy.

## ■ Discussion Questions

1. To what extent has historical criminology been concerned with public health issues?

2. Do the disciplines of criminology and public health seem compatible in terms of integration?

3. Identify and discuss the barriers that prevent researchers and policy makers from harnessing the full potential of a criminology concerned with public health issues.

4. Discuss some contemporary criminal justice policy issues that might benefit from an integration of criminology and public health perspectives.

5. Discuss the potential disadvantages of a criminology that is concerned with public health. What steps can policy makers take to address the problems identified?

## ■ References

Aguirre, A., & Turner, J. (2007). *American ethnicity: The dynamics and consequences of discrimination.* New York: McGraw-Hill.

Akers, R. (2000). *Criminological theories: Introduction, evaluation, and application* (3rd ed.). Los Angeles: Roxbury Publishing.

Bourdieu, P. (1993). Some properties of fields. In Richard Nice (Trans.), *Sociology in Question* (p. 7277). London: Sage.

Callahan, D., & Jennings, B. (2002). Ethics and public health: Forging a strong relationship. *American Journal of Public Health, 92*(2), 169–176.

Chiricos, T. (1996). Moral panic as ideology: Drugs, violence, race and punishment in America. In M. Lynch and E. B. Patterson (Eds.), *Justice with Prejudice: Race and Criminal Justice in America* (19–48). New York: Harrow & Heston.

Dowdy, E. (1994). Federal funding and its effect on criminological research: Emphasizing individualistic explanations for criminal behavior. *The American Sociologist, 25*(4), 77–89.

Foucault, Michel. (1973). *The order of things: An archaeology of the human sciences.* New York: Vintage Books.

Foucault, M. (1979). *Discipline and punish: The birth of the prison.* New York: Vintage Books.

Frenk, J. (1993). The new public health. *Annual Review of Public Health, 14,* 469–490.

Jenkins, S. (2007). *The real all Americans: The team that changed a game, a people, a nation.* New York: Random House.

Kennedy, N. (2006). Nebraska's methamphetamine treatment study: Final report. University of Nebraska at Omaha.

Kiecolt-Glaser, J. K., & Newlon, T. L. (2001). Marriage and health: His and hers. *Psychological Bulletin, 127*(4), 472–503.

Korn, D. A., & Shaffer, H. J. (1999). Gambling and the health of the public: Adopting a public health perspective. *Journal of Gambling Studies, 15*(4), 289–365.

Liberman, A. M. (2009). Advocating evidence-generating policies: A role for the ASC. *The Criminologist, 34*(1), 1–5.

Mariner, W. (2003). Public health and law: Past and future visions. Book review essay of *Public health law: Power, duty, restraint,* by Lawrence O. Gostin. *Journal of Health Politics, Policy & Law, 28*(2/3), 525–552.

McNamara, R., & Burns, R. (2009). *Multiculturalism in the criminal justice system.* New York: McGraw-Hill.

Mills, C. W. (1943). The professional ideology of social pathologists. *The American Journal of Sociology, 49, 2,* 165–180.

O'Connor, S. (2001). *Orphan trains: The story of Charles Loring Brace and the children he saved and failed.* New York: Houghton Mifflin.

Platt, A. (1977). *The child savers: The invention of delinquency.* Chicago: University of Chicago Press.

Rothman, D. (1971). *The discovery of the asylum: Social order and disorder in the new republic.* Boston: Little, Brown.

Sample, L. (2008). Final report for the evaluation of Nebraska's serious and violent offender reentry program. School of Criminology and Criminal Justice, University of Nebraska at Omaha.

Sampson, R., Morenoff, J., & Raudenbush, S. (2005). Social anatomy of racial and ethnic disparities in violence. *American Journal of Public Health, 95*(2), 224–232.

Spitzer, S. (1975). Toward a Marxian theory of delinquency. *Social Problems, 22*(5), 638–651.

Vold, G., & Bernard, T. (1986). *Theoretical criminology.* New York: Oxford University Press.

Wilson, W. J. (1996). *When work disappears: The world of the new urban poor.* New York: Knopf.

# Subject Index

*Figures and tables are indicated by f and t following the page number.*

ABA (American Bar Association), 101

Abandoned buildings, 34

Abuse
  domestic violence. *See* Domestic violence
  in prison, 150

Accountability
  confidentiality of juvenile records and, 221
  parental accountability laws, 217–218
  in problem-solving approach, 131

Accounting fraud, 294, 301–302

Accreditation of crime scene examiners, 281

Adoption and foster parenting, 174

Adoption and Safe Families Act of 1997, 174

Adult courts and justice system
  juvenile courts compared to, 214
  juvenile justice system, unification with, 218. *See also* Blended sentencing for juveniles
  transfer of juveniles to, 216, 220–221, 224

Adversarialism, 129–130

AFDC (Aid to Families with Dependent Children), 173

African Americans. *See also* Race and ethnicity; Racial discrimination
  disenfranchisement and political influence of, 152
  drug laws and, 174
  incarceration rates for, 151–152, 151f
  media portrayal of, 142
  reentry process and, 163

Age
  incarceration and, 147, 162
  juvenile court jurisdiction and, 214, 219–220, 225, 227

Agencies. *See* Federal agencies; State governments

Agenda setting
  in anti–human trafficking legislation, 319
  policy making stage, 309–311

Aggravated felony and immigration law, 251, 252

Aid to Families with Dependent Children (AFDC), 173

AIG bailout, 299

Alaska, funding for juvenile mental health treatment in, 223

Alcohol abuse
  commission of crimes and, 145
  drunk drivers, 98–101, 99t

Alcoholics Anonymous, 129

Alexandria Domestic Violence Intervention Project (Virginia), 127

Alien and Sedition Acts of 1798, 249

Al Qaeda, 55, 67

American Bar Association (ABA), 101

American Judges Association, 100, 103, 104

American Medical Association, 289

American Public Health Association (APHA) mission statement, 333

American Society of Criminology policy statements, 340

American State Administrators Project, 187

Amnesty International, 276

Anarchist bombings (1919), 52. *See also* Palmer raids (1919–1920)

Anger management programs for domestic violence offenders, 127

Anti-Drug Abuse Acts of 1986 & 1988, 169–170, 172

Anti-human trafficking. *See* Human trafficking

Antiterrorism. *See* Counterterrorism

Antiterrorism and Effective Death Penalty Act (AEDPA) of 1996, 53, 252

APHA (American Public Health Association) mission statement, 333

Arab Americans, post-September 11 treatment of, 54, 55–56, 58–61, 259

Arizona
  illegal immigrant detention services, 201
  juvenile DNA collection, 222

Arkansas, juvenile jurisdiction reform, 219–220, 222

Arrest-focused policing, 36–38

Arrests
  DNA collection and, 281, 282. See also DNA
  of illegal immigrants, 61, 62, 257–258
  for low-level offenses, 36
  for marijuana in public view (MPV), 27–28
  during Palmer raids (1919–1920), 51
  policing focused on, 36–38
  quotas for, 37
  racial inequality in, 151
  repeated. *See* Recidivism
  September 11 attacks and, 54
  violent crime rates and, 28

Arthur Andersen, 295, 296

Ashcroft, John, 60

Assessing Policy Options (APO) project (NIJ), 212, 232–234, 233*f*

Asylums, history of, 334

Asylum seekers, detainment of, 252, 261

Attentive public, 10, 13

Auditing committees, 296

Autonomy of police departments, 23

Bailout bill in financial crisis of 2007–2008, 299–300

Battered women, 125, 126, 127. *See also* Domestic violence

Bicycle patrols, 50

Biological evidence, collection of, 281. *See also* DNA

BJA. *See* Bureau of Justice Assistance

BJS (Bureau of Justice Statistics) report, 161, 162

Black Americans. *See* African Americans

Blended sentencing for juveniles, 216, 218, 225, 228

Blood and DNA analysis, 280

Bobby Ross Group, 196–197

Bombings, 52

BOP. *See* Bureau of Prisons

Border enforcement, 249, 254, 255

Border Patrol, U.S., 250, 251–252

Bracero program and Mexican immigration, 247

"Broken windows" philosophy, 36–38

Bureaucrats. *See* Federal agencies; State governments

Bureau of Immigration and Naturalization (formerly Bureau of Immigration), 248–249

Bureau of Justice Assistance (BJA)
  human trafficking grants program, 324
  "1996 Survey of State Sentencing Structures," 166
  on problem-solving courts, 119

Bureau of Justice Statistics (BJS) report, 161, 162

Bureau of Naturalization, 249

Bureau of Prisons (BOP), 189, 190, 199

California
  illegal immigrant detention services, 201
  juvenile jurisdiction reform, 221
  parole supervision, 165
  prison overcrowding, 135
  procedural fairness initiative, 95, 98, 109–110, 109*f*
  sentencing guidelines, 164
  Youthful Offender Block Grant, 223

Camden, New Jersey, police union lawsuit in, 37
Canada, crime rates compared with incarceration rates, 147
Capital Correctional Resources, Inc. (CCRI), 197
Case method for human trafficking legislation, 311–312
CBP (Customs and Border Protection), 249, 254
CCA. *See* Corrections Corporation of America
CCRI (Capital Correctional Resources, Inc.), 197
CEDs (conducted energy devices), 276–279
Censorship of counterterrorism information, 55
Center for Court Innovation on problem-solving courts, 119
Central America, immigration from, 251
Certifications of companies' financial disclosures, 296–297
Challenge of Crime in a Free Society (presidential report, 1967), 22, 34, 39
Chapin Hall Center for Children, University of Chicago, 232
Chertoff, Michael, 255
Chicago School, 336–337
Chief executive officers (CEOs), 293, 296–297
Child custody and reentry, 174–175
Children. *See also headings starting with "Juvenile"*
    conducted energy devices use on, 278–279
    early childhood intervention programs, 149
    parental incarceration and, 149, 174
    in sex tourism trade, 321, 322
Children's Aid Society, 334
"Child Savers," 335
Child sex tourism, 321, 322
China and Falun Gong, 63
Chinese Exclusion Acts of 1882–1943, 249–250
Church Committee (Senate), 51
Cities. *See headings starting with "Urban"*
Citizenship, eligibility for, 249

CiviGenics, 196
Civil rights movement, 143
Civil rights of U.S.-born citizens, compared with immigrants, 247
Classical perspective of policy-making process, 5–6
Clinton, Hillary, 312–313
Clinton welfare reform, 173
CODIS (Combined DNA Index System), 281
Cognate form of white-collar crime, 290, 292
Colorado, juvenile justice system, 222, 223
Columbine High School tragedy, 78
Combative policing, opposition to, 60
Combined DNA Index System (CODIS), 281
Commission on Security and Cooperation in Europe (CSCE), 314
Commission on Trial Court Performance, 103
Community Corrections Act of 1971 (Minnesota), 190
Community Education Centers, Inc., 196
Community Oriented Policing Services (COPS), 34, 50, 56, 67, 84, 337
Community–police collaboration, 26, 29, 34
Community policing, 33–39
    in Arab American neighborhoods after September 11 attacks, 59–60
    counterintelligence from, 60
    funding cuts after September 11 attacks, 56
    in New York City, beginnings of, 36
    principles of, 35, 50
    professional policing vs., 35
    provision of services, 34
    replaced by homeland defense as catchphrase, 55
    in San Diego, California, 37–38
    school resource officers (SROs) and, 79
    violence prevention in schools and, 84
Community programs. *See also* Community policing
    collaboration with correctional agencies, 175
    domestic violence, 127–128
    juveniles, treatment for, 222

Competency hearings, 228
Competition in private prison industry, 183, 198–202
Computerized crime statistics (CompStat), 36, 37
Concentration ratio (CR) in private prison industry, 190–198, 192–193*t*
Conditional releases, 164
Conducted energy devices (CEDs), 276–279
*Confessions of a Subprime Lender* (Bitner), 301
Confidentiality of juvenile records, 216–217, 221–222, 228–229, 234
Conflict resolution programs, 83
Conflicts of interest, 294, 296
Conflict theory and police behavior, 27
Congressional hearings
    on corporate scandals (2001–2002), 295–296
    on human trafficking, 314–316, 318, 319, 320
Connecticut, juvenile court jurisdiction in, 219
Conservatives' stance on immigration, 246
Conspiracy in human trafficking, 321
Constitutional rights
    homeland security tactics and, 57, 64
    protection of, 29
Consumer groups, 291
Contextual orientation to policy research, 6–7
Contingency theory and court reform, 102–103
Continuing Financial Crimes Enterprise Act of 1990, 301
Contract monitoring in private prisons, 187
Cooperative antiterrorism planning, 53
COPS. *See* Community Oriented Policing Services
COPS in Schools (CIS), 84–85
Cornell Corrections, Inc., 196, 200
Corporations. *See also* White-collar crime
    congressional hearings on scandals of (2001–2002), 295–296
    environmental practices of, 292
    Sarbanes-Oxley Act, impact on, 295–298
    self-policing by, 289
Correctional institutions. *See* Prisons

Correctional Services Corporation (CSC), 195
Correctional Systems, Inc., 196
Corrections Corporation of America (CCA), 189, 193–194, 201
Corrections policy, 135–157. *See also* Incarceration; Prisons; Privatization of prisons
    community impact of, 149–150
    consequences of, 148–152
    costs of incarceration and, 137, 138*f*, 149
    crime rate impact on, 139–141, 143–148
    deterrence as goal of, 144–145
    family impact of, 149–150
    fear of crime and, 141–142
    incapacitation as goal of, 146–147
    incarceration rate and, 136–139, 137–138*f*
    news media and, 142–143
    prison living conditions, 136, 150–151
    public officials and, 142–143
    racial inequality and, 151–152, 151*f*, 162
Corruption, 23, 307
Costs
    of crime reduction, 148
    of drug courts, 123–124
    of incarceration, 137, 138*f*, 149
    of problem-solving courts, 128
    of public vs. private prisons, 186, 199, 201
    of Sarbanes-Oxley Act compliance, 297
    of technology, 273, 274, 275–276
Council of State Governments, 175
Counterintelligence from community policing, 60
Counterterrorism
    combative vs. cooperative strategies, 56–58
    immigration enforcement and, 254, 255, 259–260
    law enforcement agencies, role of, 52, 53, 65
"CourTools," 107
Courts. *See also* Judges
    consolidation of, 101
    domestic violence courts, 124, 125–128, 130

evidence admissibility standards, 280
jurisdiction of, 214
juvenile. *See* Juvenile courts
nonverbal communication, 104, 106
performance of, 103, 107–108
on privatization of prison services, 189
procedural fairness and, 95–96, 99, 101–106, 108–110
public trust in, 109
reform of, 102–103, 118
specialized courts, 117–134. *See also* Problem-solving courts
staff of, 104, 107
CR. *See* Concentration ratio (CR) in private prison industry
Crack cocaine, 336
Credit rating agencies, 301
Crime rates
decline in, 28, 161
incarceration rates and, 139–140, 141, 143–148
mass media and perception of, 13
research on, 147–148
trends in, 139–141
Crime reduction
adaptation to new types of crime, 343
arrest-focused policing and, 37–39
costs of, 148
mass incarceration and, 144
parenting programs and, 149
policing policies and, 29
racial profiling and, 39
technology and, 273
Crime scene examiners, 281
Crimes of moral turpitude and immigration law, 252
Criminal aliens, 253
Criminal courts. *See* Courts; Drug courts; Juvenile courts
Criminal justice, 1–18
criminology described, 1–2
deterrence theory as dominant in, 100
initiatives in, 175
outcomes and improved technology in, 273
policy-making process of, 7–15
actors in, 9–12
forces in, 7, 8f

networks and relationships within policy community, 12–15
unique aspects of, 7–9, 8f
as subfield of criminology, 4–5
subgovernment and attentive public, 10–12
technology. *See* Technology
Criminal record checks, 174
Criminal record databases, 222
Criminology
conceptual base, 339, 340–341
conceptual overlap with public health, 338–342
defined, 1–2
integration with public health, 336–338
interdisciplinary approach to, 333
policy-making process and, 3–5
popular conceptions of right and wrong and, 13
potential for growth based on expanded research, policy and treatment, 343–345
production base, 339, 341
public health problem, crime as, 333–346. *See also* Public health problem, crime as
reproduction base, 339, 341
research in, 3, 336
utilization base, 340, 341
"Crisis" situations as drivers of public policy, 336–337, 338
CSCE (Commission on Security and Cooperation in Europe), 314
Cuban immigrants, 251
Cultural cues, 32
Customs and Border Protection (CBP), 249, 254

DARPA (Defense Advanced Research Projects Agency), 271
Daubert standard, 280
Dearborn, Michigan, residents from local Arab community, 59
Death penalty in juvenile cases, 225
Decentralization and innovation in trial courts, 103
Defense Advanced Research Projects Agency (DARPA), 271

Defense attorneys, role in problem-solving courts, 130
Defense Department
    technology research by, 272
    Threat And Local Observation Notice (TALON), 57
    white-collar crime investigations, 289
Deferred prosecution agreements, 299
Deficit Reduction Act of 2005, 177
Delaware
    juvenile jurisdiction reform, 221
    school crime reporting laws, 84
*Department of Housing and Urban Development v. Rucker* (2002), 172
Deportation, 61–62, 247, 249, 250, 252
Deregulation, role in financial crisis of 2007–2008, 300
Detention
    of asylum seekers, 252, 261
    of illegal immigrants, 189, 201, 251, 254, 255, 258, 260–261, 281
    in juvenile facilities, 213
    of noncitizens post-September 11 attacks, 60–61
    of suspected terrorists, 54
Determinate sentencing, 120, 164, 165
Deterrence. *See also* Crime reduction
    broadening of approach to, 343–344
    as corrections policy goal, 144–145, 163
    in domestic violence cases, 97
    as dominant criminal justice policy, 100
    immigration policy and, 254, 257, 258, 260
    procedural fairness vs., 97–98, 100
Development grants for technology, 269
DHS. *See* Homeland Security Department
*Dilworth, People v.* (1996), 89
Disadvantaged neighborhoods, 29, 159–160, 174. *See also* Urban policing
Discipline in schools, 89
Discretionary release, 164–165. *See also* Parole
Discrimination. *See* Racial discrimination
Disenfranchisement of convicted felons, 152
Distributive justice, 98–99, 99*t*, 101, 104
District courts, 117
Diversity. *See* Race and ethnicity

Diversity training for police, 40
DNA, 279–283
    analysis of, 280
    collection from arrestees, 282
    databases, 281–282
    handling of, 280–281
    juvenile cases, 221–222, 225, 227
Domestic Abuse Intervention Project (Minnesota), 127
Domestic violence
    courts for, 124, 125–128, 130
    offenders, 97
    procedural fairness policies and, 97–98
    recidivism rates in, 97
DoveDevelopment Corporation, 197–198
Driving while intoxicated, 98–101, 99*t*
Drug Abuse Resistance Education (DARE), 335
Drug abusers. *See* Drug offenders
Drug courts, 120–125. *See also* Drug offenders
    as alternative to traditional courts, 136, 335
    best practices, 131
    cost savings of, 123–124
    nonviolent offenders in, 129
    procedural fairness and, 107
    research on, 122–125
    supervision of drug offenders by, 121
Drug-free schools, 82
Drug-Free Schools and Communities Act of 1986, 82–83
Drug offenders. *See also* Drug courts
    commission of crimes and, 142, 145
    court supervision of, 121
    expanding outside of traditional approaches to, 343–344
    federal and state education loans for, 170
    immigrants as, deportation of, 252
    juvenile cases. *See* Juveniles
    mandatory minimum sentences and, 166
    public assistance and, 173
    racial inequality in arrests of, 151
Drug prevention, 335
Drug treatment
    as alternative programs, 120, 335
    juvenile offenders, 222, 223, 234

mandatory, under supervision of drug courts, 121–123
release planning, 167
Drunk drivers, 98–101, 99*t*
Due process, 96, 129–130

Early adopters of technology, 271–272
Early childhood intervention programs, 149
Eclectic Communications, Inc., 189
Economic factors
in financial crisis of 2007–2008, 300
incarceration's impact on economic inequality, 152
policing and, 29
of technology adoption, 272
Education. *See also* School–police partnerships
federal loans for, 170
of prisoners, 169–170
reentry process and, 167, 169–170
Eighth Amendment, 135
Electronic eavesdropping, 51
Elementary and Secondary Education Act of 1965, 82
Embassies, terrorist attacks on, 52
Emotional factors in crime policy, 8
Employment of ex-offenders, 150, 152, 163, 167, 169
Enemy combatants, 54
Enron, 293–296
Environmental movements, 291
Environmental Protection Agency, 289
Environmental racism and justice, 336, 344
Epidemiology, 340
Equal protection clause violations, 31
Esmor Correctional Services, Inc., 195
Ethnicity. *See* Race and Ethnicity
Eugenics as form of crime control, 334
Europe
juvenile justice, 213–214
restricted immigration from, 249
Evictions from public housing, 171, 172
Evidence, 279–283. *See also* DNA
Exclusion as immigration policy, 249
Ex-offenders
rearrest of. *See* Recidivism
reentry process for. *See* Reentry
Expiration release, 165–166
Expunged records, 216–217, 221–222, 228

FAFSA (Free Application for Federal Student Aid), 170
Falun Gong (a spiritual exercise) in China, 63
Familial searching in DNA databases, 282
Families
corrections policy impact on, 149–150
ex-offenders reconnecting with, 174
reentry process and, 168
Family courts. *See* Juvenile courts
Fannie Mae, 299–300, 301–302
Federal agencies. *See also specific agencies (e.g., Justice Department)*
criminal justice policy and, 9–12
human trafficking legislation
drafting and testifying about, 315, 316
implementation of, 317–319, 318*f*
immigration, 248–249, 254
public distrust of, 58
white-collar crime responses by, 289
Federal Bureau of Investigation (FBI), 51, 53–54, 55, 56
Federal Judicial Center conference on international exploitation of women and children, 313
Federal Trade Commission, 289
Felony convictions
aggravated felony and immigration law, 251, 252
reentry process and, 168
Fenton Security, Inc., 196
Financial crisis of 2007–2008, 66, 292, 299–302
Financial restatements filed in response to Sarbanes-Oxley Act, 297
Financier crime, 293, 299–302. *See also* White-collar crime
Fingerprinting, 217, 221, 222
First responders, funding for, 53
Florida
conducted energy device use, 277, 278
juvenile correctional facilities, 203
Florida State University, 232
Foot patrols in community policing, 34, 35
Forced labor. *See* Human trafficking
Ford Foundation, 338
Foreign Intelligence Surveillance Court, 51
Foreign students, interviews of, 58
Forensic DNA. *See* DNA

Forensic scientists and laboratories, accreditation, 281
Forensic technicians, 281
Foster parenting, background checks on, 174
Fourth Amendment rights, 51
Frankfurter, Felix, 96
Fraud, 294, 301–302
Freddie Mac, 299–300, 301–302
Free Application for Federal Student Aid (FAFSA), 170
Frye standard, 280
*Fuerzas Armadas de Liberación Nacional*, terrorist activities of, 52
Fugitive slave laws, 24

GAO (Government Accountability Office) report on cost-effectiveness of private vs. public prisons, 186
Gender
    incarceration and, 162
    procedural fairness and, 97
Genetic profiling, 282
Genocide, 344
GEO Group, 195–196, 197, 200
Georgia Association of Chiefs of Police, 278–279
Get-tough policies
    immigration policies, effect on, 251, 254
    incapacitation and, 145
    incarceration boom and, 139
    juvenile justice, 212, 216, 219, 223–224
    post-September 11 government policies, 63
    sentencing and, 144
Goals 2000: Educate America Act of 1994, 82
Governance of corporations and Sarbanes-Oxley Act, 295–298
Government Accountability Office (GAO), 186
Gross incapacitation, 146
Group crime, replacement effect in, 147
Group 4 Falck, 195
Guest worker legislation, 252
"Guidelines for the Interviews Regarding International Terrorism" (Office of the Deputy Attorney General), 55
Gun control, 31

H.A. Guggenheim Foundation, 338
Habeas corpus, 254
Haiti, immigration from, 251
Halfway houses, 166, 190
Hands-on tactics, injuries from, 277
Harassment by police, 25, 31
Hate crimes, 260, 336
Hawaii, juvenile justice system, 223
Hawes-Cooper Act of 1929, 185
Health and Human Services Department, 85, 317, 318*f*
Health care, 167
Hennepin County District Court (Minneapolis), 105–106
Herfindahl–Hirschman Index (HHI), 192, 193, 194*t*
Higher Education Act (HEA) of 1965, 169, 170
Higher Education Opportunity Act of 2008, 169
High Performing Courts project, 107
"Holder memo" on corporate crime prosecutions, 298
Home confinement as alternative to prison, 136
Homeland security, 52–55, 57
Homeland Security Department (DHS)
    creation of, 54
    human trafficking
        enforcement, 317, 318*f*
        policy making, role in, 315, 316
    immigration agencies as part of, 249, 254
    immigration policies and, 247
    recommendation for prior to September 11 attacks, 53
Homicides, 220–221, 280
Hoover, J. Edgar, 51
Housing and Urban Development Department, 172, 289
Housing for ex-offenders, 163, 167, 170–173
Housing initiatives of federal government, 170–171
Housing Opportunity Program Extensions (HOPE) Act of 1996, 171
Human rights violations, 336
Human trafficking, 307–331
    agenda-setting stage of policy making, 309–310, 319

background, 308
case method for analysis, 311–312
civil suits against traffickers, 321
congressional hearings on, 314–316, 318, 319, 320
criminal law provisions, 319–322
evidentiary challenges in, 321
framework of policy making in, 308–311, 309*f*
grants program from Justice Department, 323–324
implementation stage of policy making, 310–311
interest groups' role, 316–317, 316*f*, 322
legislation, 312–319
    chronology of, 312–316, 313*f*
    implementation of, 317–318, 318*f*
    limitations of 2000 law, 317
legislative stage in policy making, 310
methodology for analysis, 311–312
raising awareness of, 312–313
reauthorization stage of policy making, 311
state and local enforcement task forces, 323–324
as threat to global security, 344
victim assistance programs, 322–323
visas for victims of, 317

ICE. *See* Immigration and Customs Enforcement
Idaho
    funding for mental health treatment and facilities, 223
    juvenile justice system, 219
IIRIRA (Illegal Immigration Reform and Immigration Responsibility Act of 1996), 62, 252
Illegal immigrants. *See also* Immigration policies
    arrest of, 257–258
    criminalization of, 250, 251
    criminal vs. noncriminal, 61
    detention of. *See* Detention
    effectiveness of deterrence, 257, 259–260
    policies protecting, 256–257
    suspected, arrests of, 62
    as term depicting person's status, 253

Illegal Immigration Reform and Immigration Responsibility Act (IIRIRA) of 1996, 62, 252
Illinois, juvenile DNA collection and, 222
*Illinois v. Wardlow* (2000), 32
Immaturity of juveniles as reason for law breaking, 213–214
Immigrant Reform Act of 1990, 251
Immigration Act of 1882, 249
Immigration Act of 1917, 249
Immigration Act of 1921, 249
Immigration Act of 1924, 249
Immigration Act of 1965, 250
Immigration and Customs Enforcement (ICE), 199, 249, 254, 255
Immigration and Naturalization Service (INS), 247, 249, 251, 252, 254
    privatization of detention services, 189, 190, 195
Immigration policies, 245–264. See also Illegal immigrants
    arrest of illegal immigrants, 257–258
    background, 246–247
    contemporary challenges of, 252–257, 260
    crime control model's consequences, 259–261
    get-tough policies and, 251, 254
    historic benchmarks, 248–252
    human trafficking victims, visas for, 317
    liberal vs. conservative, 246
    national security and, 257–258
    post-September 11 policies, 60–63, 254–255
    punitive aspects of, 246
    racism, 247
    raids, 255, 256
    reform, 252–253
    sanctuaries for undocumented aliens, 256–257
    September 11 attacks and, 252–253
    social problems and, 245
    USA PATRIOT Act, effect of, 254–255
Immigration Reform and Control Act (IRCA) of 1986, 251
Implementation stage of policy making, 310–311
Imprisonment. *See* Incarceration

Improving America's Schools Act of 1994, 82
Incapacitation, 145, 146–147, 163
Incarceration. *See also* Mass incarceration; Prisons
  community impact of, 149–150
  consequences of, 148–152
  costs of, 137, 138*f*, 149
  crime rates and, 139–140, 147–148
  drug abusers and, 120
  family impact of, 149–150
  for felonies, 146
  international comparisons, 137, 138*f*
  prevalence of, 161–162
  race and ethnicity and, 151, 151*f*
  recidivism and, 150
  as rite of passage, 145
  trends in, 136–139, 137–138*f*, 143–148
  violent crime rates and, 140
  of women, parental rights and, 174
Incident-driven policing, advent of, 50
Indeterminate sentencing, 164
Indiana, funding for specialized courts, 222
Information sharing in juvenile justice, 221
Injuries in use-of-force situations, 277
*In loco parentis*, schools' role, 88
Inmates. *See* Prisoners
Innovation
  adopters of, 271–273
  in criminal justice technology, 266
  decentralization in trial courts and, 103
  in private prisons, 184, 200
Insider trading, 195–196
Institutional fragmentation, 7–8
Insurance mandates for court-ordered mental health services, 223
Intelligence agencies, 51, 57
Interdiction, efficacy of, 257
Interest groups
  case studies of, 311–312
  criminal justice policy-making role of, 11–12, 13
  human trafficking and, 316–317, 322
  human trafficking legislation and policy making, role of, 314
  policy making and, 308
Intergovernmental policy community, 6, 7
Internal Revenue Service (IRS), 289
International Association of Chiefs of Police, 275, 278

Intervention programs, 149
Investment banking houses. *See* Financier crime
Investment fraud, 294
Involuntary servitude, 308
IRS (Internal Revenue Service), 289
Issue advocacy, 340
Italy and Red Brigades, 63

Japanese-American internment during World War II, 63, 247
Job Corps, 189, 195
Judges
  mandatory sentencing laws and, 166
  policy-making role of, 11, 101
  problem-solving courts, role of, 129–130
  procedural fairness policy and, 100, 103–104
Judicial Oversight Demonstration (JOD) sites for domestic violence court services, 127–128
Jurisdiction
  of juvenile courts, 214, 219–220, 227
  of problem-solving courts, 129
Jury experience, impact of, 111
Justice Department
  conference on international exploitation of women and children, 313
  corporate crime and, 298
  criminal justice initiatives, funding for, 175
  domestic violence cases, grants for, 127
  human trafficking legislation

    enforcement role in, 317, 318*f*
    grants program, 323–324
    policy making role in, 315, 316, 318, 319, 322
Judicial Oversight Demonstration (JOD) sites for domestic violence court services, 127–128
  Office of Legal Counsel's memo on warrantless searches, 55
  problem-solving courts, support for, 102
  procedural fairness reforms and, 101
  "Stewards of the American Dream" report, 119
  white-collar crime policy of, 295–297

Juvenile courts. *See also* Juvenile justice
    compared to adult courts, 214
    get tough policies and, 212, 216, 219,
        223–224
    integration with social service agencies,
        217
    jurisdiction of, 214, 219–220, 227
    origin of, 213, 335
    process in, 214, 215*f*
    in progressive juvenile justice, 229
    research on, 232
    role of, 213
    as specialized courts, 117, 124
    treatment and adjudication in, 120
Juvenile justice, 211–241. *See also* Juvenile
        courts
    age for jurisdiction, 214, 219–220, 225,
        227
    competency proceedings and standards,
        228–229
    confidentiality and records, 216–217,
        221–222, 228
    court jurisdiction. *See* Juvenile courts
    death penalty, 225
    origin, evolution, and mission of, 213–
        215
    *parens patriae* doctrine, 214
    parental accountability laws, 217–218
    policies, 230
    practitioner perspectives, 232–234, 233*f*
    privatization of detention services, 189,
        203
    progressive indicators, 228–229
    progressive scores, 229–232, 230f
    punishment vs. rehabilitation, 224–232.
        *See also* Punitive juvenile justice
    punitive indicators, 224–225, 226*f*
    punitive scores in, 225–228, 227*f*
    purpose clause of state as guidance, 229
    recommendations for, 234
    reform, 211–212, 216–218, 223
    specialized services in, 217
    state legislation, 218–224
    state systems, 213
    transfer, sentencing, and penalties, 220–
        221, 224
    treatment and rehabilitation, 222–224
    victims' rights, 217
Juvenile Justice and Delinquency
        Prevention Act of 1974, 337

Juveniles
    detention facilities for, 189, 203, 213
    DNA collection from, prevalence of, 227
    drug offenders, 106–107, 216, 217, 222,
        223, 234
    immaturity as reason for law breaking by,
        213–214
    life without parole for, 227–228
    prostitution and trafficking, 317
    recidivism rates, 203
    sentencing guidelines for, 216
    transfer to adult courts, 216, 220–221
Juvenile tactical officers. See School
        resource officers (SROs)

Kaczynski, Theodore ("Unabomber"), 52
Kansas, juvenile arrest records and DNA
        collection, 222
*Katz v. United States* (1967), 51
Kerner Commission (1968) on racial
        ghettos, 24, 26

Labor Appropriation Act of 1924, 250
Labor Department and human trafficking
        enforcement, 317, 318*f*
Labor unions and prison privatization, 185,
        187, 188
Late adopters of technology, 271, 272
Law and order politics, 143
    emotional aspects of, 8
Law enforcement. *See also* Police;
        Post-September 11 attacks and law
        enforcement; Urban policing
    conducted energy devices (CEDs) in,
        276–279
    evolution of, 49–50
    federal, 54, 57
    forensic DNA and, 279–283
    human trafficking
        laws, implementation of, 320–321
        task forces for, 323–324
    immigration control and, 255, 257
    partnerships with academic researchers,
        275
    post-September 11, 49–76. *See also*
        Post-September 11 attacks and law
        enforcement
    school policing as fastest growing area
        of, 80
    technology and, 265–268

Law Enforcement and Corrections Technology Advisory Council (LECTAC), 268
Law Enforcement Assistance Administration (LEAA), 101–102
Law Enforcement Management and Administrative Statistics on local planning for terrorist attacks, 59
LECTAC (Law Enforcement and Corrections Technology Advisory Council), 268
Legal Action Center, 173
Legislation
  anti–human trafficking, 312–319
  juvenile justice, 218–224, 230. *See also* Juvenile justice
  for reentry process, 167–176
  sunset provisions in, 337
Lehman Brothers, 299
Leniency of courts, 164
Less lethal technologies, 276–279
Liberals' stance on immigration, 246
Life without parole for juveniles, 225, 227–228
Literacy tests for immigrants, 249
Local governments
  counterterrorism and, 56, 61, 62–63
  as immigrant sanctuaries, 256–257
Louisiana, juvenile jurisdiction reform, 221
Low-level offenses
  arrests for, 36
  private contracting of incarceration services, 190

Maine, sentencing guidelines, 164
Management and Training Corporation (MTC), 189, 190
Management information systems, Office of Court Drug Treatment Programs (New York State), 130
Mandatory arrest of domestic violence offenders, 97
Mandatory release, 166
Mandatory sentences
  deterrence impact of, 135, 139, 144
  drug offenses, 121
  juveniles, 216
  reentry process and, 165, 166, 170
Mann Act of 1910, 308
Maranatha Production Company, 198

Maricopa County, Arizona, immigration checks of suspicious persons, 62, 63
Marijuana in public view (MPV), arrests for, 27–28
Market concentration in private prison industry, 190–198, 192–193*t*
Marketing of technology, 274
Mark-to-market valuations, 294
Marriage, stable social effects of, 337
Marshals Service, U.S., 199, 200
Massachusetts, juvenile drug offenders in, 222
Mass arrests, authorized by Pentagon/Twin Towers Bombing (PENTTBOM) investigation, 61
Mass incarceration, 144, 150, 160. *See also* Incarceration
Mass media
  coverage of crime, 13–14, 287–288
  on incarceration boom, 136, 142–143
  on white-collar crime, 290
"McNulty memo" on corporate crime prosecutions, 298–299
McVeigh, Timothy, 52
Mediation, 107
Medical care in prisons, 150
Megan's law, 288
Mental health courts for juveniles, 217
Mental health treatment, 222, 223
Mentally ill and criminally insane, 335
Metal detectors in New York City schools, 77
Mexico
  immigration from, 201, 247, 251–252
  NAFTA and, 251
Michigan
  juvenile records, 222
  post-September 11 investigations of Arab population in, 58–59
Michigan State University, foreign students interviews at, 58
Military technology R&D, 272
Minnesota, procedural fairness initiative in, 100, 105–106
Minorities. *See* Race and ethnicity
Money laundering, 321
Montana, juvenile records in, 222
Mortgage market and financial crisis of 2007–2008, 300–301

MTC (Management and Training Corporation), 189, 190
Multicultural competence (MCC) model, 40–41
Multiculturalism. *See* Race and ethnicity
Murders, 220–221, 280
Muslims in U.S., post-September 11 treatment of, 54, 55–56, 58–61, 259

NAFTA (North American Free Trade Agreement), 251
Narcotics Anonymous, 129
National Advisory Commission on Civil Disorders, 22
National Association of Drug Court Professionals, 121
National Association of School Resource Officers (NASRO), 80, 86
National Center for Education Statistics' School Survey on Crime and Safety of, 78
National Center for State Courts, 102, 107
National Center on Juvenile Justice (NCJJ), 229
National Crime Victimization Survey, 145
National Criminal Justice Association on juvenile accountability, 217
National DNA Index System (NDIS), 281
National Education Goals Panel, 82
National Guard and border enforcement, 255
National Institute of Education's Safe Schools Study, 81
National Institute of Justice (NIJ)
    Assessing Policy Options (APO) project and juvenile justice, 212, 232–234, 233*f*
    COPS program. *See* Community Oriented Policing Services
    investment portfolio of, 267, 267*t*
    research, design, testing, and evaluation (RDT&E) model of, 268–271, 269*f*, 272
    technology needs identified by, 265
National Institute of Standards and Technology, 271
National Judicial College, 102
National Research Council, 28, 34

National security, 52–55
    immigration and, 252–253, 254, 257–258
National Security Presidential Directive 22 (NSPD-22) on human trafficking, 315, 323
National Sheriffs' Association, 87
NDIS (National DNA Index System), 281
Nebraska, funding for substance abuse and mental health treatment, 222
Neo-Marxist theories, 335
Net widening, described, 129
Nevada, legalized prostitution, 320
New public management (NPM) philosophies, 185
News media. *See* Mass media
New York City
    class differences of feelings of safety, 30
    crime reduction, 37–39
    immigration violations, arrests solely for, 62
    police behavior in schools, 88
    school security and safety, 77
New York City Police Department (NYPD)
    community policing as philosophy of, 36
    counterterrorism capabilities, 65
    quality-of-life (QOL) approach, 27–28
    stop, question, and frisk (SQF) policy, 30–32
New York Civil Liberties Union, 88
New York State
    juvenile justice system in, 219
    Lobbying Commission, 195
    Office of Court Drug Treatment Programs, 130
    recidivism rates and drug courts, 130
    Safe Streets, Safe City Act (1991), 36
Nightsticks, 276
NIJ. *See* National Institute of Justice
911 emergency phone number, 50
NJCC (National Center on Juvenile Justice), 229
No Child Left Behind Act (NCLB) of 2001, 83
Nongovernmental groups, human trafficking hearings and, 314, 315
Nonprofit organizations, prison operation by, 202, 203

Nonverbal communication in courtroom, 104, 106
Nonviolent offenders in drug courts, 129
North American Free Trade Agreement (NAFTA), 251
North Carolina
  juvenile justice system, 219
  juvenile record confidentiality, 222
Northern Ireland, 63

Oakland Housing Authority evictions, 172
Obama, Barack, 67
Occupational crime, 290
Offender registries, 217
Offenders. *See also* Prisoners
  domestic violence
    court system and, 125
    treatment for, 126–127
  perception of judges' fairness by, 103
  problem-solving courts and, 129
  satisfaction with sentencing of, 100–101
Office of Court Drug Treatment Programs (New York State), 130
Office of Juvenile Affairs in Oklahoma, 223
Office of National Drug Control Policy, 122
Office of Superintendent of Immigration, 248
Office of the Inspector General, Pentagon/Twin Towers Bombing (PENTTBOM) investigation, 60–61
Ohio, juvenile justice in, 222
Oklahoma
  juvenile transfers to adult court, 220–221
  Office of Juvenile Affairs, 223
Oklahoma City bombing, 52
Oleoresin capsicum (pepper spray), 276
Oligopoly in private prison industry, 191
Oligopsony in private prison industry, 200
Omnibus Appropriations Act of 2009, 176
Omnibus Crime Control and Safe Streets Act of 1968, 34, 79
One-strike policy in public housing, 171, 172
Operational pilot testing of technology, 275
Operation Blockade, 252

Operation Gatekeeper, 252
Operation Return to Sender, 61
Operation Safeguard, 252
Oregon, juvenile treatment and rehabilitation in, 223
Organizational destabilization from adoption of new technology, 273
Organized terrorism, defined, 52
Orphan Trains as public health initiative, 334

Palmer raids (1919–1920), 51, 58, 250
*Parens patriae*, 214, 335
Parental accountability laws, 217–218
Parental incarceration, 149
Parental rights, termination of, 174
Parenting programs, 149
Parole, 136, 163, 164–165, 166, 167, 213. *See also* Release
Partial privatization, 189
PATRIOT Act. *See* USA PATRIOT Act of 2001
Payoff scandals, 195
*Payton v. New York* (1980), 51
Pell Grant program, 169–170
Pentagon/Twin Towers Bombing (PENTTBOM) Investigation, 53, 58, 60–61
*People v. See name of opposing party*
Pepper spray (oleoresin capsicum), 276
Performance standards for courts, 103
Personal Responsibility and Work Opportunity Reconciliation Act (PRWORA) of 1996, 173
PHAs (Public Housing Authorities), 172, 173
Pirate Capital, LLC, 196
Plea bargains in state courts, 120
Police. *See also* Law enforcement; Policing
  behavior of, 27
  biological evidence collection and, 281
  community and, 35
  in community vs. professional policing, 35
  domestic violence situations and, 127
  functions of, 33
  juvenile units, 213

number of, 30, 35

safety of, technology improvements for, 274

school partnerships with. *See* School–police partnerships

training inadequacies of, 26

use-of-force continuum for, 276

Police batons, 276

Police brutality, 25, 29

Police–community relationships, 26, 29, 34, 38–39, 50, 63, 256

Police departments, 272

Police–federal agency cooperation, 60

Policing. *See also* Community policing; Urban policing

  aggressive vs. collaborative, 38

  combative, opposition to, 60

  constitutional rights and, 29

  crime reduction and, 29

  incident-driven, 50

  police–community relations and, 29

  professional model of, 23

  response time, 50

  zero-tolerance, 336

Policy research, 6–7

Political era policing (1840s–early 1900s), 21, 23, 34

Politicians' role in policy making, 11

Postal Inspection Service, U.S., 289

Postrelease supervision, 164, 167. *See also* Parole

Post-September 11 attacks and law enforcement, 49–76

  Arab Americans, treatment of, 54, 55–56, 58–61, 259

  arrests after, 54

  Ashcroft's and Mueller's reactions to, 53

  combative vs. cooperative strategies, 55–60

  community policing after, 56, 59–60

  comparing local law enforcement responsibilities before, 65

  detention of noncitizens after, 60

  FBI budget increase after, 56

  federal law enforcement agencies, reorganization after, 54

  financial crisis of 2007–2008, effect of, 66

  foreign student interviews and, 58

  homeland security and, 52–55

  immigration policies and, 60–63, 252–253, 254–255

  law enforcement mission after, 52

  local law enforcement agencies after, cuts in funds, 56

  political shift after 2008 elections, 66–67

  reduction of funding for Community Oriented Policing Services (COPS) after, 56

  U.S. unpreparedness for, 53

Predictive force and distributive justice, 101

President's Commission on Law Enforcement and Administration of Justice (1967), 26

Pretrial diversion in corporate cases, 298

"Principles of Federal Prosecution of Business Organizations" (Thompson memo), 298–299

Prisoners. *See also* Prisons

  education of, 169–170

  Eighth Amendment rights of, 135

  expiration release and, 166

  family roles of, 149

  Higher Education Opportunity Act and, 169

  Pell grants and, 170

  in private prisons, 186, 189

  prospects after release for, 150

  racial disparity in, 162

  rehabilitation goal for, 136

  release of. *See* Release

  vocational training for, 136

Prisons. *See also* Incarceration; Prisoners; Privatization of prisons; Reentry

  costs of operating, 186

  education in, 169–170

  ex-offenders' return to, 167. *See also* Recidivism

  history of, 184–185

  living conditions, 136, 150–151

  officer pay rates, 201

  overcrowding in, 135, 136, 139

  population growth in, 160, 161

  as public good, 184–185

  rehabilitation and treatment in, 162, 164

  release from. *See* Release

  security in, compared with schools, 78

Private vs. public interest, 185

Privatization of prisons, 183–209
   companies with success in, 193–196
   competition and innovation in, 198–202
   failed companies, 196–198
   growth rates of, 191, 191*t*
   incarceration services market, 188–190
   low-level offender services, 190
   market concentration, 190–198,
      192–193*t*
   opposition to, 188, 189
   reasons for, 184–186
   research on, 186–188
Privileged information waivers, 298–299
Probation, 104, 120, 121
   of juvenile offenders, 213
Problem-solving approach, 131
Problem-solving courts, 117–134
   adversarialism and due process, 129–130
   background of, 118–120
   best practices, 130–131
   broad mandate of, 119
   cost issues, 128
   described, 119
   domestic violence courts as, 124,
      125–128
   drug courts as, 120–125
   jurisdiction of, 129
   juvenile courts as, 117
   procedural fairness and, 106, 107
   provision of services by, 130
   societal problems resolved in, 129–130
   traditional courts vs., 111
   treatment as priority for, 119
Procedural fairness, 95–116
   court staff, policies for, 103–104
   defined, 96–101
   deterrence model vs., 97–98
   for domestic violence offenses, 97–98
   for drunk driving offenses, 98–101, 99*t*
   due process vs., 96
   judges, policies for, 103–104
   policy design factors, 106–108
   policy making for, 101–103, 108–110
   public input on policies, 110–112
   statewide court policy making and,
      108–110, 109*f*
   trial court policies for, 104–106
Professional model of policing, 23, 35
Profiling, 29, 39, 56, 344

Progressive juvenile justice
   indicators, 228–229
   scores, 229–232, 230*f*
Property crimes, DNA evidence and,
      281
Prosecution
   of corporate crime, 298–299
   of financier crime, 295
   as goal in domestic violence, 127
   in problem-solving courts, 130
   of white-collar crime vs. conventional
      crime, 291
Prostitution. *See also* Human trafficking
   Nevada law on, 320
PRWORA (Personal Responsibility and
      Work Opportunity Reconciliation
      Act) of 1996, 173
Public assistance, 173–174
Public choice perspective, 185
Public Company Accounting Reform and
      Investor Protection Act of 2002.
      *See* Sarbanes–Oxley Act of 2002
Public distrust
   of government agencies, 58
   of police, 88
Public good, correctional services as,
      184–185
Public health problem, crime as, 333–346
   conceptual overlap of criminology and
      public health, 338–342
   contemporary convergence of theory and
      policy, 335–336
   education and, 341
   historic convergence of theory and policy,
      334–335
   integration of criminology and public
      health, 336–338
   interdisciplinary approach to
      criminology, 333
   research, policy, and treatment potential
      from approach of, 343–346
Public hostility toward police, 50, 58
Public housing, 38, 168, 170–173
Public Housing Authorities (PHAs), 172,
      173
Public interest vs. private interest in public
      choice perspective, 185
Public investments in technology R&D,
      271

Public opinion
  on conducted energy devices, 278
  corrections policy and fear of crime,
      141–142
  crack cocaine and, 142
  procedural fairness policies and, 102, 104,
      110–112
  reentry process and, 162
  technology adoption and, 271, 274
  on white-collar crime, 291–292
Public policy
  agenda-setting stage, 309–310
  in California's judicial branch, 109*f*
  classical perspective view of policy
      making, 5–6
  contextual orientation to, 6–7
  on corporate crime, 298
  corrections. *See* Corrections policy
  court managers and, 101
  criminal justice. *See* Criminal justice
  criminologists' lack of influence on, 4
  "crisis" situations as drivers of, 336–337,
      338
  definition and scope of, 1–2
  federal role in, 10–12, 101
  human trafficking framework, 308–311,
      309*f*, 321
  implementation stage of policy making,
      310–311
  judges and, 101
  legislative stage, 310
  mass media influence on, 13
  policy-making process, 5–7
  political process' influence on
      policy-making process, 3
  prison privatization, 199–200
  procedural fairness, 101–103, 108–110
  public health research and, 341–342
  reauthorization stage of policy making,
      311
  reentry process, 168–175, 177
  stages of policy making, 308
  state role in, 10–12, 101
  white-collar crime, 289–291, 298
Public safety, 29
Pump-and-dump schemes, 294
Punitive juvenile justice, 119, 224–228
  analysis of, 225–228, 226*f*
  components of, 225

indicators in U.S. juvenile justice policy,
      224–225
  states with, 225–228, 227*f*

Quality-of-life (QOL) approach of NYPD,
      27–28
Quota systems in arrest-focused policing,
      37

Race and ethnicity, 21–48. *See also* Racial
      discrimination
  broken windows and arrest-focused
      policy, 36–38
  community policing and, 33–39
  disproportionate incarceration, 259
  effectiveness of urban policing, 29–32
  exclusion of migrants based on, 250
  fairness of urban policing, 27–29
  fear of crime and fear of police, 38–39
  history of policing, 23–27
  immigration policies and, 247, 249, 250,
      259
  incarceration rates and, 142, 151–152,
      151*f*, 162, 169
  of majority of police, 40
  multicultural competency as part of
      policing, 39–41
  positive multiculturalism, 345
  procedural fairness and, 97, 110
  profiling. *See* Profiling
  in questionable deaths, 38
  reentry process and, 169
  as social indicators for urban violence,
      337–338
  white-collar crime opinions and, 291
Race riots of 1960s, 25, 26, 50
Racial discrimination, 24
  of NYPD, 30–32
Racketeering predicate offenses, 321
RAND Corporation reports
  on NYC stop, question, and frisk
      practices, 22–23, 31
  on SDFSCA, 83
RCA Corporation, 189
Reauthorization stage of policy making, 311
Recidivism
  corrections policy impact on, 150
  in domestic violence, 97, 126
  drug courts and, 121, 123

Recidivism (continued)
  Judicial Oversight Demonstration (JOD)
    sites, 128
  juveniles, 203
  procedural fairness and, 104
  Reentry Policy Council (RPC) and,
    175
  reoffending by new crimes distinguished
    from, 167
  Second Chance Act and, 176
Red Hook Community Justice Center, 107,
  110
Reentry, 159–181. See also Parole; Release
  challenges of, 163–167
  child custody policies and, 174–175
  context for, 161–162, 161f
  discretionary release and, 164–165
  of drug offenders, 335
  education access issues, 169–170
  expiration release and, 165–166
  future trends in policies for, 175–176
  legislative barriers for, 167–176
  mandatory release and, 165
  policies inhibiting, 168–175
  postrelease supervision, 167
  public assistance eligibility issues, 173–
    174
  public housing access issues, 170–173
  release mechanisms and, 163–167
Reentry Policy Council (RPC), 175–176
Reform era of policing, 21, 23, 24, 25
Rehabilitation
  corrections policy and, 136–137
  federal education loans and, 170
  indeterminate sentencing and, 164
  juvenile justice policy and, 224–232
  reentry process and, 162
  treatment and, 159, 162, 164
Reintegration. See Reentry
Reinventing government philosophy, 185,
  186
Reinvesting in Youth program
  (Washington), 223
Release. See also Parole; Reentry
  benefits for minorities after, 169
  challenges after, 163–167
  community impact of, 168
  conditional, 164
  context for, 159, 161–162, 161f, 164, 166

  discretionary, 164–165
  employment issues after, 150, 152, 159,
    163, 167, 169
  expiration, 165–166
  mandatory, 165
  mechanisms for, 163–164
  poverty in prisoners' communities and,
    159
  supervision after, 164, 167
  unconditional, 164
Religious persecution and immigration, 248
Rendition, 54
Repeat offenders, mandatory minimum
    sentences for, 166
Replacement effect in group crime, 147
Reporting requirements for schools, 84
Republican party and corrections policy,
    142, 143
Research, development, testing and
    evaluation (RDT&E) model,
    268–270, 269f
Research priorities in criminology, 336,
    337–338. See also Criminology
  conflicts between public health and
    criminology, 341
Restorative justice programs, 107, 118
Restriction as immigration policy, 249
Reverse mortgages, 300
Rhode Island
  juvenile court jurisdiction, 219
  parole supervision, 165
RICO predicate offenses, 321
Risk identification, 344
Risk tolerance in technology adoption,
    273
Rite of passage, incarceration as, 145
Rockefeller Foundation, 338
Roper v. Simmons (2005), 225
RPC (Reentry Policy Council), 175–176
Rucker v. Davis (2001), 171–172
Rudolph, Eric Robert, 52
Russia, human trafficking in, 313

Safe and Drug-Free Schools and
    Communities Act (SDFSCA) of
    1994, 82–84
Safe Schools/Healthy Students initiative, 85
Safe Schools Study (National Institute of
    Education), 81

Safe Streets, Safe City Act (New York State, 1991), 36
San Diego, California, crime reduction in, 37–38
Sarbanes-Oxley Act of 2002 (SOX), 295–298
School discipline, 86, 89
School liaison officers (UK), 88
School–police partnerships, 77–94
    challenges and successes of, 85
    COPS in Schools (CIS) and, 84–85
    crime reporting laws, 84
    effectiveness of, 86, 90
    effects on schools and students, 87–89
    implementation of, 85–86
    *in loco parentis* role of schools, 88
    National Education Goals and, 82
    policies supporting, 81–90
    safe schools acts and, 82–84
    Safe Schools/Healthy Students initiative and, 85
    school crime reporting laws and, 84
    school resource officers. *See* School resource officers (SROs)
    school security and safety, 77
    school violence and, 80–81
    searching students by school police officers, 89
School resource officers (SROs), 78–80
    benefits of, 87
    COPS in Schools (CIS) funding for, 85
    discipline by, 89
    effectiveness of, 86
    grants for hiring, 84
    police–youth relationships and, 88
    questioning of students by, 88–89
    Safe Schools/Healthy Students initiative funding for, 85
School Survey on Crime and Safety (National Center for Education Statistics), 78
School violence, 77–78
    federal understanding of, 80–81
    public perception of, 78
Sealed records, 216–217, 221–222, 228
Searches
    of students by school resource officers, 89
    warrantless, 51, 54, 55

Second Chance Act of 2008, 176
Securities and Exchange Commission (SEC), 195–196, 289
Securities Exchange Act of 1934, 297
Selective incapacitation, 146
Self-policing by corporations and professions, 289
Semen and DNA analysis, 280
Senate Select Committee to Study Governmental Operations with Respect to Intelligence Activities, 51
Sentencing
    deterrence effect of, 145
    human trafficking crimes, 321
    integrating treatment alternatives into, 117. *See also* Problem-solving courts
    for juveniles, 216, 220–221
    mandatory, 121, 135, 139, 144, 165, 166, 170, 216
    reentry process and, 160, 163
    satisfaction of offenders with, 100–101
Sentencing Project, 136
September 11 attacks. *See* Post-September 11 attacks and law enforcement
Serial killers, 343
Service provision
    in community policing, 34
    as police function, 33
    in private vs. public prisons, 187
Services-Training-Officers-Prosecutors (STOP), 127
Sex offenders, DNA evidence and, 281
Sex offenses, mandatory minimum sentences for, 166
Sexual assaults, DNA evaluation in, 280
Slavery, 24
Smugglers of illegal immigrants, 260
Social conditions, contributing to crime, 50
Social ecology theories, 335, 336–337
Social justice research, funding for, 338
Social problems
    addressing, as means of crime reduction, 50
    problem-solving courts' ability to deal with, 129–130

Socioeconomic status and white-collar crime, 291

South Carolina
conducted energy device use, 277
lack of counterterrorism measures, 59

Specialized courts. *See also* Problem-solving courts
domestic violence courts, 124, 125–128, 130

Specialized dockets. *See* Problem-solving courts

State courts and procedural fairness, 96, 101–102, 108–110

State Department
conference on international exploitation of women and children, 313
human trafficking enforcement, 317, 318*f*

State governments
agencies, 9–12, 186, 323–324
budgets, 149, 187
judicial system, 101
juvenile justice systems, 213, 229
legislation. See Legislation
mental health treatment funding, 222
prison privatization and, 199–200
public assistance ban on drug offenders and, 173

STOP (Services-Training-Officers-Prosecutors), 127

Stop, question, and frisk (SQF) policy of NYPD, 30–32

Students. *See* School–police partnerships

Subprime mortgage market, 299, 300–301

Substance abuse. *See* Alcohol abuse; *headings starting with "Drug"*

Substance Abuse and Mental Health Services Administration (SAMHSA), 85

Sunset provisions in laws, 337

Superior courts, 117

Supervision after reentry, 164

Supreme Court, U.S.
Arthur Andersen verdict overturned, 295
on one-strike policy, 171
*Rucker v. Davis* overturned, 172
on student searches, 89

"Survey of State Sentencing Structures" (Bureau of Justice Assistance), 166

Suspects
identification of, DNA collection and, 281
illegal immigrants. *See* Illegal immigrants

Sutherland, Edwin H., 290

Targeted enforcement, police legitimacy and, 29

TASER International, 277

Task forces for human trafficking, 323–324

Technology, 265–285
acquisition decisions, 273–274
adopter categories, 271–273
in Chicago police departments, 273
of conducted energy devices, 276–279
evaluation of, 266, 270, 274–276
of forensic DNA evidence, 279–283
police work using, 38
research, development, testing, and evaluation model (RDT&E), 268–271, 269*f*
review committees for, 274
stages of development and adoption, 268–276

Technology working groups (TWGs), 268, 269, 269*f*

Temporary Assistance to Needy Families (TANF), 173

Tennessee
funding for specialized courts, 222
juvenile drug treatment programs, 223

Terrorism
attacks before September 11, 2001, 52
dealing with new types of crime associated with, 336, 343
embassies as focus of attacks, 52
"Guidelines for the Interviews Regarding International Terrorism" (Office of the Deputy Attorney General), 55
organized, defined, 52
persistence of terrorist organizations, 67
political agendas of organized terrorists, 52

Testing of technology, 275

Theft in schools, 81

Therapeutic jurisprudence (TJ), 118–119

Therapeutic lawyering, 118
Third-party policing, 34
"Thompson memo" on corporate crime
    prosecutions, 298–299
Threat And Local Observation Notice
    (TALON, Defense Department),
    57
Three-strikes laws, 135, 144, 288
Total quality management, 103
Trafficking in persons. See Human
    trafficking
Trafficking Victims Protection Act (TVPA)
    of 2000, 308, 314–322
Trafficking Victims Protection
    Reauthorization Act (TVPRA) of
    2003, 316, 321, 322
Training
    for conducted energy devices, 278
    for court employees, 104
    for police officers, 26
    technology, 274
Transfers of juveniles to adult court
    systems, 216, 220–221, 224
Transnational crime. See Human trafficking
Transportation Security Administration,
    54, 64
Treatment
    for drug offenders. See Drug treatment
    due process and, trade-off for, 130
    incarceration for drug abusers vs., 120
    in juvenile courts, 120
    offenders' rights vs., 119–120
    as probation condition, 120
    problem-solving courts and, 119
    as sentencing option, 117
Triad Concept of SROs, 80
Trial courts, 117
    in California, 108
    court staff and, 102
    decentralization and innovation in, 103
    procedural fairness and, 104–106
Trusted Criminals (Friedrichs), 290
TWGs. See Technology working groups

Ukraine, human trafficking in, 313
Unconditional releases, 164
Undocumented aliens. See Illegal
    immigrants

Unemployment and released prisoners, 159
Uniform Crime Report data, 340
Unions and prison privatization, 185, 187,
    188
United Kingdom and school liaison officers,
    88
United Nations
    Fourth World Conference on Women
        (1995), 313
    human trafficking protocol, 314
    Universal Declaration of Human Rights,
        64
University of Florida, conducted energy
    device incident at, 278
University of Michigan, foreign students
    interviews at, 58
University of Nebraska Medical Center,
    College of Public Health, 341
Unmanned aerial vehicles (UAVs), 266
Urban ghettos
    creation of, 24–27
Urban Institute, 176, 232
Urban policing, 21–48
    arrest quotas, 37
    broken windows and arrest-focused
        policing and, 36–38
    effectiveness of, 29–32
    efficacy of community policing and
        broken windows approach, 33–39
    fairness of, 27–29
    fear of crime and fear of police, 38–39
    history of policing race and place, 23–27
    political and social power and, 27
    political era vs. reform era vs. community
        era, 21, 23, 24, 25
U.S. Citizen and Immigration Service
    (USCIS), 249
U.S. Corrections Corporation, 190,
    193–195
USA PATRIOT Act of 2001, 54, 57, 61, 64,
    254
Use-of-force continuum, 276

Vandalism in schools, 81
Vera Institute of Justice, 176
Verbal abuse, 127
Victim assistance programs for human
    trafficking, 322–323

Victims
  domestic violence courts and, 125, 126
  human trafficking, 319
    assistance for, 322–323
    cooperation with law enforcement, 321
    identification of, 323–324
    visas for, 317
  juvenile justice proceedings, rights of, 217
Violence
  in schools, 84. *See also* School–police
    partnerships; School violence
  social predictors of, 337–338
Violence Against Women Act (VAWA) of
  1994, 127, 281
Violent Crime Control and Law
  Enforcement Act of 1994, 34, 170,
  337
Violent crimes
  emotional aspects of, 144
  juvenile transfers to adult court and,
    220–221
  political agendas and, 140
  racial discrimination and, 142
  rates, 28
Virginia, juvenile jurisdiction reform,
  221
Vital Voices, 313
Vocational training for inmates, 136

Wackenhut Corrections Corporation, 189,
  195, 197
Wackenhut Services, Inc., 189
Wall Street and subprime mortgage market
  collapse, 301
War on Drugs, 120, 170, 174, 255, 288,
  335
War on terror, as de facto war on
  immigrants, 254
Warrantless wiretaps and searches, 51, 54,
  55
Washington State
  Institute for Public Policy, 276
  Reinvesting in Youth program, 223

Weapons offenses
  juvenile transfers for, 220
  mandatory minimum sentences for,
    166
Welfare, 173–174
West Virginia, juvenile drug offenders in,
  222–223
White-collar crime, 287–306
  conventional crime compared to,
    288–289
  Enron case, 293–296
  financial crisis of 2007–2008 and,
    299–302
  financial losses from, 292
  financier crime, 293, 299–302
  Justice Department policy on, 298–299
  perceptions of, 291–292
  as public policy focus, 289–291,
    292
  Sarbanes-Oxley Act and, 295–298
White Collar Crime Penalty Enhancement
  Act of 2002, 297
William Wilberforce Trafficking Victims
  Protection Reauthorization Act of
  2008, 316, 321–322, 323
Women
  incarceration rates, 162
  parental rights and incarceration of,
    174–175
Work-assistance programs, 170
WorldCom, 295
World Trade Center
  bombing of (1993), 52
  September 11, 2001, attacks on. *See*
    Post-September 11 attacks and law
    enforcement
World War II, immigration policies during,
  247

Youthful Offender Block Grant (California),
  223

Zero tolerance policy, 8, 36, 336

# Author Index

Albini, J. L., 58
Alpert, G., 24
Aos, S., 276
Armstrong, S., 203
Atkinson, M., 10
Austin, J., 162

Balmer, S., 89
Bayer, P., 203
Beckett, K., 13, 141, 143
Berman, G., 120
Bishop, D. M., 232
Bonczar, T. P., 161, 162
Bond, J. W., 280
Bourdieu, P., 338–340
Brady, K. P., 89
Brownstein, H., 4
Brudney, J. L., 187
Bucqueroux, B., 37

Callahan, D., 341
Carey, S. M., 124
Cavender, G., 13
Clear, T. R., 150
Coderoni, G. R., 41
Coleman, W. D., 10
Connor, G., 278
Cooper, W. E., 30
Cornelius, W., 257
Crawford, S., 186
Crepet, T., 281

Daicoff, S, 118, 119
DiIulio, J. J., Jr., 200
Donahue J., 183
Dorf, M. C., 119, 120, 126
Dowdy, E., 338
Dunham, R., 24

Fagan, J., 119, 120, 126
Fairchild, E. S., 13
Feller, I., 271
Fernandez, S., 187
Fischer, F., 6
Forester, F., 6
Fournier, E., 186
Fox, A., 122, 124, 274
Frenk, J., 334, 336, 338, 340–341, 343

Geis, G., 288
Genders, E., 200
Goldstein, H., 50

Hammond, C., 280
Hartnett, S. M., 273
Hendricks, N. J., 59
Heuer, L., 105
Hickman, M. J., 59
Hoffman, M., 119, 120
Hopkins, N., 88

Irwin, J., 162

Jackson, A., 88
Jacobs, J., 281
Jennings, B., 341
Johnson, I. M., 86, 87
Jones, T., 3, 8

Kanstroom, D., 246, 248
Kappeler, V. E., 65
Kelling, G. L., 33
King, R. S., 151
Kingdon, J. W., 309
Klockars, C. B., 35
Korn, D. A., 338
Krumholz, S. T., 118

Laswell, H., 6, 15
Laub, J. H., 4
Lee, E., 250
Lewis, J., 257
Liberman, A. M., 340

Maahs, J., 186
Mariner, W., 333
Marion, N., 10, 13
Martinson, R., 164
Mauer, M., 151
McCoy, C., 120
McDevitt, J., 87, 88
McDonald, D., 186
Michalowski, R., 140
Miller, K. S., 65
Miller, T., 246
Mirchandani, R., 125
Moore, M. H., 33
Morenoff, J., 337
Morgan, J., 277
Mukamal, D., 173
Munsterman, T., 111

Nagel, S., 7, 8, 12
Nagin, D., 147
Newburn, T., 3, 8
Nolan, L. J. L, 121, 123

Oliver, W. M., 65
Ortiz, C. W., 59

Panniello, J., 87, 88
Paulanka, B., 40
Pelfrey, W. V., 59, 66
Perrone, D., 187
Phenix, D., 89
Pigou, A. C., 184, 185
Platt, A., 335
Pozen, D. E., 203
Pratt, T.C., 186, 187
Purnell, L., 40

Quinn, M. C., 130

Raphael, S., 140
Raudenbush, S., 337
Rawls, J., 27

Reaves, B. A., 59
Reitz, K. R., 4, 165
Reuter, P. H., 83
Roderick, D., 118
Rogers, E. M., 271, 272
Rubenstein, G., 173
Ruegg, R., 271
Russell-Einhourn, M., 186
Ruth, H., 4
Ryu, J. E., 187

Sampson, R., 337, 338
Sasson, T., 141, 143
Schuiteman, J. G., 86
Shaffer, H. J., 338
Sickmund, M., 220
Silverman, E. B., 38
Simoncelli, T., 282
Skogan, W. G., 273
Skolnick, J. H., 32
Sloan, A., 300
Slobogin, C., 118
Snyder, H. N., 220
Spelman, W., 141
Spitzer, S., 335, 336
Staton, J., 279
Steinhardt, B., 282
Stolz, B. A., 11
Stone, D., 6
Sugie, N. F., 59
Sutherland, E. H., 290, 291, 292
Sykes, G. M., 150

Timpane, P. M., 86
Tonry, M., 140, 160
Trojanowicz, R., 37
Tyler, T. R., 96, 98, 100, 105, 106, 111, 112

Van Slylce, D. M., 187
Visher, C. A., 147

Walker, S., 146
Welch, M., 246, 248
Western, B., 140, 148
Wexler, D. B., 118, 119
Wilson, J. Q., 33
Wolf, R. V., 122, 124
Wright, D. S., 187